River and Lake Ice Processes

River and Lake Ice Processes

Impacts of Freshwater Ice on Aquatic Ecosystems in a Changing Globe

Special Issue Editors

Karl-Erich Lindenschmidt
Helen M. Baulch

MDPI • Basel • Beijing • Wuhan • Barcelona • Belgrade

MDPI

Special Issue Editors

Karl-Erich Lindenschmidt
University of Saskatchewan
Canada

Helen M. Baulch
University of Saskatchewan
Canada

Editorial Office
MDPI
St. Alban-Anlage 66
4052 Basel, Switzerland

This is a reprint of articles from the Special Issue published online in the open access journal *Water* (ISSN 2073-4441) from 2017 to 2018 (available at: https://www.mdpi.com/journal/water/special_issues/RaLIP)

For citation purposes, cite each article independently as indicated on the article page online and as indicated below:

LastName, A.A.; LastName, B.B.; LastName, C.C. Article Title. *Journal Name* **Year**, *Article Number*, *Page Range*.

ISBN 978-3-03897-388-1 (Pbk)
ISBN 978-3-03897-389-8 (PDF)

Contents

About the Special Issue Editors

Karl-Erich Lindenschmidt's, Ph.D., P.Eng., key research focus is the development of models designed to better forecast and understand flooding as related to ice-jams, an increasingly important risk to our prairie and other communities. Over the course of his career, Dr. Lindenschmidt has developed experience in water quality and quantity modelling of rivers and their basins internationally, including the Elbe, Saale, Havel, and Spree Rivers in Germany. His research interests in modelling river ice freeze-up and ice-jamming have also brought him to Manitoba, Saskatchewan, Alberta and the Northwest Territories. As his projects require intensive field research, he has significantly involved local community members in his work, and much of his research has been directed by community partners.

Helen Baulch, Dr., is an aquatic ecosystem biogeochemist with a research focus at the interface of field research and modelling. Her research interests include water quality, aquatic ecology, global change, biogeochemical cycles, greenhouse gas emissions, and eutrophication. The overall goal of her research program is to determine how to maintain crucial freshwater ecosystem services in the face of climate change and nutrient pollution.

Preface to "River and Lake Ice Processes"

Water quality studies of aquatic ecosystems in both rivers and lakes have often been falsely assumed to be dormant under ice-covered conditions compared to open-water seasons. This book encompasses 12 articles that comprise the Special Issue of the MDPI journal Water, entitled 'River and lake ice processes—impacts of freshwater ice on aquatic ecosystems in a changing globe'. Many topics are presented by world-class scientists and engineers that cover topics in river and lake ice processes and the water quality of riverine and lacustrine ecosystems under ice-covered conditions. The book also explores new research avenues applied to this specialized field, such as remote sensing and computer modelling, to help grapple with under-ice aquatic environmental research.

Karl-Erich Lindenschmidt, Helen M. Baulch
Special Issue Editors

water MDPI

Editorial

River and Lake Ice Processes—Impacts of Freshwater Ice on Aquatic Ecosystems in a Changing Globe

Karl-Erich Lindenschmidt *,† , **Helen M. Baulch** † and **Emily Cavaliere**

Global Institute for Water Security, 11 Innovation Blvd., Saskatoon, SK S7N 3H5, Canada;
helen.baulch@usask.ca (H.M.B.); emily.cavaliere@usask.ca (E.C.)
* Correspondence: karl-erich.lindenschmidt@usask.ca; Tel.: +1-306-966-6174
† These authors contributed equally to this work.

Received: 29 September 2018; Accepted: 31 October 2018; Published: 6 November 2018

Abstract: This special issue focuses on the effects of ice cover on surface water bodies, specifically rivers and lakes. Background information on the motivation of addressing this topic is first introduced with some selected references highlighting key points in this research field. A summary and synthesis of the eleven contributions is then provided, focusing on three aspects that provide the structure of the special issue: Physical processes, water quality, and sustainability. We have placed these contributions in the broader context of the field and identified selected knowledge gaps which impede our ability both to understand current conditions, and to understand the likely consequences of changing winters to the diversity of freshwater ecosystems subject to seasonal ice cover.

Keywords: aquatic ecosystems; field sampling; ice regime; modelling; northern lakes and rivers; water quality

1. Introduction

Most freshwater aquatic ecosystems have focused on open-water conditions, during spring, summer, and autumn. Studies in winter during ice-covered conditions are relatively sparse in many regions [1] due to the logistical challenges, issues of safety, and the historic assumption made by some that these ecosystems are biologically inactive during winter. Despite this, there is growing evidence that the winter period is not quiescent [2,3]—and indeed, many researchers have known this for decades, helping to build a year-round perspective of lakes and rivers in some regions [4].

Instead of winter being a quiescent period, it can be a period of rapid change. In some ecosystems, large under-ice blooms may occur [5,6] while in others anoxia is common, governed in large part by ice-cover duration [7,8]. Key biogeochemical processes continue through winter, and, as such, this can lead to significant greenhouse gas fluxes at ice out [9], and can alter the speciation and concentration of nutrients through winter and early spring [10–14]. Ice conditions can have strong impacts on the physical environment of lakes and rivers, impacting light, temperature, and mixing, and strongly influencing sediment transport. As such, ice can have both direct, and indirect effects on the flora, fauna, and water quality of freshwater systems [2,3,6]. High spatial variability in physical and chemical conditions is the norm. Within lakes vertical zonation, with regions of high and low light, and high and low oxygen can impact biota and geochemistry [2]. Horizontal patchiness in light is influenced by snow and ice conditions. Within rivers, the light environment changes in winter, but ice-induced changes in flow and habitat have received greater study due to the marked impacts on biotic communities. Variability within rivers may even be more extreme than lakes, with frazil ice, anchor ice, and surface ice affecting different aspects of the ecosystem, altering the habitat for fish, macrophytes, and invertebrates [15–17].

We still lack a complete understanding of how winter conditions affect aquatic ecosystems. While this is an important challenge, given the millions of lakes and rivers which are subject to ice cover, this also represents a vast challenge, given the multitude of factors that differ including geomorphology, morphometry, hydrology, trophic status, anoxia risk, carbon concentrations, and of course, the constituent biotic communities. Winter conditions set the stage for biological succession, and seasonal trajectories of change in water quality. As such, changes such as earlier ice-off may affect the duration and magnitude of the spring bloom in lakes [18], and changing winters may alter summer blooms [19,20]. In rivers, shorter ice periods may mean changes in anoxia risk, as many rivers show similar progressive declines in oxygen through winter to lakes [21,22]. However, changes in breakup intensity may represent the greatest impact of changing winter conditions on river ecosystems [23] due to the effects on channel morphology, along with direct impacts on sediment transport, water chemistry, and biota.

Observations of rapid declines in the duration of ice cover for lakes and rivers, (e.g., [24,25]) and evidence of changing timing and risk associated with ice-jam floods [26], combined with concern about implications of declining ice cover for water quality and biota (e.g., [3,27]), create a sense of urgency for winter aquatic research. This is because without an understanding of current conditions and processes governing winter and ice conditions, and with a limited understanding of how winter changes influence the open water season, we cannot effectively predict climate-related changes in seasonally ice-covered lakes and rivers. Likewise, without developing and applying modelling tools to understand physical processes and to anticipate how these processes will change, we cannot predict resultant changes in ecosystems, or risks to ecosystem services. This Special Issue provides a venue to report new findings in field-based and modelling research to highlight the importance of the ice regime within rivers and lakes, building a stronger understanding of current conditions, and ultimately helping to frame our understanding of future change.

2. Contributions and Current State of Knowledge

2.1. Physical Processes and Ice Phenology

Lakes and rivers are already experiencing decreased ice-cover duration in many regions [24,28,29], with the largest impact of climate change on ice-cover duration expected to be a much earlier spring melt, up to four-weeks earlier by 2050 [30,31]. Work by Hewitt et al. [32] presented here compares long-term breakup and freezeup records over nine lakes in Wisconsin USA, and Ontario, Canada. They demonstrate that over 35 years, there has been a shift to loss of ice cover five days earlier in spring, and freeze up eight days later, associated with warmer fall, winter, and spring temperatures. By 2070, they predict much more dramatic changes may occur, but these changes will be strongly impacted by the degree of warming observed, with the warmest scenario suggesting breakup could be 43 days earlier. Dramatic decreases in ice-cover duration could have negative impacts on aquatic ecosystems that are already vulnerable to regime shifts [33]. However, there are many unknowns about changes during winter, and how winter duration will affect aquatic ecosystems.

A precursor for understanding change within ice-covered rivers is characterization of its flow and ice regime. Alfredsen [34] investigated the effects of ice on the flow conditions of rivers in relation to their hydrological variability during winter. He used the index of hydrological alteration (IHA) to characterize the interaction between the rivers' flow and ice regimes. Ice processes considered in the IHA include frazil and anchor ice formation and the formation and breakup of ice covers. The modelling tool introduced by Lindenschmidt [35] also includes the process of ice jamming, an important process leading to flooding in many communities along northern rivers. The tool has been successfully implemented to simulate flood hazard and risk induced by ice-jam flooding [36,37], climate change [38], and ice-jam flood forecasting [39]. Zhang et al. [40] introduced a novel space-borne remote sensing method for determining ice volume that breaks up along a river that forms an ice jam. The ice volume constituting an ice jam can be a very sensitive parameter for the accuracy of backwater levels induced from jams [41]. An interesting approach to predict ice-cover failure is presented by

Zhang et al. [42] who used an extension of the smooth particle hydrodynamics method, which is based on a simplified finite difference interpolation scheme.

The physical processes of greatest interest in lakes tend to center around dynamics of temperature, light, and mixing. Temperature, light, and mixing tend to be reduced with the onset of ice cover in lentic systems. As the ice forms, the denser, warmer water in the water column sinks and is further heated by the sediments [43], while the colder, less dense water stays near the surface as the ice forms [44]. Snow accumulation on the surface of ice-covered lakes, reservoirs, and ponds decreases light penetration [44], hence snowfall, blowing snow, and snowmelt can alter light inputs over winter [44,45]. When light penetration increases, due to melting of snow and ice, water below the ice–water interface warms, causing mixing [2,45], and altering the light environment for phytoplankton. The type of ice can also alter the light environment, with more opaque, or white ice blocking much of the light while black or crystal ice can be nearly transparent [4,46]. Changes in ice and snow through the season have a critical impact on rates of primary productivity, and the biomass of phytoplankton under ice [45]. Although low temperatures decrease microbial activity [47], important under-ice blooms can occur when light conditions are suitable (e.g., see [2,6]).

2.2. Water Quality

Changes in winter oxygen represent one of the most significant areas of scientific interest, and the longest-studied aspects of winter limnology, in part due to risk of winter fish kills. Within rivers and lakes, dissolved oxygen depletion can occur due to continued respiration, coupled with often low primary productivity, and the prevention of reaeration due to ice cover [28,48]. Oxygen depletion can be exacerbated due to municipal and industrial emissions, exfiltration of oxygen-depleted groundwater, and can be strongly influenced by the light environment, and connectivity of river–lake networks [16,22,49–51].

Akomeah et al. [52] and Terry et al. [53] both used surface water quality models to investigate the water quality of ice-covered water bodies, the former concentrating on a river and the latter on a lake within the Qu'Appelle river–lake system on the Canadian Prairies. Both studies are initial steps in larger projects to investigate the dynamics of dissolved oxygen concentrations and sediment oxygen demand during the winter seasons. The strain on oxygen concentrations increases as the ice cover persists longer at the end of the winter season—consistent with work suggesting that shorter periods of ice cover will lead to reduced anoxia risk, particularly in shallow ecosystems [22,26]. While aerobic respiration, particularly benthic respiration, are considered the dominant drivers of anoxia risk, there is growing interest in the role of methanotrophy [54] and nitrification [10] in mediating differences among lakes in their trajectories of oxygen decline.

An extensive examination of the effects of ice covers on river ecology is provided by Prowse [16,17] with a review on the effects of ice-cover breakup on riverine aquatic systems provided by Scrimgeour et al. [55]. Recent literature also focuses on nutrient dynamics in the water column [56] and dissolved oxygen in the river bed sediments [57] of frozen-over rivers. In addition, there has been significant interest, and progress in the ecology of ice-covered rivers and effects of ice on fish behavior and habitat (e.g., [58–61]).

Within lakes, impacts of ice phenology (ice duration due to the timing of ice-on and ice-off dates) can impact the following spring and summer seasons' water quality, as indicated by the work presented by Warner et al. [62] in this special issue. These trends are consistent with work in Sweden [63] and Lake Erie [18] showing increased diatom biomass with shorter or no-ice cover years. This area of understanding winter impacts on the spring bloom, and on summer water quality is a key area where more work is required to understand different responses across lakes, and to characterize interactions between physical, chemical, and biotic drivers of change.

Winter changes span many more variables. Turcotte and Morse [64] studied the impact of ice breakup on peaks in specific conductivity and turbidity using a conceptual model, the winter environment continuum, to help relate the winter conditions in headwater streams to the water quality state along the river network during the winter. This linking of physical and chemical change is a key area for additional work across lentic and lotic environments. Physical effects of freeze-out may have important impacts upon solute chemistry, particularly in shallow ecosystems [65–67]. Low temperature will affect microbial processes, with impacts on , oxygen and biogeochemical processes [8,11,12,47]. More broadly, the full integration of lakes and their watersheds, necessary to understand likely effects of climatic change represents a vast challenge—as illustrated by the work presented here.

The potential impact of changing ice phenology on a river's future (2050s and 2080s) water quality was modelled by Hosseini et al. [68]. They found that thin ice and shorter ice periods may lead to a decrease in winter nutrient concentrations in the winter, but this could be offset by increased flow if more water is passed through the river system to meet increased water demand. Warner et al. [62] also found that changes in the ice-off dates can have marked effects on the thermal regimes, mixing depths and length of turnover in the spring, which may further influence spring and summer water quality and algal species succession. These changes have already been observed in some areas of the world, including in Lake Erie, and Sweden [18,63,69], and in some cases, resulting in increased spring bloom diatom biomass causing water quality problems for local municipalities [63]. Warner et al. have provided important insights that will help build an understanding of the different responses we might anticipate across lakes of different regions, building on some of our knowledge of different physical responses, for example among lakes with varied mean depths [32]. Parks et al. [70,71] also found that river ice in the arctic is lessening due to regional climate warming, as are other components of the cryosphere such as permafrost, glacier ice, and sea ice.

2.3. Sustainability

Rokaya et al. [71] have criticized how little attention has been given to questions of sustainability within ice research. They noted that there is limited cross-disciplinary collaboration and integration of social sciences despite the massive socio-economic consequences of ice-jam flooding. They provide a scheme to sustainably manage a regulated river system to both (i) mitigate ice-jam flood risk at communities and (ii) promote ice-jam flooding to help replenish moisture and sediment supply of an inland delta's aquatic and terrestrial ecosystem [71]. We are fortunate to have a paper in this issue that delves, on a conceptual level, on the management of ice-jam floods to better sustain socio-economic and socio-ecological systems [72]. They provide a framework for such a management strategy that calls for a more interdisciplinary approach to ice management, which integrates social, economic, and ecological perspectives.

This need for a more interdisciplinary approach within ice science holds more broadly. Ultimately, the ice-cover period, and climate-related changes in the ice-cover period is strongly linked to ecosystem services provided by freshwater. Communities depend upon seasonally ice-covered rivers and lakes for necessities like water, fish, and transportation, and for recreational opportunities. While some of the effects of ice cover on ecology, ecosystem services, and economics have been known for decades (e.g., anoxia risk, fish kills, and ice-jam floods), there is growing awareness of broader effects of changing winters and associated risks and opportunities. More communities are grappling with transportation challenges associated with threats to ice roads, while others may see opportunities to increase transport via barge, or the need to grapple with the high cost of building all weather roads [73] and there is growing awareness of the impacts of ice, and changing ice on industries such as hydroelectric power generation [74]. Changing ice cover may have important economic consequences in some areas [73], and effects on fisheries may have profound cultural impacts, particularly in Indigenous communities. More work is required to better understand how loss of ice will impact people and communities, and support building the adaptive capacity required to respond to changing freshwater ice [75].

3. Conclusions

Winter ice cover is changing rapidly, and with it, many changes within lentic and lotic ecosystems will result, impacting key ecosystem services, and more broadly, affecting sustainability of key resources, livelihoods, and traditional practices. Climate change effects must be viewed through a regional lens, understanding the specific types of changes in precipitation, temperature, wind, and other variables that may occur, and building an understanding of how that will affect the ecosystems in a region. To help build the necessary local–regional–global understanding, synthetic work to understand differences across watersheds, across lakes, across rivers, and across regions such as those already presented here [32,62,64] and elsewhere [3,9] is very valuable—identifying common responses of ecosystems through winter, and differences across regions, and across types of lakes and rivers. This type of comparative approach will help to build a more diverse, regionally-informed understanding of physical, hydrological, hydraulic, biogeochemical, and ecological change, particularly when coupled with work to inform process-based understanding of changes through winter (e.g., [10,12]), and to integrate our growing understanding of winter changes into model-based frameworks (e.g., [76–78]).

As we work to anticipate future climatic changes, and effects upon aquatic ecosystems, much of our work has focused on understanding changes within the ice-free season, yet the ice cover and ice-free season must be understood jointly. Despite a rapid growth of interest in this area, key gaps remain. Building the scientific basis to assemble and integrate an understanding of physical, biogeochemical, ecological, and socio-economic change across regions is a vast challenge that will take decades of integrative research to address. Where possible, this represents work that should progress in tandem with community-based approaches to foster climate adaptation.

Author Contributions: K.-E.L. developed the idea for the special issue and introductory manuscript. K.-E.L. and H.M.B. wrote much of the text, with major contributions from E.C.

Funding: Funding sources for this work include the University of Saskatchewan's School of Environment and Sustainability and Global Institute for Water Security and a NSERC Discovery Grant to H.M.B.

Acknowledgments: We thank all authors for their contributions to this special issue.

Conflicts of Interest: The authors declare no conflict of interest.

References

1. Powers, S.M.; Hampton, S.E. Winter Limnology as a New Frontier. *Limnol. Oceanogr. Bull.* **2016**, *25*, 103–108. [CrossRef]
2. Bertilsson, S.; Burgin, A.; Carey, C.C.; Fey, S.B.; Grossart, H.-P.; Grubisic, L.M.; Jones, I.D.; Kirillin, G.; Lennon, J.T.; Shade, A.; et al. The under-ice microbiome of seasonally frozen lakes. *Limnol. Oceanogr.* **2013**, *58*, 1998–2012. [CrossRef]
3. Hampton, S.E.; Galloway, A.W.E.; Powers, S.M.; Ozersky, T.; Woo, K.H.; Batt, R.D.; Labou, S.G.; O'Reilly, C.M.; Sharma, S.; Lottig, N.R.; et al. Ecology under lake ice. *Ecol. Lett.* **2017**, *20*, 98–111. [CrossRef] [PubMed]
4. Schindler, D.W.; Welch, H.E.; Kalff, J.; Brunskill, G.J.; Kritsch, N. Physical and Chemical Limnology of Char Lake, Cornwallis Island (75° N Lat.). *J. Fish. Res. Board Can.* **1974**, *31*, 585–607. [CrossRef]
5. Vehmaa, A.; Salonen, K. Development of phytoplankton in Lake Paajarvi (Finland) during under-ice convective mixing period. *Aquat. Ecol.* **2009**, *43*, 693–705. [CrossRef]
6. Katz, S.L.; Izmest'eva, L.R.; Hampton, S.E.; Ozersky, T.; Shchapov, K.; Moore, M.V.; Shimaraeva, S.V.; Silow, E.A.; Izmest'eva, L.R.; Hampton, S.E.; et al. The "Melosira years" of Lake Baikal: Winter environmental conditions at ice onset predict under-ice algal blooms in spring. *Limnol. Oceanogr.* **2015**, *60*, 1950–1964. [CrossRef]
7. Meding, M.E.; Jackson, L.J. Biological implications of empirical models of winter oxygen depletion. *Can. J. Fish. Aquat. Sci.* **2001**, *58*, 1727–1736. [CrossRef]
8. Barica, J.; Mathias, J.A. Oxygen Depletion and Winterkill Risk in Small Prairie Lakes under Extended Ice Cover. *J. Fish. Res. Board Can.* **1979**, *36*, 980–986. [CrossRef]

9. Denfeld, B.A.; Baulch, H.M.; Giorgio, P.A.; Hampton, S.E.; Karlsson, J. A synthesis of carbon dioxide and methane dynamics during the ice-covered period of northern lakes. *Limnol. Oceanogr. Lett.* **2018**, *3*, 117–131. [CrossRef]

10. Powers, S.M.; Baulch, H.M.; Hampton, S.E.; Labou, S.G.; Lottig, N.R.; Stanley, E.H. Nitrification contributes to winter oxygen depletion in seasonally frozen forested lakes. *Biogeochemistry* **2017**, *136*, 119–129. [CrossRef]

11. Powers, S.M.; Labou, S.G.; Baulch, H.M.; Hunt, R.J.; Lottig, N.R.; Hampton, S.E.; Stanley, E.H. Ice duration drives winter nitrate accumulation in north temperate lakes. *Limnol. Oceanogr. Lett.* **2017**, 177–186. [CrossRef]

12. Cavaliere, E.; Baulch, H.M. Denitrification under lake ice. *Biogeochem. Lett.* **2018**, *137*, 285–295. [CrossRef]

13. Orihel, D.M.; Baulch, H.M.; Casson, N.J.; North, R.L.; Parsons, C.T.; Seckar, D.C.M.; Venkiteswaran, J.J. Internal phosphorus loading in Canadian fresh waters: A critical review and data analysis. *Can. J. Fish. Aquat. Sci.* **2017**, *74*, 2005–2029. [CrossRef]

14. Palacin-Lizarbe, C.; Camarero, L.; Catalan, J. Denitrification Temperature Dependence in Remote, Cold and N-Poor Lake Sediments. *Water Resour. Res.* **2018**, *2*. [CrossRef]

15. Linnansaari, T.; Alfredsen, K.; Stickler, M.; Arnekleiv, J.V.; Harby, A.; Cunjak, R.A. Does ice matter? Site fidelity and movements by Atlantic salmon (Salmo salar L.) parr duirng winter in a substrate enhanced river reach. *River Res. Appl.* **2009**, *25*, 773–787. [CrossRef]

16. Prowse, T.D. River-ice ecology. I.: Hydrologic, geomorphic, and water-quality aspects. *J. Cold Reg. Eng.* **2001**, *15*, 1–16. [CrossRef]

17. Prowse, T.D. River-ice ecology. II. Biological aspects. *J. Cold Reg. Eng.* **2001**, *15*, 17–33. [CrossRef]

18. Twiss, M.R.; McKay, R.M.L.; Bourbonniere, R.A.; Bullerjahn, G.S.; Carrick, H.J.; Smith, R.E.H.; Winter, J.G.; D'souza, N.A.; Furey, P.C.; Lashaway, A.R.; et al. Diatoms abound in ice-covered Lake Erie: An investigation of offshore winter limnology in Lake Erie over the period 2007 to 2010. *J. Gt. Lakes Res.* **2012**, *38*, 18–30. [CrossRef]

19. Weyhenmeyer, G.A. Rates of change in physical and chemical lake variables—Are they comparable between large and small lakes? *Hydrobiologia* **2008**, *599*, 105–110. [CrossRef]

20. Reavie, E.D.; Cai, M.; Twiss, M.R.; Carrick, H.J.; Davis, T.W.; Johengen, T.H.; Gossiaux, D.; Smith, D.E.; Palladino, D.; Burtner, A.; et al. Winter-spring diatom production in Lake Erie is an important driver of summer hypoxia. *J. Gt. Lakes Res.* **2016**, *42*, 608–618. [CrossRef]

21. Prowse, T.D. Environmental significance of ice to streamflow in cold regions. *Freshw. Biol.* **1994**, *32*, 241–259. [CrossRef]

22. Chambers, P.A.; Scrimgeour, G.J.; Pietroniro, A.; Culp, J.M.; Loughran, I. Oxygen modelling under river ice covers. In Proceedings of the Workshop on Environmental Aspects of River Ice; Prowse, T.D., Ed.; NHRI Symposium: Saskatoon, SK, Canada, 1993; pp. 235–260.

23. Prowse, T.D.; Beltaos, S. Climatic control of river-ice hydrology: A review. *Hydrol. Process.* **2002**, *16*, 805–822. [CrossRef]

24. Magnuson, J.J.; Robertson, D.M.; Benson, B.J.; Wynne, R.H.; Livingstone, D.M.; Arai, T.; Assel, R.A.; Barry, R.G.; Card, V.; Kuusisto, E.; et al. Historical trends in lake and river ice cover in the Northern Hemisphere. *Science* **2000**, *289*, 1743–1746. [CrossRef] [PubMed]

25. Sharma, S.; Magnuson, J.J.; Batt, R.D.; Winslow, L.A.; Korhonen, J.; Aono, Y. Direct observations of ice seasonality reveal changes in climate over the past 320–570 years. *Sci. Rep.* **2016**, *6*, 25061. [CrossRef] [PubMed]

26. Rokaya, P.; Budhathoki, S.; Lindenschmidt, K.-E. Trends in the Timing and Magnitude of Ice-Jam Floods in Canada. *Sci. Rep.* **2018**, *8*, 5834. [CrossRef] [PubMed]

27. Hampton, S.E.; Moore, M.V.; Ozersky, T.; Stanley, E.H.; Polashenski, C.M.; Galloway, A.W.E. Heating up a cold subject: Prospects for under-ice plankton research in lakes. *J. Plankton Res.* **2015**, *37*, 277–284. [CrossRef]

28. Fang, X.; Stefan, H.G. Simulations of climate effects on water temperature, dissolved oxygen, and ice and snow covers in lakes of the contiguous United States under past and future climate scenarios. *Limnol. Oceanogr.* **2009**, *54*, 2359–2370. [CrossRef]

29. Magee, M.R.; Wu, C.H. Effects of changing climate on ice cover in three morphometrically different lakes. *Hydrol. Process.* **2017**, *31*, 308–323. [CrossRef]

30. Butcher, J.B.; Nover, D.; Johnson, T.E.; Clark, C.M. Sensitivity of lake thermal and mixing dynamics to climate change. *Clim. Chang.* **2015**, *129*, 295–305. [CrossRef]

31. Beltaos, S.; Prowse, T. River-ice hydrology in a shrinking cryosphere. *Hydrol. Process.* **2009**, *23*, 122–144. [CrossRef]

32. Hewitt, A.; Lopez, L.S.; Gaibisels, K.M.; Murdoch, A.; Higgins, S.N.; Magnuson, J.J.; Paterson, A.M.; Rusak, J.A.; Yao, H.; Sharma, S. Historical Trends, Drivers, and Future Projections of Ice Phenology in Small North Temperate Lakes in the Laurentian Great Lakes Region. *Water* **2018**, *10*, 70. [CrossRef]

33. Scheffer, M.; Hosper, S.H.; Meijer, M.L.; Moss, B.; Jeppesen, E. Alternative equilibria in shallow lakes. *Trends Ecol. Evol.* **1993**, *8*, 275–279. [CrossRef]

34. Alfredsen, K. An Assessment of Ice Effects on Indices for Hydrological Alteration in Flow Regimes. *Water* **2017**, *9*, 914. [CrossRef]

35. Lindenschmidt, K.-E. RIVICE—A Non-Proprietary, Open-Source, One-Dimensional River-Ice Model. *Water* **2017**, *9*, 314. [CrossRef]

36. Lindenschmidt, K.-E.; Das, A.; Rokaya, P.; Chun, K.P.; Chu, T. Ice jam flood hazard assessment and mapping of the Peace River at the Town of Peace River. In Proceedings of the CRIPE 18th Workshop on the Hydraulics of Ice Covered Rivers, Quebec City, QC, Canada, 18–20 August 2015.

37. Lindenschmidt, K.-E.; Das, A.; Rokaya, P.; Chu, T. Ice jam flood risk assessment and mapping. *Hydrol. Process.* **2016**, *30*, 3754–3769. [CrossRef]

38. Das, A.; Rokaya, P.; Lindenschmidt, K.-E. Impacts of climate change on ice-jam flooding along a northern river, Canada. *Clim. Chang.* **2018**, submitted.

39. Lindenschmidt, K.-E.; Rokaya, P.; Das, A.; Li, Z.; Richard, D. A novel stochastic modelling approach for operational real-time ice-jam flood forecasting. *J. Hydrol.* **2018**, submitted.

40. Zhang, F.; Mosaffa, M.; Chu, T.; Lindenschmidt, K.-E. Using Remote Sensing Data to Parameterize Ice Jam Modeling for a Northern Inland Delta. *Water* **2017**, *9*, 306. [CrossRef]

41. Lindenschmidt, K.-E. Using stage frequency distributions as objective functions for model calibration and global sensitivity analyses. *Environ. Model. Softw.* **2017**, *92*, 169–175. [CrossRef]

42. Zhang, N.; Zheng, X.; Ma, Q. Updated Smoothed Particle Hydrodynamics for Simulating Bending and Compression Failure Progress of Ice. *Water* **2017**, *9*, 882. [CrossRef]

43. Bengtsson, L. Ice-covered lakes: Environment and climate-required research. *Hydrol. Process.* **2011**, *25*, 2767–2769. [CrossRef]

44. Catalan, J. Evolution of dissolved and particulate matter during the ice-covered period in a deep, high-mountain lake. *Can. J. Fish. Aquat. Sci.* **1992**, *49*, 945–955. [CrossRef]

45. Pernica, P.; North, R.L.; Baulch, H.M. In the cold light of day: The potential importance of under-ice convective mixed layers to primary producers. *Inland Waters* **2017**, *7*, 138–150. [CrossRef]

46. Petrov, M.P.; Terzhevik, A.Y.; Palshin, N.I.; Zdorovennov, R.E.; Zdorovennova, G.E. Absorption of Solar Radiation by Snow-and-Ice Cover of Lakes. *Water Resour.* **2005**, *32*, 546–554. [CrossRef]

47. Søndergaard, M.; Bjerring, R.; Jeppesen, E. Persistent internal phosphorus loading during summer in shallow eutrophic lakes. *Hydrobiologia* **2013**, *710*, 95–107. [CrossRef]

48. McBean, E.; Farquhar, G.; Kouwen, N.; Dubek, O. Predictions of ice-cover development in streams and its effect on dissolved oxygen modelling. *Can. J. Civ. Eng.* **1979**, *6*, 197–207. [CrossRef]

49. Wharton, R.A.; Simmons, G.M.; McKay, C.P. Perennially ice-covered Lake Hoare, Antarctica: Physical environment, biology and sedimentation. *Hydrobiologia* **1989**, *172*, 305–320. [CrossRef] [PubMed]

50. Jakkila, J.; Lepparanta, M.; Kawamura, T.; Shirasawa, K.; Salonen, K. Radiation transfer and heat budget during the ice season in Lake Paajarvi, Finland. *Aquat. Ecol.* **2009**, *43*, 681–692. [CrossRef]

51. Mackinnon, B.D.; Sagin, J.; Baulch, H.M.; Lindenschmidt, K.-E.; Jardine, T.D. Influence of hydrological connectivity on winter limnology in floodplain lakes of the Saskatchewan River Delta, Saskatchewan. *Can. J. Fish. Aquat. Sci.* **2016**, *73*, 140–152. [CrossRef]

52. Akomeah, E.; Lindenschmidt, K.E. Seasonal variation in sediment oxygen demand in a Northern chained River-lake system. *Water* **2017**, *9*, 254. [CrossRef]

53. Terry, J.A.; Sadeghian, A.; Lindenschmidt, K.E. Modelling dissolved oxygen/sediment oxygen demand under ice in a shallow eutrophic prairie reservoir. *Water* **2017**, *9*, 131. [CrossRef]

54. Denfeld, B.A.; Canelhas, M.R.; Weyhenmeyer, G.A.; Bertilsson, S.; Eiler, A.; Bastviken, D. Constraints on methane oxidation in ice-covered boreal lakes. *J. Geophys. Res. Biogeosci.* **2016**, *121*, 1924–1933. [CrossRef]

55. Scrimgeour, G.J.; Prowse, T.D.; Culp, J.M.; Chambers, P.A. Ecological Effects of River Ice Break-Up—A Review and Perspective. *Freshw. Biol.* **1994**, *32*, 261–275. [CrossRef]
56. Shakibaeinia, A.; Kashyap, S.; Dibike, Y.B.; Prowse, T.D. An integrated numerical framework for water quality modelling in cold-region rivers: A case of the lower Athabasca River. *Sci. Total Environ.* **2016**, *569–570*, 634–646. [CrossRef] [PubMed]
57. Sharma, K. Factors Affecting Sediment Oxygen Demand of the Athabasca River Sediment under Ice Cover. Ph.D. Thesis, University of Alberta, Edmonton, AB, Canada, 2012.
58. Bergeron, N.E.; Enders, E.C. Fish response to freeze up. In *River Ice Formation*; Beltaos, S., Ed.; Committee on River Ice Processes and Environment: Edmonton, AB, Canada, 2013; ISBN 978-0-9920022-0-6.
59. Brown, R.S.; Duguay, C.R.; Mueller, R.P.; Moulton, L.L.; Doucette, P.I.; Tagestad, J.D. Use of Synthetic Aperture Radar (SAR) to Identify and Characterize Overwintering Areas of Fish in Ice-Covered Arctic Rivers: A Demonstration with Broad Whitefish and Their Habitats in the Sagavanirktok River, Alaska. *Trans. Am. Fish. Soc.* **2010**, *139*, 1711–1722. [CrossRef]
60. Carr, M.; Lacho, C.; Pollock, M.; Watkinson, D.; Lindenschmidt, K.-E. Development of geomorphic typologies for identifying Lake Sturgeon (*Acipenser fulvescens*) habitat in the Saskatchewan River System. *River Syst.* **2015**, *21*, 215–227. [CrossRef]
61. Linnansaari, T.; Cunjak, R.A. Effects of ice on behavior of juvenile Atlantic salmon (*Salmo salar*). *Can. J. Fish. Aquat. Sci.* **2013**, *70*, 1488–1497. [CrossRef]
62. Warner, K.; Fowler, R.; Northington, R.; Malik, H.; McCue, J.; Saros, J. How Does Changing Ice-Out Affect Arctic versus Boreal Lakes? A Comparison Using Two Years with Ice-Out that Differed by More Than Three Weeks. *Water* **2018**, *10*, 78. [CrossRef]
63. Weyhenmeyer, G.A.; Westoo, A.K.; Willen, E. Increasingly ice-free winters and their effects on water quality in Sweden's largest lakes. *Hydrobiologia* **2008**, *599*, 111–118. [CrossRef]
64. Turcotte, B.; Morse, B. The Winter Environmental Continuum of Two Watersheds. *Water* **2017**, *9*, 337. [CrossRef]
65. Schmidt, S.; Moskal, W.; De Mora, S.J.; Howard-Williams, C.; Vincent, W.F. Limnological properties of antarctic ponds during winter freezing. *Antarct. Sci.* **1991**, *3*, 379–388. [CrossRef]
66. Dugan, H.A.; Helmueller, G.; Magnuson, J.J. Ice formation and the risk of chloride toxicity in shallow wetlands and lakes. *Limnol. Oceanogr. Lett.* **2017**, *2*, 150–158. [CrossRef]
67. Chambers, M.K.; White, D.M.; Lilly, M.R.; Hinzman, L.D.; Hilton, K.M.; Busey, R.C. Exploratory analysis of the winter chemistry of five lakes on the North Slope of Alaska. *J. Am. Water Resour. Assoc.* **2008**, *44*, 316–327. [CrossRef]
68. Hosseini, N.; Johnston, J.; Lindenschmidt, K.-E. Impacts of Climate Change on the Water Quality of a Regulated Prairie River. *Water* **2017**, *9*, 199. [CrossRef]
69. Weyhenmeyer, G.A.; Livingstone, D.M.; Meili, M.; Jensen, O.; Benson, B.; Magnuson, J.J. Large geographical differences in the sensitivity of ice-covered lakes and rivers in the Northern Hemisphere to temperature changes. *Glob. Chang. Biol.* **2011**, *17*, 268–275. [CrossRef]
70. Park, H.; Yoshikawa, Y.; Oshima, K.; Kim, Y.; Ngo-Duc, T.; Kimball, J.S.; Yang, D. Quantification of Warming Climate-Induced Changes in Terrestrial Arctic River Ice Thickness and Phenology. *J. Clim.* **2016**, *29*, 1733–1754. [CrossRef]
71. Park, H.; Yoshikawa, Y.; Yang, D.; Oshima, K. Warming Water in Arctic Terrestrial Rivers under Climate Change. *J. Hydrometeorol.* **2017**, *18*, 1983–1995. [CrossRef]
72. Rokaya, P.; Budhathoki, S.; Lindenschmidt, K.-E. Ice-jam flood research: A scoping review. *Nat. Hazards* **2018**. [CrossRef]
73. Das, A.; Reed, M.; Lindenschmidt, K.-E. Sustainable Ice-Jam Flood Management for Socio-Economic and Socio-Ecological Systems. *Water* **2018**, *10*, 135. [CrossRef]
74. Prowse, T. Introduction: Hydrologic effects of a shrinking cryosphere. *Hydrol. Process.* **2009**, *23*, 1–6. [CrossRef]
75. Prowse, T.; Alfredsen, K.; Beltaos, S.; Bonsal, B.R.; Bowden, W.B.; Duguay, C.R.; Korhola, A.; McNamara, J.; Vincent, W.F.; Vuglinsky, V.; et al. Effects of changes in arctic lake and river ice. *Ambio* **2011**, *40*, 63–74. [CrossRef]
76. Olsson, P.; Folke, C.; Berkes, F. Adaptive comanagement for building resilience in social-ecological systems. *Environ. Manag.* **2004**, *34*, 75–90. [CrossRef] [PubMed]

77. Hosseini, N.; Akomeah, E.; Davies, J.-M.; Baulch, H.; Lindenschmidt, K.-E. Water quality modelling of a prairie river-lake system. *Environ. Sci. Pollut. Res.* **2018**, *25*, 1–15. [CrossRef] [PubMed]
78. Hosseini, N.; Chun, K.P.; Wheater, H.; Lindenschmidt, K.E. Parameter Sensitivity of a Surface Water Quality Model of the Lower South Saskatchewan River—Comparison Between Ice-On and Ice-Off Periods. *Environ. Model. Assess.* **2017**, *22*, 291–307. [CrossRef]

water

MDPI

Article

Historical Trends, Drivers, and Future Projections of Ice Phenology in Small North Temperate Lakes in the Laurentian Great Lakes Region

Bailey A. Hewitt [1,†], Lianna S. Lopez [1,†], Katrina M. Gaibisels [1], Alyssa Murdoch [1], Scott N. Higgins [2], John J. Magnuson [3], Andrew M. Paterson [4], James A. Rusak [4] [iD], Huaxia Yao [4] and Sapna Sharma [1,*]

[1] Department of Biology, York University, 4700 Keele Street, Toronto, ON M3J 1P3, Canada; bailey24@yorku.ca (B.A.H.); liannalopez18@gmail.com (L.S.L.); k.gaibisels@gmail.com (K.M.G.); alyssamurdoch@gmail.com (A.M.)
[2] IISD Experimental Lakes Area Inc., Winnipeg, MB R3B 0T4, Canada; shiggins@iisd-ela.org
[3] Center for Limnology, University of Wisconsin, Madison, WI 53706, USA; john.magnuson@wisc.edu
[4] Dorset Environmental Science Centre, Ontario Ministry of the Environment and Climate Change, Dorset, ON P0A 1E0, Canada; Andrew.Paterson@ontario.ca (A.M.P.); Jim.Rusak@ontario.ca (J.A.R.); Huaxia.Yao@ontario.ca (H.Y.)
* Correspondence: sharma11@yorku.ca; Tel.: +1-416-736-2100 (ext. 33756)
† These authors contributed equally to this work.

Received: 31 October 2017; Accepted: 10 January 2018; Published: 15 January 2018

Abstract: Lake ice phenology (timing of ice breakup and freeze up) is a sensitive indicator of climate. We acquired time series of lake ice breakup and freeze up, local weather conditions, and large-scale climate oscillations from 1981–2015 for seven lakes in northern Wisconsin, USA, and two lakes in Ontario, Canada. Multiple linear regression models were developed to understand the drivers of lake ice phenology. We used projected air temperature and precipitation from 126 climate change scenarios to forecast the day of year of ice breakup and freeze up in 2050 and 2070. Lake ice melted 5 days earlier and froze 8 days later over the past 35 years. Warmer spring and winter air temperatures contributed to earlier ice breakup; whereas warmer November temperatures delayed lake freeze. Lake ice breakup is projected to be 13 days earlier on average by 2070, but could vary by 3 days later to 43 days earlier depending upon the degree of climatic warming by late century. Similarly, the timing of lake freeze up is projected to be delayed by 11 days on average by 2070, but could be 1 to 28 days later. Shortened seasonality of ice cover by 24 days could increase risk of algal blooms, reduce habitat for coldwater fisheries, and jeopardize survival of northern communities reliant on ice roads.

Keywords: climate change; lake ice phenology; weather; climate oscillations; climate change projections; ice breakup; ice freeze; ice loss

1. Introduction

Temperate regions of the Northern Hemisphere have undergone faster warming trends in the past three to four decades than over the last 1300 years [1]. Lake ice phenology (the timing of ice breakup, freeze up and duration) is highly sensitive to changes in climate [2,3] and therefore, long-term ice phenological records can serve as indicators of climate dynamics over time, both in the past and into the future. Over a 150-year period, ice has melted earlier, frozen later, and ice duration has become shorter in lakes and rivers across the Northern Hemisphere [2,4]. Specifically within the Great Lakes region, Jensen et al. [5] found that on average, lake ice melted 6.3 days earlier (*n* = 64 lakes and 1 river) and froze 9.9 days later (*n* = 33 lakes) from 1975 to 2004. Shorter periods of lake ice cover can lead to earlier stratification and warmer summer surface water temperatures [6,7], earlier spring phytoplankton blooms [8], and alterations in fish feeding behaviour such that in warmer years lake trout eat smaller prey from deeper, offshore regions [9]. Ice phenology is also important to terrestrial mammals; such as the Isle Royale wolves that require lake ice for gene flow into their population [10].

Observed historical trends in lake ice phenology have been associated with changes in local weather and large-scale climate oscillations [11–14]. For example, air temperature, precipitation, wind, cloud cover, and solar radiation have been correlated with ice phenology [4,14–20]. Air temperature has consistently been found to be the most important driver of lake ice phenology [4,15,16,21–25]. For example, Assel and Robertson [22] found that a 1 °C change in air temperatures resulted in ice breakup occurring 8.4 days earlier and ice freeze up occurring 7.1 days later in Grand Traverse Bay, Michigan. Interestingly, air temperature has been found to be a more important driver of ice phenology in lakes south of 61° N, whereas solar radiation is a more influential driver than air temperatures at latitudes north of 61° N [19]. A decrease in snowfall by 50% corresponded to breakup dates that were 4 days earlier in Southern Wisconsin, whereas a 50% increase in snowfall resulted in ice breakup occurring six days later [23]. However, spring rainfall can either accelerate the physical process of ice melting or delay ice breakup by decreasing the amount of solar radiation input to a lake's surface [16,21,23,26].

In addition to relatively long-term changes in climate and weather, large-scale climate oscillations, including the Quasi-biennial Oscillation (QBO), El Nino Southern Oscillation (ENSO), North Atlantic Oscillation (NAO), Pacific Decadal Oscillation (PDO), and the solar sunspot cycle, have been shown to explain variation in lake ice phenology [4,11–13,15,16,18,27–33]. For example, Anderson et al. [27] found significantly earlier breakup dates during the mature warm phase of the ENSO than the average breakup dates in Wisconsin lakes. Further, NAO's influence on winter air temperature [34], snowfall [15], and southerly and westerly wind strength [12] may affect ice breakup dates. In Lake Mendota, Wisconsin, for example, ice duration and breakup were primarily affected by NAO and PDO; NAO influenced lake ice dynamics through snowfall rates and PDO through local air temperatures [15]. In south-central Ontario, Canada, ice breakup dates were affected by solar activity, ENSO, NAO and the Arctic Oscillation [32].

Few studies have explored the impact of future climatic change on lake ice phenology and duration of ice cover in the winter. For example, in Dickie Lake, Ontario, warmer air temperatures, increased snowfall, and reduced wind speed were important drivers of earlier lake ice breakup, whereas warmer air temperatures, reduced wind speed, and increased heat storage corresponded to delayed lake freeze up [17]. Projections on Dickie Lake using regression and physically-based models suggested that lake ice duration may decrease by 50 days, from approximately 130 days in 2010 to 80 days by the year 2100 [17]. There appear to be differences in lake ice response to future climate change, owing to lake type, surface area, depth or volume [35]. For example, a study on three lakes in southern Wisconsin suggested that deep lakes, both small (Fish Lake) and large (Lake Mendota), could experience no lake ice cover in multiple years with increases in daily mean air temperature as little as 4 °C [36]. However, a small, shallow lake would continue to freeze with increases in daily mean air temperatures up to 10 °C, suggesting that ice cover in shallow lakes may be more resilient to climatic change [36].

The overall goal of our study is to expand our understanding of the impacts of future climatic changes on lake ice phenology for north temperate lakes in the Laurentian Great Lakes region of North America. The Laurentian Great Lakes watershed is home to tens of thousands of small north temperate lakes similar to the nine lakes that we studied over the past 35 years. Specifically, we are interested in addressing the following questions: (1) What are the historical trends in the timing of lake ice breakup and freeze up in nine small north temperate lakes in the Laurentian Great Lakes region of Wisconsin, USA and Ontario, Canada between 1981 and 2015? (2) What are the local weather and large-scale climate drivers of lake ice breakup and freeze up over this time period based on multiple regression models? and (3) What is the projected timing of lake ice breakup and freeze up in 2050 and 2070 based on coupling regression models with the suite of downscaled Global Circulation Models

(GCM) projections across a range of greenhouse gas emission (RCP) scenarios? We aim to contribute to the scant literature on the effects of future climatic change on lake ice phenology by further exploring the influence of climatic projections on future predictions of lake ice.

2. Materials and Methods

2.1. Data Acquisition

2.1.1. Ice Breakup and Freeze up Dates

Lake ice breakup and freeze up dates for nine north temperate lakes in Wisconsin, United States and Ontario, Canada, were acquired for the period between 1981/1982 and 2014/2015 (Figure 1). Lake ice data for seven northern Wisconsin lakes (Allequash Lake, Big Muskellunge Lake, Crystal Bog, Crystal Lake, Sparkling Lake, Trout Bog, and Trout Lake) were acquired from the North Temperate Lakes Long Term Ecological Research Program (NTL-LTER; Table 1) [37,38]. The timing of lake ice breakup for the northern Wisconsin lakes was defined as the day a boat could be driven from the dock to the deepest point of the lake without encountering ice. The day the lake froze was defined as the day the deepest point of the lake was ice covered.

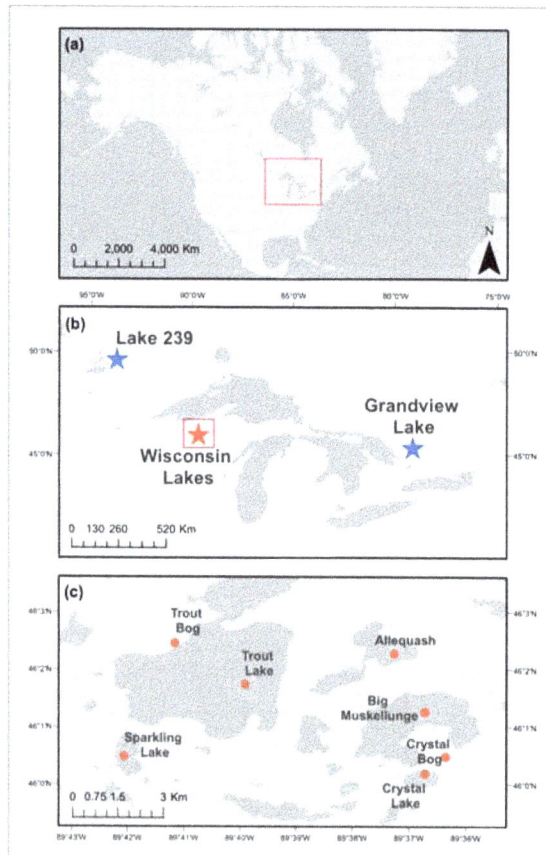

Figure 1. Maps of (**a**) North America (the red box indicates the location of the study regions); (**b**) the study regions in Ontario, Canada (blue stars) and Wisconsin, USA (orange star); and (**c**) a close up of the seven study lakes in northern Wisconsin.

We obtained lake ice phenological data for Grandview Lake in south-central Ontario from the Ontario Ministry of Environment and Climate Change and Lake 239 in north-western Ontario from the IISD Experimental Lakes Area. Lake ice breakup date in Grandview Lake was defined as the date it was less than ~15% ice covered and frozen when it was more than 85% ice covered. Lake 239 was considered thawed when 90% of the lake was ice-free and considered frozen when 90% of the lake was ice covered. Importantly, each site defined ice breakup and freeze up in the same manner every year, although each source of data defined ice breakup and freeze up slightly differently. Trends analyses were conducted on each lake separately and therefore consistency in data measurements between years within a lake is imperative.

Table 1. Morphometric and geographic characteristics of the nine north temperate study lakes.

Region	Lake	Latitude (°)	Longitude (°)	Elevation (m)	Surface Area (km^2)	Mean Depth (m)	Maximum Depth (m)
Wisconsin	Allequash Lake	46.04	−89.62	494	1.64	2.9	8.0
Wisconsin	Big Muskellunge Lake	46.02	−89.61	500	3.63	7.5	21.3
Wisconsin	Crystal Bog	46.01	−89.61	503	0.01	1.7	2.5
Wisconsin	Crystal Lake	46.00	−89.61	502	0.38	10.4	20.4
Wisconsin	Sparkling Lake	46.01	−89.70	495	0.64	10.9	20.0
Wisconsin	Trout Bog	46.04	−89.69	499	0.01	5.6	7.9
Wisconsin	Trout Lake	46.03	−89.67	492	15.65	14.6	35.7
Ontario	Grandview Lake	45.20	−79.05	335	0.74	10.0	28.0
Ontario	Lake 239 (Rawson Lake)	49.66	−93.72	387	0.54	10.5	30.4

2.1.2. Historical Meteorological and Large-Scale Climate Oscillation Data

We obtained monthly weather data for the historical period (1981–2015) in the form of air temperature, precipitation, and cloud cover from the University of East Anglia's Climatic Research Unit. The weather data were derived from meteorological station measurements that were interpolated into 0.5° latitude/longitude gridded datasets [39]. Seasonal averages of fall, winter, and spring were calculated using monthly values. We defined fall as September, October, and November; winter as December plus January and February of the following year; and spring as March, April, and May. As lake ice breakup in the nine lakes ranged from 18 to 28 April on average, we also calculated the average of March and April temperatures and precipitation, to include as predictor variables. Large-scale climate oscillations including monthly and annual index values of the North Atlantic Oscillation (NAO), El Niño Southern Oscillation (ENSO), Arctic Oscillation (AO), and Quasi-biennial Oscillation (QBO), as well as sunspot numbers were obtained from online open source databases (Table 2). In the case of climate drivers with monthly index values, an annual average was calculated.

Table 2. Large-scale climate oscillations and local weather data used to identify drivers of lake ice phenology.

Climate Variable	Source	Length of Record	Scale
Total Sunspot Number (SS)	Sunspot Index and Long-term Solar Observations (SILSO) http://www.sidc.be/silso/	1700–2015	Annual
North Atlantic Oscillation Index (NAO)	National Center for Atmospheric Research (NCAR) https://climatedataguide.ucar.edu/climate-data/hurrell-north-atlantic-oscillation-nao-index-station-based	1865–2015	Annual
El Nino Southern Oscillation (ENSO)-(SOI)	National Climate Center, Australia (Bureau of Meteorology) http://www.bom.gov.au/climate/enso/#tabs=SOI	1876–2016	Monthly
Quasi-Biennial Oscillation Index (QBO)	National Oceanic and Atmospheric Administration (NOAA) http://www.esrl.noaa.gov/psd/data/climateindices/list/	1948–2016	Monthly
Arctic Oscillation (AO)	National Oceanic and Atmospheric Administration (NOAA) http://www.esrl.noaa.gov/psd/data/climateindices/list/	1950–2016	Monthly
Local Air Temperature and Precipitation	University of East Anglia's Climatic Research Unit (CRU) https://crudata.uea.ac.uk/cru/data/hrg/	1901–2015	Monthly

2.1.3. Projected Climate Data

We acquired projected climate data for mid-century (2050; average of 2041–2060) and late-century (2070; average of 2061–2080) from the Intergovernmental Panel on Climate Change 2013 fifth assessment report [40]. We extracted projected monthly air temperature and precipitation from all 19 general circulation models (GCMs) for both 2050 and 2070 (Supplementary Table S1). Each GCM consisted of one to a maximum of four representative concentration pathways (RCP) of greenhouse gas emissions including RCP 2.6, 4.5, 6.0 and 8.5. RCP 2.6 represents the most conservative estimate of forecasted greenhouse gas concentrations, in which an aggressive mitigation strategy is implemented and temperatures are kept below 2 °C above pre-industrial temperatures [40]. In contrast, RCP 8.5 represents the "business-as-usual" scenario and forecasts the highest emissions of greenhouse gases. RCP 4.5 and RCP 6.0 are greenhouse gas emissions scenarios which forecast intermediate increases in greenhouse gas emissions [40]. The north temperate region is projected to become warmer and wetter (Supplementary Table S1).

We used the full suite of 19 GCMs and corresponding 4 RCPs for mid and late century totalling 126 climate change scenarios in our projections of climate change on lake ice phenology. We used all scenarios available to incorporate the uncertainty and variability in forecasted air temperatures and precipitation among the GCMs and RCPs. Differences in projections of future air temperature and precipitation stem from variations in spatial and vertical resolution of GCMs, modelling of several processes such as ocean mixing and terrestrial processes, and climate feedback mechanisms [41]. Incorporating all of the climate change scenarios has been suggested to account for this variability and uncertainties among GCMs [40].

2.2. Data Analyses

2.2.1. Trends in Lake Ice Phenology

We used Sen's slopes to calculate trends in lake ice breakup and freeze up between 1981 and 2015 using the "openair" package in R [42]. Sen's slopes are a nonparametric method of statistically testing trends. The Sen's slope is the median of the slopes calculated between each pair of points [43,44]. This analysis has previously been used to discern temporal trends in ice phenology [4,45].

2.2.2. Drivers of the Timing of Lake Ice Breakup and Freeze up

We used multiple linear regression models on the time series of lake ice phenology, local weather, and large-scale climate oscillations, to identify significant local weather and large-scale climate oscillations explaining the timing of lake ice breakup and freeze up. We ran a forward selection procedure with dual criterion, such that each predictor variable was potentially included in the model if it was significant at $\alpha = 0.05$ and explained significant amounts of variation (R^2_{adj}) using the "packfor" package in R [46]. We assessed multicollinearity among predictor variables using Spearman correlations. Correlations between predictor variables that had a rho value greater than 0.70 and with a *p*-value less than 0.05 were considered multicollinear and removed from the models. For lake ice breakup, we developed a linear regression model for all lakes in our dataset using year as a covariate in the model. For lake ice freeze up, we developed individual linear regression models for each lake. The freeze up process is more heavily influenced by individual lake characteristics such as mean depth, than climate drivers [36,47,48]. Therefore, we found that developing individual models for lake ice freeze up explained substantially more variation than a generalized model. In addition, we ran linear regressions to examine the relationships between ice breakup and freeze up (trends and average day of breakup/freeze up) and lake morphometric characteristics including volume, surface area, and mean depth. Models were selected using the Akaike Information Criterion (AIC), such that the most parsimonious model yielded the lowest AIC value [49].

2.2.3. Projections in Lake Ice Phenology

We forecasted the timing of lake ice breakup and freeze up date for 2050 and 2070 under all 126 climate change scenarios for 9 north temperate lakes (Supplementary Table S1). The aforementioned linear models were extrapolated using projected air temperatures and precipitation to forecast the day of year (DOY) the ice would breakup or freeze in 2050 (2041–2060) and 2070 (2061–2080). The change in the timing of lake ice breakup and freeze from forecasted to historical was calculated by subtracting the forecasted average DOY of 126 climate change scenarios from the historical average DOY (1981–2015).

3. Results

3.1. Trends in Lake Ice Phenology

Lake ice breakup was 5 days earlier between 1981/2 and 2014/5. The average rate was 1.5 days per decade in northern Wisconsin lakes. There were no trends in ice breakup in the Ontario lakes (Figure 2; Supplementary Figures S1–S9). All trends for lake ice breakup in both regions were nonsignificant ($p > 0.05$), perhaps because of high inter-annual variation and shorter nature of the time series. Lake ice freeze up was 7.8 days later between 1981/2 and 2014/5. The average change was 2.2 days per decade in all lakes. Only the two Ontario Lakes, Grandview Lake and Lake 239, had significant trends in lake ice freeze. Notably, Grandview Lake froze 12 days later and experienced the greatest rate of change in the timing of freeze during the study period (Figure 2; Supplementary Figures S1–S9).

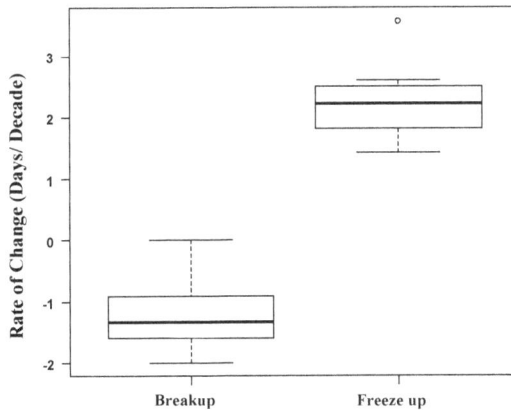

Figure 2. Rate of change of lake ice breakup and freeze up (day of year) in nine north temperate lakes between 1981/2 and 2014/5.

3.2. Drivers of the Timing of Lake Ice Breakup and Freeze up

The most important predictor variables of the timing of lake ice breakup in all study lakes between 1981/2 and 2014/5 were the combined mean of March and April air temperature, winter air temperature, and winter precipitation. March and April were the months including and preceding the timing of lake ice breakup. We found that with increases in spring and winter air temperatures, lake ice broke earlier in the year. Increases in winter precipitation led to later ice breakup date. No large-scale climate oscillation was significant. The model explained 91% variation and was significant at $p < 0.05$ (Table 3).

Table 3. Multiple linear regression model results for the timing of lake ice breakup and freeze up. The most parsimonious models with their respective R^2_{adj}, AIC, and *p*-values are displayed.

Response Variable	Region	Lake	Model Equation [1]	R^2_{adj}	AIC	*p*-Value
Break-up Day of Year	All	All lakes	$DOY_b = 99.28 - 2.79$ (MarAprTemp) $- 1.13$ (WinTemp) $+ 0.06$ (WinPrecip)	0.91	1643.22	<0.001
Freeze-up Day of Year	Wisconsin	Allequash Lake	$DOY_f = 344.90 + 2.85$ (NovTemp)	0.60	226.85	<0.001
Freeze-up Day of Year	Wisconsin	Big Muskellunge Lake	$DOY_f = 344.11 + 3.42$ (NovTemp)	0.70	223.60	<0.001
Freeze-up Day of Year	Wisconsin	Crystal Bog	$DOY_f = 327.14 + 2.75$ (NovTemp)	0.63	220.52	<0.001
Freeze-up Day of Year	Wisconsin	Crystal Lake	$DOY_f = 343.63 + 3.06$ (NovTemp)	0.69	218.02	<0.001
Freeze-up Day of Year	Wisconsin	Sparkling Lake	$DOY_f = 345.66 + 2.88$ (NovTemp)	0.58	230.42	<0.001
Freeze-up Day of Year	Wisconsin	Trout Bog	$DOY_f = 328.31 + 2.65$ (NovTemp)	0.66	212.26	<0.001
Freeze-up Day of Year	Wisconsin	Trout Lake	$DOY_f = 352.61 + 3.24$ (NovTemp)	0.61	233.86	<0.001
Freeze-up Day of Year	Ontario	Grandview Lake	$DOY_f = 338.57 + 3.22$ (NovTemp)	0.39	242.32	<0.001
Freeze-up Day of Year	Ontario	Lake 239	$DOY_f = 308.67 + 3.93$ (FallTemp)	0.63	209.38	<0.001

Notes: [1] Model variables include DOY_b = breakup day of year, MarAprTemp = mean air temperature during the March–April period, WinTemp = mean air temperature from December to February, WinPrecip = mean precipitation from December to February, DOY_f = freeze day of year, NovTemp = mean November air temperature, and FallTemp = mean air temperature from September to November.

Mean November air temperature (i.e., the month including and preceding lake freeze up) was the most important predictor variable explaining the timing of lake ice freeze up for eight of the nine lakes in our study. The only exception was Lake 239, which was influenced by fall air temperature instead of November air temperature. No large-scale climate oscillations were significant for any lake. The mean variation explained for all models was 61% with a range of 39–70% variation explained (Table 3).

We found a significant linear relationship between lake ice freeze up date and mean depth ($p < 0.05$), such that deeper lakes froze later. However, there were no other significant relationships between lake ice phenology and lake morphology within our study sites (Supplementary Table S2).

3.3. Forecasted Lake Ice Loss

Mean ice duration is forecasted to decrease by 20 days in northern Wisconsin lakes, 15 days in Grandview Lake in south-central Ontario, and 19 days in Lake 239 in northwestern Ontario by 2050 (Figure 3a). By 2070, ice duration is projected to decrease even further by a total of 25 days on average in northern Wisconsin lakes, 21 days in Grandview Lake, and 25 days in Lake 239 (Figure 3b). Concurrently, mean annual air temperatures are forecasted to increase between 1.6 and 2.9 °C in mid century, and by 1.5–4.6 °C in late century. Mean annual precipitation is projected to increase by 1 mm to 2 mm by 2050 and from 1.5 mm to 3.5 mm by 2070 (Supplementary Table S1). We forecast that this will result in, on average, 15 to 23 days shorter ice duration by 2050, and 14 to 34 days shorter ice duration by 2070 (Supplementary Table S1).

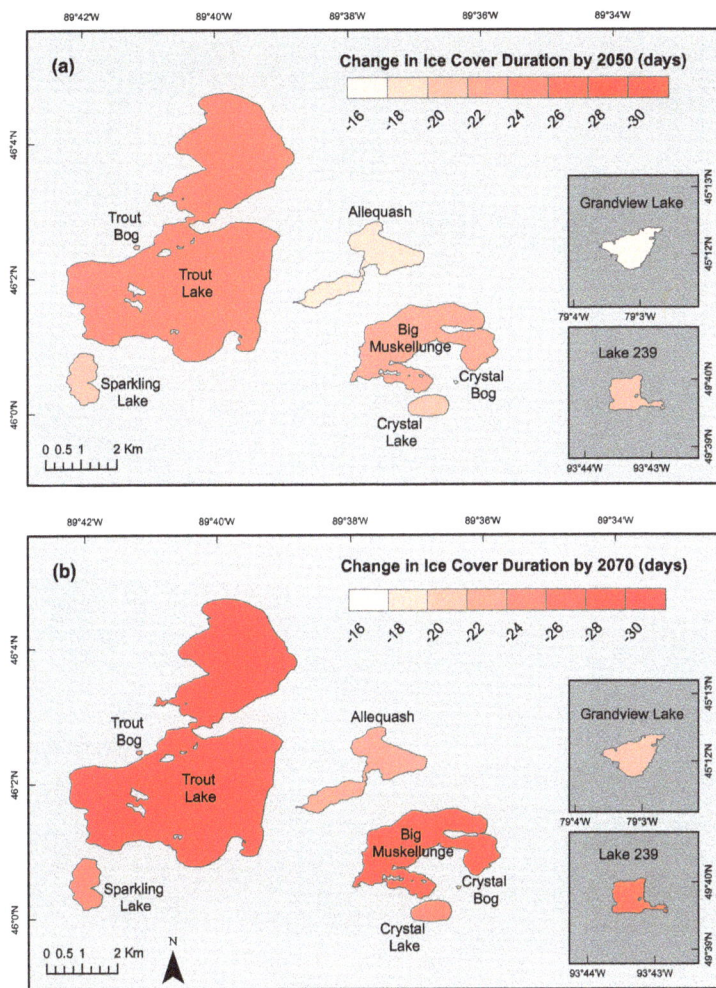

Figure 3. Projected mean loss of ice duration in nine north temperate study lakes by the year (**a**) 2050 and (**b**) 2070. The seven northern Wisconsin lakes are featured in the main map layout; Grandview Lake and Lake 239 in Ontario are featured in the darker insets.

We predict that lake ice breakup will be on average 10 days earlier by 2050 and 13 days by 2070 in these nine north temperate lakes (Supplementary Table S1). In the past 34 years, lake ice breakup occurred between 21 March to 18 May. However, by 2050, lake ice breakup is projected to occur earlier between 20 March and 2 May and between 13 March and 30 April by 2070 (Figure 4a). With a 1 °C increase in forecasted spring air temperature we calculated earlier ice breakup by 2.5 days (Equation (1); $R^2 = 0.93$; $p < 0.05$; Figure 4b).

Change in ice breakup date = $0.97 - 3.45 *$ Forecasted mean March and April air temperature   (1)

For example, an increase in spring air temperatures by 2 °C could translate to ice breakup occurring between 0 and 12 days earlier. An increase in spring air temperatures by 5 °C could correspond to earlier ice breakup by 9 and 24 days (Figure 4b).

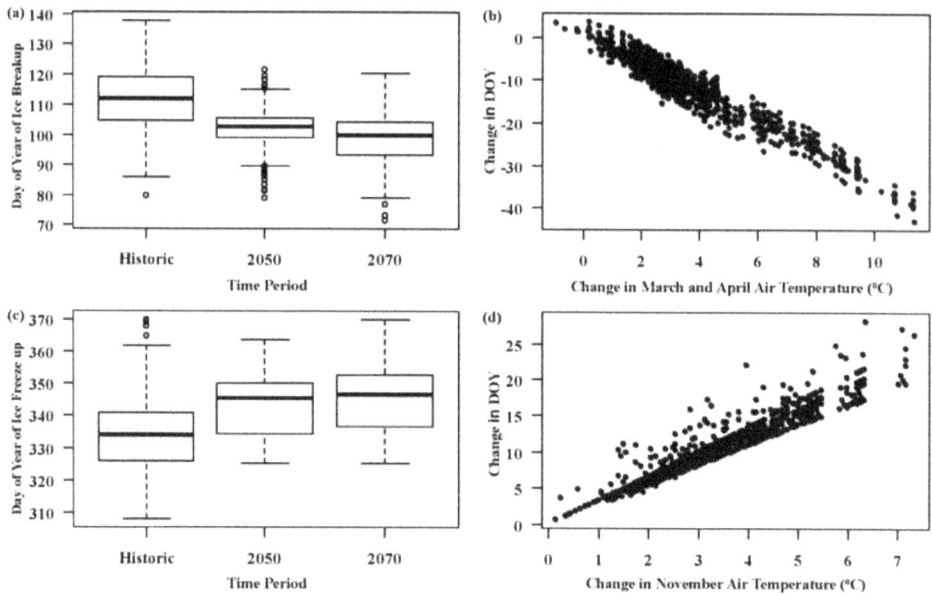

Figure 4. (**a**) The timing of lake ice breakup (day of year) for the historic period (1981/2–2014/5), and forecasted in 2050, and 2070; (**b**) Forecasted change in the day of ice breakup with the corresponding change in mean March–April air temperature under 126 projected climate scenarios; (**c**) The timing of lake ice freeze up (day of year) for the historic period (1981–2015), 2050, and 2070; (**d**) Forecasted change in the day of ice freeze up with the corresponding change in mean November air temperature under 126 projected climate scenarios.

We forecast that lake ice freeze up will be 9 days later by 2050 and 11 days later by 2070 (Supplementary Table S1). Over the past 35 years, lake ice freeze up occurred between 4 November and 5 January. However, by 2050, lake ice freeze up is projected to occur between 21 November and 30 December and between 21 November and 5 January by 2070 (Figure 4c). With a 1 °C increase in forecasted November air temperature, we calculated later ice freeze up by 3.3 days (Equation (2); $R^2 = 0.89$; $p < 0.05$; Figure 4d). An increase in November air temperatures by 2 °C could translate to ice freeze up occurring between 4 and 11 days later. An increase in November air temperatures by 6 °C could correspond to later ice freeze up by 16 to 28 days (Figure 4d).

$$\text{Change in ice freeze up date} = 0.28 + 3.02 * \text{Forecasted mean November air temperature} \qquad (2)$$

The variability in forecasted breakup and freeze up dates arises from the assumptions of varying Global Circulation Models (GCMs) and corresponding greenhouse gas emissions scenarios (RCPs). For example, the business-as-usual greenhouse gas emissions scenario (RCP 8.5) forecasted that by 2070, lake ice breakup could occur 18 days earlier with a range of 4 to 41 days earlier. Lake ice freeze up could be 16 days later (6 to 28 days later), depending upon the GCM (Supplementary Table S1). Intermediate greenhouse gas emissions scenarios (e.g., RCP 4.5) project that lake ice breakup could occur 12.5 days earlier on average, with a range of 0.5 to 33.5 days earlier by 2070 and lake ice freeze up could be delayed by 11 days on average, ranging between 1 and 23 days later (Supplementary Table S1). The best case greenhouse gas emissions scenario, which assumes stabilization of greenhouse gases by mid-century (RCP 2.6), forecasts ice breakup to be 1 week earlier on average with a range of 2 days later to 24 days earlier, and ice freeze up to be on average 1 week later with a range of 2 to 14 days later by 2070 (Supplementary Table S1).

4. Discussion

4.1. Trends in Lake Ice Phenology

In northern Wisconsin, lake ice breakup became earlier at a rate of 1.5 days per decade between 1981/2 and 2014/5. There were no trends in ice breakup in Grandview Lake and Lake 239. Unsurprisingly, none of the trends were significant, at the $p < 0.05$ level. This is likely attributed to the high inter-annual variation and shorter nature of the time series as longer ice records have shown significant trends (e.g., [2,4,44,45]). For example, Hodgkins [50] calculated trends in ice breakup for lakes in New England for varying record lengths from 25 to 150 years. He found nonsignificant trends in the shorter 25-year period, although trends were significant for the same lakes with records extending 50 to 150 years [50]. A second possible explanation for the nonsignificant trends in ice breakup might be an off-set or compensation among several drivers; the role of increased air temperatures may be off-set by the effects of increased snowfall and reduced wind locally [17]. However, for lakes across the Northern Hemisphere, lake ice trends are becoming faster in recent decades [4,16]. Ice melted 0.88 days per decade earlier over a 150-year period spanning 1854 to 2004 for lakes across the Northern Hemisphere. In the most recent 30-year time period (1974–2004), ice melted twice as fast at a rate of 1.86 days per decade earlier [4].

All nine study lakes showed a trend towards later freeze up over the past 35 years. Rates of warming in recent decades are much higher than what has been recorded in the North America historically [5,17]. For example, Jensen et al. [5] found that the lakes froze an average of 3.3 days per decade later, concomitantly with an increase of average fall-spring air temperature of 0.7 °C per decade in 65 waterbodies in the Great Lakes Region recording ice phenology from 1975–2004. The nine lakes we studied in Wisconsin and Ontario have been freezing at a rate approximately 4 times faster than rates of lakes across the Northern Hemisphere over a 150-year period between 1846 and 1995, where the average freeze up date warmed by 0.58 days per decade [2]. Dickie Lake and Lake Utopia, both within the Great Lakes region, have been warming especially fast [17,45]. Freeze up date was delayed in Dickie Lake (close in proximity and similar characteristics to Grandview Lake) by 4.9 days per decade between 1975 and 2009 [17] and 12.3 days per decade later between 1971 and 2000 in Lake Utopia [45].

4.2. Drivers of the Timing of Lake Ice Breakup and Freeze up

The most important predictors for lake ice breakup were weather variables, specifically spring and winter air temperatures, and winter precipitation. Air temperature has been suggested to be the most prominent driver of lake ice breakup timing in lakes and rivers across the Northern Hemisphere [4,15,16,21–23]. For example, in Lake Mendota in Wisconsin, a 1 °C increase in early spring and winter temperatures resulted in ice break-up occurring 6.4 days earlier [51], at a rate much faster than projected for the nine study lakes here under future climatic change. Warming of early spring temperatures may result in the premature arrival of the 0 °C isotherm and thereby earlier ice breakup date [45]. Likewise, warmer winter temperatures can limit ice growth throughout the winter and therefore ice may be more easily melted in the spring [52]. In contrast, increased winter snowfall has been associated with later ice breakup dates monotonically as greater snow cover on lake ice can increase the albedo and generally results in thicker lake ice [23]. However, a nonlinear relationship exists between snowfall decreases and ice decay partly in response to a positive feedback because of decreased albedo and increased solar penetration [23].

Air temperature was also the most important driver of lake ice freeze up in these nine north temperate lakes in the Laurentian Great Lakes watershed over the past 35 years. We found that November or fall air temperature was the only significant predictor of lake ice freeze date, explaining up to 70% of the variation in freeze date across all nine lakes. Air temperature during the fall is consistently one of the most important influences on freeze up date [4,17,53,54], because warmer temperatures prevent the lake from releasing sensible heat and dropping to a temperature where it can freeze [53]. For example, over a 150-year period, fall air temperatures were correlated strongly (r = 0.6) with freeze up date in lakes across the Northern Hemisphere [4].

We did not find any significant relationships between lake ice phenology and large-scale climate oscillations in our lakes between 1981/2 and 2014/5, although many previous studies have suggested the importance of climate oscillations on lake ice phenology and ice cover across the Northern Hemisphere [11–13,33,55]. However, our study is consistent with findings from Dickie Lake, south-central Ontario, for which NAO and ENSO did not explain significant variation in freeze up date [17]. There are several reasons large-scale climate oscillations may not have a direct influence on ice breakup and freeze up in our study lakes. First, several climate indices have been shown to affect temperature and precipitation across the Northern Hemisphere [11,33,56–58] and these relationships may have already been embedded in our models by the inclusion of temperature and precipitation variables. Second, although climate oscillations may play an important role in explaining temporal fluctuations (i.e., ice, local climate, water quality), their contribution to overall trends may be weak within our study period. Third, the influence of large-scale climate oscillations with longer cycle lengths, such as NAO [59], may be underestimated because these cycles would not have occurred repeatedly within our study period [16].

Morphometric characteristics of lakes such as volume, surface area, and depth are known to impact lake ice phenology [53,60]. We found that deeper lakes tend to freeze later, but no other morphometric characteristics were significantly related to lake ice breakup or freeze up trends. However, mean depth is known to be an important physical characteristic of a lake, specifically in relation to lake ice formation [60]. Deeper lakes can store more heat and will take longer to cool to a temperature where it can freeze [61]. In contrast, lake morphometry has been shown to have little effect on lake ice breakup as it is more influenced by climatic and geographic variables such as air temperature and latitude [62].

4.3. Forecasted Lake Ice Loss

The seasonal duration of lake ice cover is projected to decline in north temperate lakes on average by 24 days, but estimates of ice loss range between 0 to 63 days in late century depending upon the degree of climatic warming. Several studies have predicted similar reductions in ice cover days under future climate change. For example, Yao et al. [17,63] predicted a 50-day decline in the ice duration of Dickie and Harp Lakes located in south-central Ontario between 2010 and 2100 under a single climate projection estimated by the Canadian Regional Climate Model (CRCM V4.2) (The Ouranos Consortium, Montreal, QC, Canada). Shuter et al. [53] also expected similar changes for 19 lakes across Canada where ice breakup was estimated to occur 0–20 days earlier and freeze up was projected to be 4–23 days later by the years 2041–2070.

Although the seasonality of ice cover is projected to decline by an average of 24 days under mean climatic projections, there have already been extreme warm years over the past 34 years that may foreshadow ice seasonality in the future. For example, the earliest date lake ice melted within our study region was 21 March in 2012 within the past 34 years. By 2050, the earliest date of ice breakup is projected to be 20 March and 13 March by 2070 under projected changes in mean climatic conditions. Extreme warm events in the future may contribute to even shorter periods of ice cover on lakes in the north temperate region of North America. With breakup dates becoming earlier and freeze up dates becoming later under future climate change some studies have suggested that not only will the ice cover season shorten but there will likely be more ice free years. Magee and Wu [36] simulated future changes in daily air temperatures and lake ice thickness for 3 lakes in Madison, Wisconsin. Over the simulated 100-year period an increase in air temperatures by 4 °C to 10 °C would lead to several no-freeze years for these lakes. Similarly, Robertson et al. [51] predicted that increases in daily air temperatures by 5 °C would result in two no-freeze years in a 30-year period for Lake Mendota in Wisconsin.

4.4. Implications for Losing Lake Ice

Projected loss of lake ice in north temperate lakes by an average of 24 days, ranging from 0–63 days, by 2070 under scenarios of climate change will have far-reaching ecological and socio-economic implications for north temperate lakes. As ice cover duration declines, summer thermal habitat will be greatly altered including a longer thermal stratification period and warmer surface water temperatures [7]. The longer open water season may increase evaporation, resulting in lower lake levels with negative consequences for water quality and littoral habitat availability [4]. Earlier spring lake ice breakup has been shown to shift the timing and abundance of plankton [64,65], promoting a higher risk of toxic algal blooms in nutrient-rich lakes [66]. As many species rely on a combination of photoperiod and thermal cues as triggers for critical life history events (e.g., spawning, larval emergence), changes in ice cover phenology may produce detrimental ecological mismatches [65]. For example, fall spawning fish species may be vulnerable to a warmer incubation period, promoting earlier spring hatching and potential starvation if the spring production pulse is not similarly responsive [67]. During warmer, longer summers, cold-water species will be increasingly squeezed between warming surface waters and deep anoxic habitats [67]. As winter conditions become less severe, aquatic communities will shift from being dominated by winter specialists to species that thrive in warmer, brighter, and more productive environments [4,67].

In addition to its ecological importance, consistent year-to-year lake ice cover has extensive socio-economic implications. More frequent algal blooms and the loss of large-bodied cold-water fishes will negatively impact important ecosystem services such as clean drinking water, fisheries, and summer recreational activities. In addition, lake ice supports multi-billion-dollar recreation and tourism opportunities in north temperate regions including ice fishing, snowmobiling, ice skating, and associated winter festivals [63,68–70]. Northern transportation is predicted to be heavily impacted by climate, as ice roads spanning frozen waterways are relied upon as lifelines to remote northern communities and industrial sites [71]. The decreasing predictability of lake ice already has shown signs of undermining food security, human safety, and economic vitality in northern regions [71,72]. Results from this study suggest an alarming risk to north temperate regions within this century and stress the importance of mitigating greenhouse gas emissions to curb the ecological and socio-economic impacts of climate change in response to reduced seasonality of ice cover.

Supplementary Materials: The following are available online at www.mdpi.com/2073-4441/10/01/70/s1, Tables S1 and S2, Figures S1–S9. Table S1: The change in climatic variables (mean annual temperature and mean annual precipitation), day of ice breakup, and day of ice freeze up under each climate change scenario for 2050 and 2070. Table S2: Slope, explained variation, and significance of linear regressions examining the relationship between lake ice breakup and freeze up and lake morphometric characteristics, including volume (m^3), surface area (km^2), and depth (m). Figures S1–S9: Lake ice (**a**) breakup and (**b**) freeze up trends for each lake during the study period.

Acknowledgments: We thank Ron Ingram for collecting and providing ice phenology data of Grandview Lake. Funding for this research was provided by a Natural Sciences and Engineering Research Council (NSERC) Discovery Grant, Early Researcher Award, and Ministry of Environment and Climate Change award to S.S.; Ontario Graduate Scholarship (OGS) to B.A.H. and L.S.L., Enbridge Graduate Student Award to B.A.H., K.M.G., and A.M., C.D. Fowle Scholarship in Ecology to B.A.H.; NSERC Postgraduate Scholarship Doctoral Award (PGS-D) to A.M., and NSERC Collaborative Research and Training Experience Program to S.S. and L.S.L. We thank the associate editors and two anonymous reviewers who greatly helped improve the manuscript.

Author Contributions: S.S. conceived and designed the study; B.A.H., L.S.L., K.M.G., S.N.H., J.J.M., A.P., J.A.R., and H.Y. collected data; B.A.H., L.S.L., K.M.G., A.M., and S.S. analyzed the data; B.A.H., L.S.L., K.M.G., A.M. and S.S. wrote the paper; and B.A.H., L.S.L., K.M.G., A.M., S.N.H., J.J.M., A.P., J.A.R., H.Y. and S.S. revised the paper.

Conflicts of Interest: The authors declare no conflict of interest. The founding sponsors had no role in the design of the study; in the collection, analyses, or interpretation of data; in the writing of the manuscript, and in the decision to publish the results.

References

1. Jansen, E.; Overpeck, J.; Briffa, K.R.; Duplessy, J.-C.; Joos, F.; Masson-Delmotte, V.; Olago, D.; Otto-Bliesner, B.; Peltier, W.R.; Rahmstorf, S.; et al. *Climate Change 2007: The Physical Science Basis. Contribution of Working Group I to the Fourth Assessment Report of the Intergovernmental Panel on Climate Change*; Solomon, S., Qin, D., Manning, M., Chen, Z., Marquis, M., Averyt, K., Tignor, M., Miller, H., Eds.; Cambridge University Press: Cambridge, UK, 2007; pp. 433–497.

2. Magnuson, J.J.; Robertson, D.M.; Benson, B.J.; Wynne, R.H.; Livingstone, D.M.; Arai, T.; Assel, R.A.; Barry, R.G.; Card, V.; Kuusisto, E.; et al. Historical Trends in Lake and River Ice Cover in the Northern Hemisphere. *Science* **2000**, *289*, 1743–1746. [CrossRef] [PubMed]

3. Adrian, R.; O'Reilly, C.M.; Zagarese, H.; Baines, S.B.; Hessen, D.O.; Keller, W.; Livingstone, D.M.; Sommaruga, R.; Straile, D.; Van Donk, E.; et al. Lakes as sentinels of climate change. *Limnol. Oceanogr.* **2009**, *54*, 2283–2297. [CrossRef] [PubMed]

4. Benson, B.J.; Magnuson, J.J.; Jensen, O.P.; Card, V.M.; Hodgkins, G.; Korhonen, J.; Livingstone, D.M.; Stewart, K.M.; Weyhenmeyer, G.A.; Granin, N.G. Extreme events, trends, and variability in Northern Hemisphere lake-ice phenology (1855–2005). *Clim. Chang.* **2012**, *112*, 299–323. [CrossRef]

5. Jensen, O.P.; Benson, B.J.; Magnuson, J.J.; Card, V.M.; Futter, M.N.; Soranno, P.A.; Stewart, K.M. Spatial analysis of ice phenology trends across the Laurentian Great Lakes region during a recent warming period. *Limnol. Oceanogr.* **2007**, *52*, 2013–2026. [CrossRef]

6. Austin, J.A.; Colman, S.M. Lake Superior summer water temperatures are increasing more rapidly than regional air temperatures: A positive ice-albedo feedback. *Geophys. Res. Lett.* **2007**, *34*, L06604. [CrossRef]

7. O'Reilly, C.M.; Sharma, S.; Gray, D.K.; Hampton, S.E.; Read, J.S.; Rowley, R.J.; Schneider, P.; Lenters, J.D.; McIntyre, P.B.; Kraemer, B.M.; et al. Rapid and highly variable warming of lake surface waters around the globe. *Geophys. Res. Lett.* **2015**, *42*, 10773–10781. [CrossRef]

8. Weyhenmeyer, G.A. Warmer winters: Are planktonic algal populations in Sweden's largest lakes affected? *AMBIO A J. Hum. Environ.* **2001**, *30*, 565–571. [CrossRef]

9. Guzzo, M.M.; Blanchfield, P.J.; Rennie, M.D. Behavioral responses to annual temperature variation alter the dominant energy pathway, growth, and condition of a cold-water predator. *Proc. Natl. Acad. Sci. USA* **2017**, *114*, 9912–9917. [CrossRef] [PubMed]

10. Hedrick, P.W.; Peterson, R.O.; Vucetich, L.M.; Adams, J.R.; Vucetich, J.A. Genetic rescue in Isle Royale wolves: Genetic analysis and the collapse of the population. *Conserv. Genet.* **2014**, *15*, 1111–1121. [CrossRef]

11. Sánchez-López, G.; Hernández, A.; Pla-Rabes, S.; Toro, M.; Granados, I.; Sigró, J.; Trigo, R.M.; Rubio-Inglés, M.J.; Camarero, L.; Valero-Garcés, B.; et al. The effects of the NAO on the ice phenology of Spanish alpine lakes. *Clim. Chang.* **2015**, *130*, 101–113. [CrossRef]

12. Bai, X.; Wang, J.; Sellinger, C.; Clites, A.; Assel, R. Interannual variability of Great Lakes ice cover and its relationship to NAO and ENSO. *J. Geophys. Res. Ocean* **2012**, *117*, C03002. [CrossRef]

13. George, D.G. The Impact of the North Atlantic Oscillation on the development of ice on Lake Windermere. *Clim. Chang.* **2007**, *81*, 455–468. [CrossRef]

14. Gebre, S.; Boissy, T.; Alfredsen, K. Sensitivity of lake ice regimes to climate change in the Nordic region. *Cryosphere Dicuss* **2014**, *8*, 1589–1605. [CrossRef]

15. Ghanbari, R.N.; Bravo, H.R.; Magnuson, J.J.; Hyzer, W.G.; Benson, B.J. Coherence between lake ice cover, local climate and teleconnections (Lake Mendota, Wisconsin). *J. Hydrol.* **2009**, *374*, 282–293. [CrossRef]

16. Sharma, S.; Magnuson, J.J.; Mendoza, G.; Carpenter, S.R. Influences of local weather, large-scale climatic drivers, and the ca. 11 year solar cycle on lake ice breakup dates; 1905–2004. *Clim. Chang.* **2013**, *118*, 857–870. [CrossRef]

17. Yao, H.; Rusak, J.A.; Paterson, A.M.; Somers, K.M.; Mackay, M.; Girard, R.; Ingram, R.; McConnell, C. The interplay of local and regional factors in generating temporal changes in the ice phenology of Dickie Lake, south-central Ontario, Canada. *Inland Waters* **2013**, *3*, 1–14. [CrossRef]

18. Sharma, S.; Magnuson, J.J. Oscillatory dynamics do not mask linear trends in the timing of ice breakup for Northern Hemisphere lakes from 1855 to 2004. *Clim. Chang.* **2014**, *124*, 835–847. [CrossRef]

19. Weyhenmeyer, G.A.; Meili, M.; Livingstone, D.M. Nonlinear temperature response of lake ice breakup. *Geophys. Res. Lett.* **2004**, *31*, L07203. [CrossRef]

20. Soja, A.-M.; Kutics, K.; Maracek, K.; Molnár, G.; Soja, G. Changes in ice phenology characteristics of two Central European steppe lakes from 1926 to 2012—Influences of local weather and large scale oscillation patterns. *Clim. Chang.* **2014**, *126*, 119–133. [CrossRef]

21. Palecki, M.A.; Barry, R.G.; Palecki, M.A.; Barry, R.G. Freeze-up and break-up of lakes as an index of temperature changes during the transition seasons: A case study for Finland. *J. Clim. Appl. Meteorol.* **1986**, *25*, 893–902. [CrossRef]

22. Assel, R.A.; Robertson, D.M. Changes in winter air temperatures near Lake Michigan, 1851–1993, as determined from regional lake-ice records. *Limnol. Oceanogr.* **1995**, *40*, 165–176. [CrossRef]

23. Vavrus, S.J.; Wynne, R.H.; Foley, J.A. Measuring the sensitivity of southern Wisconsin lake ice to climate variations and lake depth using a numerical model. *Limnol. Oceanogr.* **1996**, *41*, 822–831. [CrossRef]

24. Livingstone, D.M. Break-up dates of alpine lakes as proxy data for local and regional mean surface air temperatures. *Clim. Chang.* **1997**, *37*, 407–439. [CrossRef]

25. Williams, G.P. Predicting the date of lake ice breakup. *Water Resour. Res.* **1971**, *7*, 323–333. [CrossRef]

26. Jakkila, J.; Leppäranta, M.; Kawamura, T.; Shirasawa, K.; Salonen, K. Radiation transfer and heat budget during the ice season in Lake Pääjärvi, Finland. *Aquat. Ecol.* **2009**, *43*, 681–692. [CrossRef]

27. Anderson, W.L.; Robertson, D.M.; Magnuson, J.J. Evidence of recent warming and El Nino-related variations in ice breakup of Wisconsin Lakes. *Limnol. Oceanogr.* **1996**, *41*, 815–821. [CrossRef]

28. Livingstone, D.M. Large-scale climatic forcing detected in historical observations of lake ice break-up. *Int. Ver. Theor. Angew. Limnol. Verh.* **2001**, *27*, 2775–2783. [CrossRef]

29. Robertson, D.M.; Wynne, R.H.; Change, W.Y.B. Influence of El Nino on lake and river ice cover in the Northern Hemisphere from 1900 to 1995. *Verh. Int. Ver. Limnol.* **2000**, *27*, 2784–2788. [CrossRef]

30. Bonsal, B.R.; Prowse, T.D.; Duguay, C.R.; Lacroix, M.P. Impacts of large-scale teleconnections on freshwater-ice break/freeze-up dates over Canada. *J. Hydrol.* **2006**, *330*, 340–353. [CrossRef]

31. Mudelsee, M. A proxy record of winter temperatures since 1836 from ice freeze-up/breakup in lake Näsijärvi, Finland. *Clim. Dyn.* **2012**, *38*, 1413–1420. [CrossRef]

32. Fu, C.; Yao, H. Trends of ice breakup date in south-central Ontario. *J. Geophys. Res. Atmos.* **2015**, *120*, 9220–9236. [CrossRef]

33. Yoo, J.; D'Odorico, P. Trends and fluctuations in the dates of ice break-up of lakes and rivers in Northern Europe: The effect of the North Atlantic Oscillation. *J. Hydrol.* **2002**, *268*, 100–112. [CrossRef]

34. Blenckner, T.; Adrian, R.; Livingstone, D.M.; Jennings, E.; Weyhenmeyer, G.A.; George, D.G.; Jankowski, T.; Järvinen, M.; Aonghusa, C.N.; Nõges, T.; et al. Large-scale climatic signatures in lakes across Europe: A meta-analysis. *Glob. Chang. Biol.* **2007**, *13*, 1314–1326. [CrossRef]

35. Crossman, J.; Eimers, M.C.; Kerr, J.G.; Yao, H. Sensitivity of physical lake processes to climate change within a large Precambrian Shield catchment. *Hydrol. Process.* **2016**, *30*, 4353–4366. [CrossRef]

36. Magee, M.R.; Wu, C.H. Effects of changing climate on ice cover in three morphometrically different lakes. *Hydrol. Process.* **2017**, *31*, 308–323. [CrossRef]

37. Magnuson, J.; Carpenter, S.; Stanley, E. North Temperate Lakes LTER: Ice Duration-Trout Lake Area 1981—Current. Available online: http://dx.doi.org/10.6073/pasta/48f12ec9c24e5b8fef692b2bb7aaa5ad (accessed on 1 January 2017).

38. Robertson, D.; Magnuson, J.; Carpenter, S.; Stanley, E. North Temperate Lakes LTER Morphometry and Hypsometry Data for Core Study Lake. Available online: http://dx.doi.org/10.6073/pasta/1d15f38aaf14110714add6230ef78bd8 (accessed on 1 January 2017).

39. Harris, I.; Jones, P.D.; Osborn, T.J.; Lister, D.H. Updated high-resolution grids of monthly climatic observations—The CRU TS3.10 Dataset. *Int. J. Climatol.* **2014**, *34*, 623–642. [CrossRef]

40. Intergovernmental Panel on Climate Change (IPCC). Climate change 2014: Synthesis report. In *Contribution of Working Groups I, II and III to the Fifth Assessment Report of the Intergovernmental Panel on Climate Change*; Core Writing Team, Pachauri, R.K., Meyer, L.A., Eds.; IPCC: Geneva, Switzerland, 2014; p. 151.

41. Beaumont, L.J.; Hughes, L.; Pitman, A.J. Why is the choice of future climate scenarios for species distribution modelling important? *Ecol. Lett.* **2008**, *11*, 1135–1146. [CrossRef] [PubMed]

42. R Development Core Team. *R: A Language and Environment for Statistical Computing. R Foundation for Statistical Computing*; R Core Team: Vienna, Austria, 2017. Available online: http://www.R-project.org (accessed on 1 January 2018).

43. Theil, H. A rank-invariant method of linear and polynomial regression analysis. In *Proceedings of Koninalijke Nederlandse Akademie van Weinenschatpen A*; Springer: Dordrecht, The Netherlands, 1992; pp. 345–381.

44. Sen, P.K. Estimates of the regression coefficient based on Kendall's Tau. *J. Am. Stat. Assoc.* **1968**, *63*, 1379–1389. [CrossRef]

45. Duguay, C.R.; Prowse, T.D.; Bonsal, B.R.; Brown, R.D.; Lacroix, M.P.; Ménard, P. Recent trends in Canadian lake ice cover. *Hydrol. Process.* **2006**, *20*, 781–801. [CrossRef]

46. Blanchet, F.G.; Legendre, P.; Borcard, D. Forward selection of explanatory variables. *Ecology* **2008**, *89*, 2623–2632. [CrossRef] [PubMed]

47. Assel, R.; Herche, L. Coherence of long-term lake ice records. *Verh. Int. Ver. Limnol.* **2000**, *27*, 2789–2792. [CrossRef]

48. Efremova, T.V.; Pal'shin, N.I. Ice phenomena terms on the water bodies of Northwestern Russia. *Russ. Meteorol. Hydrol.* **2011**, *36*, 559–565. [CrossRef]

49. Burnham, K.P.; Anderson, D.R. *Model Selection and Multimodel Inference*; Springer: New York, NY, USA, 2002, ISBN 978-0-387-95364-9.

50. Hodgkins, G.A. The importance of record length in estimating the magnitude of climatic changes: An example using 175 years of lake ice-out dates in New England. *Clim. Chang.* **2013**, *119*, 705–718. [CrossRef]

51. Robertson, D.M.; Ragotzkie, R.A.; Magnuson, J.J. Lake ice records used to detect historical and future climatic changes. *Clim. Chang.* **1992**, *21*, 407–427. [CrossRef]

52. Leppäranta, M. Modelling the formation and decay of lake ice. In *The Impact of Climate Change on European Lakes*; Springer: Dordrecht, The Netherlands, 2010; pp. 63–83.

53. Shuter, B.J.; Minns, C.K.; Fung, S.R. Empirical models for forecasting changes in the phenology of ice cover for Canadian lakes. *Can. J. Fish. Aquat. Sci.* **2013**, *70*, 982–991. [CrossRef]

54. Lei, R.; Leppäranta, M.; Cheng, B.; Heil, P.; Li, Z. Changes in ice-season characteristics of a European Arctic lake from 1964 to 2008. *Clim. Chang.* **2012**, *115*, 725–739. [CrossRef]

55. Livingstone, D.M. Ice break-up on southern Lake Baikal and its relationship to local and regional air temperatures in Siberia and to the North Atlantic Oscillation. *Limnol. Oceanogr.* **1999**, *44*, 1486–1497. [CrossRef]

56. Ropelewski, C.F.; Halpert, M.S. North American Precipitation and Temperature Patterns Associated with the El Niño/Southern Oscillation (ENSO). *Mon. Weather Rev.* **1986**, *114*, 2352–2362. [CrossRef]

57. George, D.G.; Jarvinen, M.; Arvola, L. The influence of the North Atlantic Oscillation on the winter characteristics of Windermere (UK) and Paajarvi (Finland). *Boreal Environ. Res.* **2004**, *9*, 389–400.

58. López-Moreno, J.I.; Vicente-Serrano, S.M.; Morán-Tejeda, E.; Lorenzo-Lacruz, J.; Kenawy, A.; Beniston, M. Effects of the North Atlantic Oscillation (NAO) on combined temperature and precipitation winter modes in the Mediterranean mountains: Observed relationships and projections for the 21st century. *Glob. Planet. Chang.* **2011**, *77*, 62–76. [CrossRef]

59. Trenberth, K.E.; Jones, P.D.; Ambenje, P.; Bojariu, R.; Easterling, D.; Tank, A.K.; Parker, D.; Rahimzadeh, F.; Renwick, J.A.; Rusticucci, M.; et al. Observations: Surface and atmospheric climate change. In *Climate Change 2007: The Physical Science Basis. Contribution of Working Group I to the Fourth Assessment Report of the Intergovernmental Panel on Climate Change*; Solomon, S., Qin, D., Manning, M., Chen, Z., Marquis, M., Averyt, K.B., Tignor, M., Miller, H.L., Eds.; Cambridge University Press: Cambridge, UK; New York, NY, USA, 2007.

60. Korhonen, J. Long-term changes in lake ice cover in Finland. *Nord. Hydrol.* **2006**, *37*, 347–363. [CrossRef]

61. Williams, G.; Layman, K.L.; Stefan, H.G. Dependence of lake ice covers on climatic, geographic and bathymetric variables. *Cold Reg. Sci. Technol.* **2004**, *40*, 145–164. [CrossRef]

62. Williams, S.G.; Stefan, H.G. Modeling of lake ice characteristics in North America Using climate, geography, and lake bathymetry. *J. Cold Reg. Eng.* **2006**, *20*, 140–167. [CrossRef]

63. Yao, H.; Samal, N.R.; Joehnk, K.D.; Fang, X.; Bruce, L.C.; Pierson, D.C.; Rusak, J.A.; James, A. Comparing ice and temperature simulations by four dynamic lake models in Harp Lake: Past performance and future predictions. *Hydrol. Process.* **2014**, *28*, 4587–4601. [CrossRef]

64. Ruhland, K.; Paterson, A.M.; Smol, J.P. Hemispheric-scale patterns of climate-related shifts in planktonic diatoms from North American and European lakes. *Glob. Chang. Biol.* **2008**, *14*, 2740–2754. [CrossRef]

65. Preston, N.D.; Rusak, J.A. Homage to Hutchinson: Does inter-annual climate variability affect zooplankton density and diversity? *Hydrobiologia* **2010**, *653*, 165–177. [CrossRef]
66. Winder, M.; Schindler, D.E. Climatic effects on the phenology of lake processes. *Glob. Chang. Biol.* **2004**, *10*, 1844–1856. [CrossRef]
67. Shuter, B.J.; Finstad, A.G.; Helland, I.P.; Zweimüller, I.; Hölker, F. The role of winter phenology in shaping the ecology of freshwater fish and their sensitivities to climate change. *Aquat. Sci.* **2012**, *74*, 637–657. [CrossRef]
68. Scott, D.; Dawson, J.; Jones, B. Climate change vulnerability of the US Northeast winter recreation-tourism sector. *Mitig. Adapt. Strateg. Glob. Chang.* **2008**, *13*, 577–596. [CrossRef]
69. Damyanov, N.N.; Damon Matthews, H.; Mysak, L.A. Observed decreases in the Canadian outdoor skating season due to recent winter warming. *Environ. Res. Lett.* **2012**, *7*, 14028. [CrossRef]
70. Magnuson, J.J.; Lathrop, R.C. Lake Ice: Winter, Beauty, Value, Changes and a Threatened Future. *LAKELINE*. 2014, pp. 18–27. Available online: http://z0ku333mvy924cayk1kta4r1-wpengine.netdna-ssl.com/wp-content/uploads/LakeLine/34-4/Articles/34-4-7.pdf (accessed on 1 January 2018).
71. Prowse, T.; Alfredsen, K.; Beltaos, S.; Bonsal, B.R.; Bowden, W.B.; Duguay, C.R.; Korhola, A.; McNamara, J.; Vincent, W.F.; Vuglinsky, V.; et al. Effects of changes in Arctic lake and river ice. *Ambio* **2011**, *40*, 63–74. [CrossRef]
72. Vincent, W.F.; Laurion, I.; Pienitz, R.; Walter Anthony, K.M. Climate impacts on Arctic lake ecosystems. In *Climatic Change and Global Warming of Inland Waters: Impacts and Mitigation for Ecosystems and Societies*; Goldman, C.R., Kumagai, M., Robarts, R.D., Eds.; John Wiley & Sons, Ltd.: Chichester, UK, 2012; pp. 27–42, ISBN 9781118470596.

water

MDPI

Article

An Assessment of Ice Effects on Indices for Hydrological Alteration in Flow Regimes

Knut Alfredsen [ID]

Department of Civil Engineering, Norwegian University of Science and Technology, 7491 Trondheim, Norway;
knut.alfredsen@ntnu.no; Tel.: +47-73-594-757

Received: 31 October 2017; Accepted: 21 November 2017; Published: 23 November 2017

Abstract: Preserving hydrological variability is important when developing environmental flow
regimes, and a number of tools have been developed to support this process. A commonly applied
method is the index of hydrological alteration (IHA), which describes a set of indices that can be
used to assess changes in flow regimes. In cold climate regions, river ice can have large effects on
flow regimes through frazil and anchor ice formation, ice cover formation, and ice break-up, and the
impact of this is usually not included in the commonly used indexes. However, to understand the
effect of ice formation and the break-up on the flow regime, the ice effects on the hydrology should
be considered when assessing winter alteration indexes. This paper looks at the effects of river ice
on winter flow conditions using data from Norwegian rivers, and discusses these effects in relation
to hydrological variability. This paper also shows how indexes can be used to classify ice-induced
variability, how this should be used to avoid ice-induced effects in the current analysis, and how
this can be combined with the current indices to improve the winter flow regime classification.
The findings from this paper show that frazil- and anchor-induced raises of the water level have a
large impact on the perceived flow in winter, producing higher flow and deeper water than what
the open water conditions discharge could do. Corresponding to this, winter lows connected to
ice-induced high flows at other locations are also common. Finally, issues related to the assessment of
the temporal and spatial effects of ice formation are discussed.

Keywords: river ice; index of hydrological alteration; river regulation; environmental flow

1. Introduction

The importance of preserving hydrological variability in environmental flow regimes has
been described by several authors, e.g., [1,2], and methods to assess and preserve flow regime
variability in environmental flows are suggested [3]. The described problems with preserving
hydrological variability in environmental flow regimes are valid for all kinds of encroachments
on river regimes. Norway produces more than 96% of its electrical energy from hydropower and
has several large-scale hydropower developments that have wide ranging effects in the affected river
systems. Processes related to the implementation of the European Union Water Framework Directive
and the upcoming relicensing of existing hydropower systems drive the need for better information
on the impact of regulation and the development of better methods for mitigating changes in flow
regimes. Numerous methods of different types and on different scales are developed for assessing
impacts and designing environmental flows, e.g., [4,5]. One method used is the computation of
hydrological indices that describes variability in the hydrological regime [2]. The various indices can
be linked to environmental performance [6], and effects of changes in flow on the environment can be
assessed. Applying these indices before (natural conditions) and then after flow changes provides a
measure of change and a foundation for finding mitigation, and a basis for further and more detailed
analysis of the impacts. Probably the best known method is the Index of Hydrological Alteration (IHA)

described by Richter et al. [7], a total of 33 indexes describing components of the flow regime relevant for ecological assessment. Table 1 shows the category of indexes suggested by Richter. This method has gained some use and has been developed into a software tool for assessment and evaluation of the before and after situation in regulated rivers. An adapted version of this method is used as a component of the environmental flow methodology recently developed in Norway [8], and also in the methodology for hydro-morphological assessment with regard to the implementation of the water framework directive [9].

Table 1. The index of hydrological alteration (IHA) parameters used in the Norwegian assessment method (adapted from [7] with permission from John Wiley and Sons).

IHA-Group	Regime Characteristics	Parameter
Magnitude of monthly flow	Magnitude Timing	Monthly means
Magnitude and duration of annual extremes	Magnitude Duration	Annual 1-,3-,7-, and 30-day minimum and maximum. Seasonal minimums and maximums
Timing of annual extremes	Timing	Julian day of minimum and maximum events
Frequency and duration of high and low pulses	Frequency Magnitude Duration	Number of high pulses, number of low pulses, mean duration of pulses
Rate and frequency of changes	Frequency Rate of change	Number of rises, number of falls, means of changes

For assessment in cold regions, it is important to note that the common description of hydrological indexes of alteration [7,10] does not consider the significant effect formation of river ice on the flow conditions during winter. The formation and break-up of river ice can have a significant effect on the water level and flow regime in rivers due to the storage and release of water when ice forms and breaks up. Particularly, during freeze-up, ice can lead to large variability in water levels along the river under stable discharge conditions. A consequence of this is that the in-stream habitat can differ between summer and winter for the same discharge. For cold climate regions, knowledge of ice should therefore be combined with the open water hydrological variability analysis to ensure that relevant ice effects on the flow regime are taken into account in environmental flow analysis. This is particularly the case for rivers regulated for hydropower, since production releases have a significant effect on the ice regime after regulation [11]. Research and data on discharge and river ice is most common in larger rivers from Canada and USA, but recent research has provided insights on the formation of ice and the effects ice has on the flow and related hydraulic parameters in rivers of the size most commonly found in Norway [12–16]. Stickler, Alfredsen, Linnansaari, and Fjeldstad [14] show how formation of anchor ice dams in the river Sokna in Norway transform the hydraulic conditions, and thereby the available habitat, under relatively constant discharge conditions. Shallow, fast riffle areas were transformed into a succession of deeper pools that eventually formed a continuous ice cover over the reach. During the formation process, the water level in this reach was considerably higher than what it would have been with the same discharge under open water conditions, and as a consequence anchor ice damming at this site would lead to ice-induced low flows further downstream. Similar results were found by Turcotte and Morse [15], who also compared three sections of a different size and documented the similarities and differences in ice formation between scales from river to creek. Turcotte, Morse, Dube, and Anctil [16] quantified the ice formation in small rivers combining an energy balance model with river hydraulics providing a method to extend the assessment to sections where direct observations don't exist. Lind, Alfredsen, Kuglerova, and Nilsson [12] used a statistical modelling approach to link ice formation on 25 river-reaches in Sweden to parameters easily obtainable from meteorological and geographical information, thereby providing tools for assessing ice formation over larger spatial scales. As discussed above, ice can induce high water conditions and even floods during freeze-up.

In addition, ice can also induce flood levels during break-up, particularly in association with ice jamming [17]. For a further discussion on low flows and the corresponding channel storage effects, see [18].

There are both direct and indirect effects on the environment from ice formation which should be incorporated in an environmental assessment. It is therefore necessary to link the relevant hydrological indices to the response of the environment. For salmonid fish, ice formation and break-up can influence habitat availability, behaviour, and apparent survival, while a stable ice cover is often considered beneficial. A reduced or removed surface ice cover due to hydropower regulation in rivers where a stable ice cover is the normal can have a negative impact on Atlantic salmon [19]. Summaries of the general impact of ice on fish can be found in reviews by Brown et al. [20,21], and Heggenes et al. The authors of [22] recently summarized the knowledge on the relations between fish, hydropower development, and ice conditions. Formation of ice may also have effects on the terrestrial environment and the river floodplains. Lind and Nilsson [23] found a higher diversity of riparian plants in rivers with flooding related to anchor ice formation, and also a difference in aquatic plants where anchor ice formed. Lind et al. [24] review the relationship between ice dynamics on riparian and aquatic vegetation and find evidence of both a direct physical impact and the physiological effects of ice on plant communities. Lesack and Marsh [25] show the relationship between flood peaks and water renewal in adjacent lakes on the Mackenzie delta, and how ice jams during the spring flood influence this process. Ice-induced effects on the flow regime can have a negative impact on the river environment, but also beneficial effects, which should be addressed in an environmental assessment. There is, therefore, a need to develop tools to do this, since omitting ice effects would eliminate effects like ice-induced flooding during freeze up, flooding during break up, the duration of the ice-covered period, and ice-induced low flows during freeze up, which are known to appear during the winter period in cold climates. In a study of the Peace–Athabasca delta, Canada, Monk et al. [26] added cold climate relevant parameters to a subset of the original IHA parameters to assess environmental change also related to the ice processes. This work was later developed into the Cold-regions Hydrological Indicators of Change (CHIC) [27]. This is a suite of indexes that characterizes the impact of ice on the flow regime, and, combined with a selection of IHA indexes, it covers the important features of a cold climate flow regime. This development work used the Mackenzie river, Canada as a case study. Table 2 shows the CHIC revised ice indexes.

Table 2. Ice influenced indices in Cold-regions Hydrological Indicators of Change (CHIC), after [27].

Ice Feature	Regime Characteristics	Parameter
Freeze-up Break-up	Timing Magnitude	Date of break-up and freeze-up, magnitude of flow at freeze-up and break-up.
Ice-induced high and low flow	Timing Magnitude	1-day minimum ice-induced flow, 1-day maximum ice-induced flow, date of 1-day max and min, peak water level during ice-influenced period, date of peak water level.
Ice season	Duration	Duration of ice-influenced period.

To compute the ice-related indexes, Peters, Monk, and Baird [27] used data from Environment Canada that carried a flag in the meta data, which showed if the measurement was influenced by ice or not. For Norway, such data is not available from the national discharge database, and data on ice duration is just available for a few rivers and has uncertainties that are difficult to assess [28]. On the other hand, ice-corrected data based on manual correction procedures are available from the national database on a daily scale and provide an alternative to data marked as ice-influenced.

The objective of this paper is to describe the effect of river ice on hydrological flow indices and suggest additional indices to measure ice effects, particularly those that focus on smaller rivers with short duration ice events. This paper also describes how indices can be used to both describe ice effects on flow in unregulated gauges and how to compare ice-induced hydrological variability before and after flow regulation. Further, this paper discusses temporal and spatial issues related to ice indices and how indices can be used to improve our understanding of winter flow regimes. Lack of data is a common issue related to winter and ice, and this paper discusses strategies for evaluating ice using air temperature, together with the flow record, to assess periods where ice effects on the flow can occur. This paper is an extended version of initial work presented at the CRIPE 2017 conference [29].

2. Materials and Methods

2.1. Study Sites and Data

Flow data is collected from the HYDRA II database at the Norwegian Water Resources and Energy Directorate (NVE). Ice-on and ice-off data is not available for the gauges used in this study, but NVE provides ice-adjusted daily data for each of the gauges. Ice adjustments are carried out retroactively based on air temperature and neighboring gauges if possible, in which flow events in cold periods are evaluated and removed if they cannot be explained by changes in natural inflow. For each gauge, both the ice adjusted data set and the raw data set (before adjustment) was downloaded and combined to detect periods where ice have been considered to be present in the data. From this combined data set, the ice season and ice effects on the flow can be derived and the indexes computed. For small rivers, ice processes may best be detected on a sub-daily time step, but only daily data have been corrected for ice. For periods where hourly data are analyzed, the daily corrected data are disaggregated into hourly resolution before the analysis is carried out. The corrections in the data are checked against air temperature, and air temperature is further evaluated as a tool for detecting ice periods in the case for which no flagged or corrected data is available for the site. The flow data used in the development and analysis is collected from the unregulated gauges at Hugdal Bru in Sokna (10.24° E, 62.99° N) and Eggafoss in Gaula (11.18° E, 62.89° N), and for the regulated gauge at Syrstad/Bjørset in Orkla (9.73° E, 63.03° N), all gauges are located in middle Norway. Hugdal bru covers a catchment area of 545 km^2 and has a mean flow of 12.7 m^3·s^{-1}. Eggafoss has a catchment area of 655 km^2 and an annual mean flow of 17.1 m^3·s^{-1}. Syrstad has a catchment area of 2280 km^2 and an annual mean flow of 48.6 m^3·s^{-1}. Figure 1 A shows the location of the study sites. Syrstad is regulated by several reservoirs and power plants upstream and has a redistributed flow regime over the year with higher winter flow and reduced spring and early summer flow. The data from Syrstad/Bjørset was used to evaluate regulation impacts on the flow regime. The winter period used in all analysis is defined from 1st of October to the 31 March. Data for air temperature is collected from the Norwegian Meteorological Institute. For Syrstad/Bjørset, data from 2000 to 2016 is used as the regulated period and 1970–1980 as the unregulated period. For the two other unregulated sites, data from 2000 to 2016 is used.

2.2. Analysis and Extended Indices

Looking at the open-water indexes, e.g., the Index of Hydrological Alteration (IHA), analysis of ice reduced data alone would underestimate winter high flow periods and probably also miss potential short low flow periods. Similarly, using uncorrected data without allowing for ice effects, ice-driven high flows would bias the computation of winter low flow and the duration of winter low flow periods. The issue of winter high flow periods is addressed by Peters, Monk, and Baird [27] as timing and magnitude. For smaller rivers and, particularly, regulated rivers with an unstable ice cover, indices describing the frequency and duration of such episodes are necessary, since an ice-driven variability in the flow regime could influence a number of ecological processes. The frequency of such events would also be an indicator of the stability of the ice cover in the reach.

During freeze-up, periods of low flow can be experienced, as ice stores water upstream both as ice and in temporary ice dams [30]. This effect is also identified by Peters, Monk, and Baird [27], and indexes for magnitude and timing are described. Here, indexes for frequency and duration are added, particularly to handle the unstable ice regime experienced in many regulated rivers. Similarly to the high flow, ice-induced low flows are mainly estimated by comparing the raw and corrected data, and, in cases where there are no available air temperature changes in the gradient of the recession, hydrograph is used to detect drops in the flow caused by ice. For small streams, the rate of change of the water level related to such low flows can be significant, and the maximum rate of change is computed following similar procedures, as for hydropeaking analysis [31].

Freeze-up and break-up indices are computed similarly to Peters, Monk, and Baird [27] (see Table 2). For many Norwegian rivers, this is a difficult parameter to compute, since data for ice formation and breakup is not available, and this parameter is not easily derived from the corrected data or from temperature either. An alternative method is to use the zero isotherm using the method described in Gebre and Alfredsen [32]. There are significant uncertainties in this approach that must be considered when data is used.

In rivers regulated for hydropower, particularly high-head systems [22], the stability of the ice cover could be a very important parameter for assessing ecological impacts [33]. The assessment of ice cover stability is an interaction between winter flow (magnitude and stability), local climate, and water temperature. This would need a more comprehensive assessment procedure, e.g., by modelling, than the other ice related indexes. However, indications of the stability of the ice cover can be derived from the frequency and magnitude of ice-induced high flows during the winter period.

The computation of indices is mainly based on utilizing the raw data series, with ice effects present at an hourly time resolution and the ice-reduced data series at a daily resolution. To be identified as a peak, the duration of the peak must be long enough so that the daily average of the hourly time series exceeds the ice corrected series for that specific day. Thereby, we avoid classifying natural peaks that are removed by averaging the hourly series into a daily series. To further control the detection of a peak and avoid noise from minor fluctuations, a user-defined threshold for when a deviation between the ice-reduced and ice-influenced series is considered a peak is defined, see Figure 1B. This ensures that only peaks above a certain size are counted. For the comparison, in which both series are on a daily resolution, this is not an issue and all; differences between the raw and the corrected data are assumed to be ice-induced.

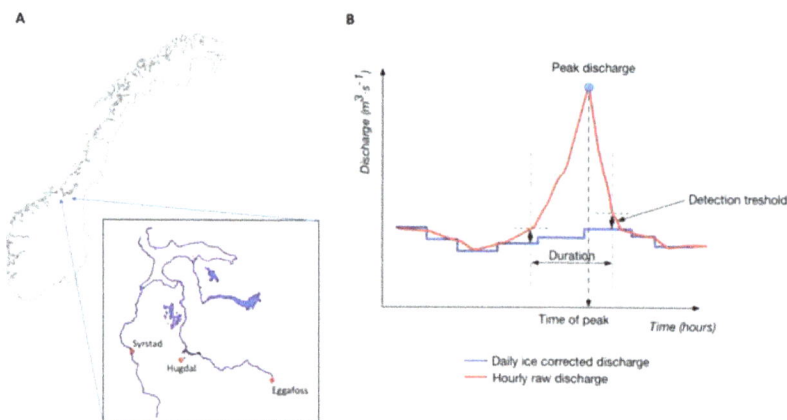

Figure 1. Study sites (panel (**A**)), and the principle of detecting ice-induced peaks and their characteristic features from a combination of a daily ice-reduced series (blue) and a raw data series (red) (panel (**B**)). For each year, the maximum peak level and the time of the maximum peak are derived from all yearly events. The detection threshold marked on the figure eliminates effects of small fluctuations on the detected peaks. This is a schematic figure only; therefore, there are no values on the axes.

The air temperature is used as a control to check that ice-induced peaks appear in periods in which the temperature is at zero or lower; however, such periods could also be used to directly detect ice-induced peaks in the case in which no meta data on the ice season or ice-corrected data is available.

As mentioned, in the case in which hourly raw data is combined with daily resolution-corrected data, it is important to take into consideration that the averaging of hourly data into the daily series that is ice-corrected might mask instantaneous, short-duration natural peaks found in the hourly data set. These deviations may not be captured by the detection threshold described above. To account for this, in situations in which the average of the hourly values in a day and the constant daily value is similar, peaks are not classified as ice-induced.

The analysis of the data is done using the R software (version 3.3.1) [34].

3. Results

Figure 2 shows a hydrograph with ice-reduced data (blue), with a time resolution of one day and raw data (red) with an hourly resolution taken from the regulated Syrstad gauge. The large variability of flow over the winter period is caused by the operation of the hydropower system. Key indices are marked on the figure, blue circles indicate high peaks, green circles indicate low peaks, and open circles indicate peaks in the hourly series that are considered to be instantaneous and subsequently lost in the daily data series because of averaging. These are not included in the following analysis. The duration is computed as the width of each peak within the detection threshold, and counting the high and low peaks provides the frequency of events.

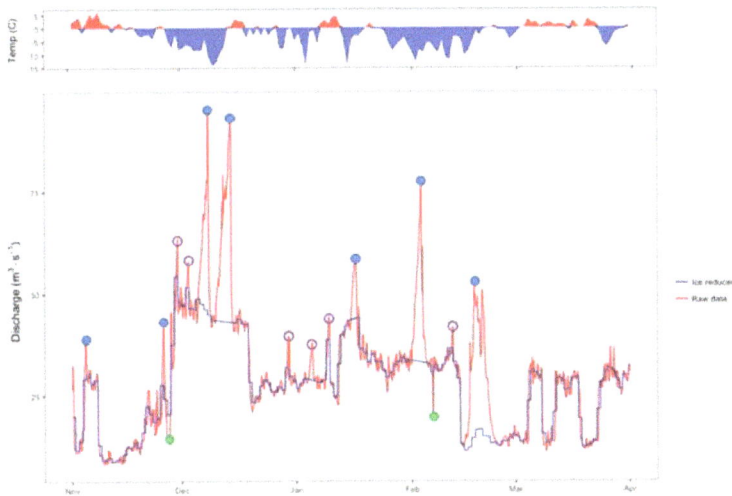

Figure 2. Air temperature (top panel) and identified peaks for one year at Syrstad (lower panel). Open circles are peaks that are considered natural, closed blue circles high peaks due to ice, and closed green circles are low peaks. Note that the variability of flow is driven by hydropower releases.

The formation of the identified ice-induced peaks on Figure 2 coincides with periods of cold weather and shows that these peaks are driven by frazil and anchor ice accumulation at the gauge site. It is also worth noting that the peaks seem to appear independent of the volume flow in the river, which, for this winter, clearly shows the effect of hydropower operation. The indices for the high events registered in Figure 2 (marked by blue circles) are summarized in Table 3. All event data is presented in $m^3 \cdot s^{-1}$; this includes the rise and fall of the hydrograph (in $m^3 \cdot s^{-1} \cdot h^{-1}$), in order to be consistent with the indices hydrological alteration commonly used, and since discharge is usually more readily available for these sites than water level. Data can easily be converted to water level in meters above sea level by utilizing the stage–discharge curve for the gauge. For evaluation purposes, e.g., purposes related to rises and falls, a conversion to water level can be a useful for computing, e.g., loss of wetted area or loss of habitat at the site.

Table 3. Summary of characteristic data for the event shown in Figure 2. A threshold of $10 \ m^3 \cdot s^{-1}$ is used, and only events with a duration of more than 5 h are shown in the table.

Start Event	Peak Date	Peak $(m^3 \cdot s^{-1})$	Duration (h)	Max Rise $(m^3 \cdot s^{-1} \cdot h^{-1})$	Max Drop $(m^3 \cdot s^{-1} \cdot h^{-1})$
25.11 19:00	26.11 01:00	43.2	10	0.97	4.33
06.12 05:00	08.12 05:00	95.2	54	14.5	19.1
11.12 03:00	14.12 03:00	93.0	80	7.7	25.9
16.01 08:00	16.01 19:00	59.1	22	0.95	2.2
01.02 16:00	03.02 12:00	78.1	57	10.8	14.8
16.02 10:00	17.02 17:00	53.7	124	9.8	3.3

Over the winters from 2000/2001 to 2015/2016, the maximum ice-induced peak was $289.7 \ m^3 \cdot s^{-1}$, while the maximum difference between ice corrected discharge and the ice-induced peak was $238.9 \ m^3 \cdot s^{-1}$, with an average of $35.3 \ m^3 \cdot s^{-1}$. The median duration of all events longer than 5 h was 11 h. The average duration was 84 h, a number strongly affected by two cold years in which nearly the entire record was corrected. In such cases, the raw flow data would significantly overestimate the inflow to the river, while on the other hand the corrected flow record would underestimate the actual environmental conditions. It is particularly for events of long duration that an evaluation of the ice effects is important.

Figure 3 shows a comparison between flow records from the Syrstad gauge in panel A and the Hugdal bru gauge in panel B. In the Syrstad case, several peaks are seen forming in cold periods, raising the discharge above the correct discharge in the river, which is controlled by releases from the power plant. For the Hugdal Bru case (panel B), two freeze-up events are detected in November and December, separated by what is, most likely, drainage of the anchor ice dam formed at the gauge site (indicated by increase in temperature and fast drainage). The two other peaks occur during milder periods and could be effects of temporary ice jams at the gauge site, showing another type of event in which ice processes alter the discharge in the river outside of the natural stage–discharge relationship. The ice-controlled data for Hugdal bru shows a natural recession into a low winter flow.

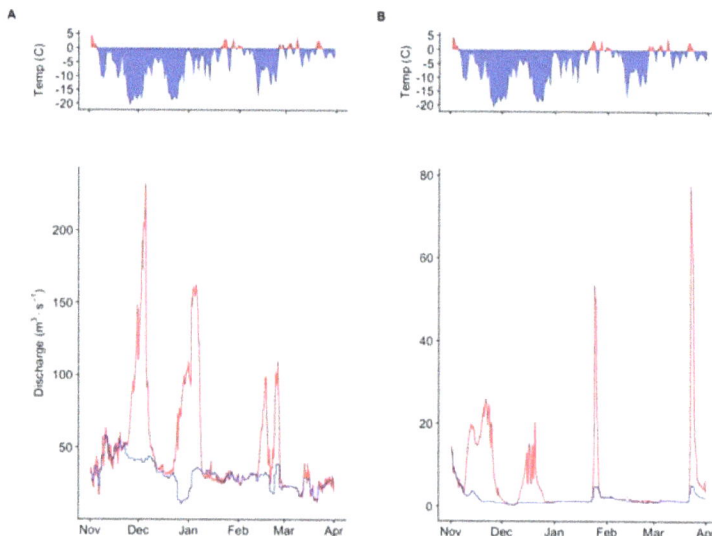

Figure 3. Comparison of flow records from the regulated Syrstad gauge (**A**) and the unregulated Hugdal Bru gauge (**B**) for the winter 2010/2011.

For many practical comparisons, hourly data is not available, particularly in cases in which older discharge records are needed for the evaluation of pre- and post-regulation effects on the flow. It is, therefore, necessary to use ice adjusted daily data for the assessment. Figure 4 shows a comparison for the Syrstad gauge between daily and hourly data.

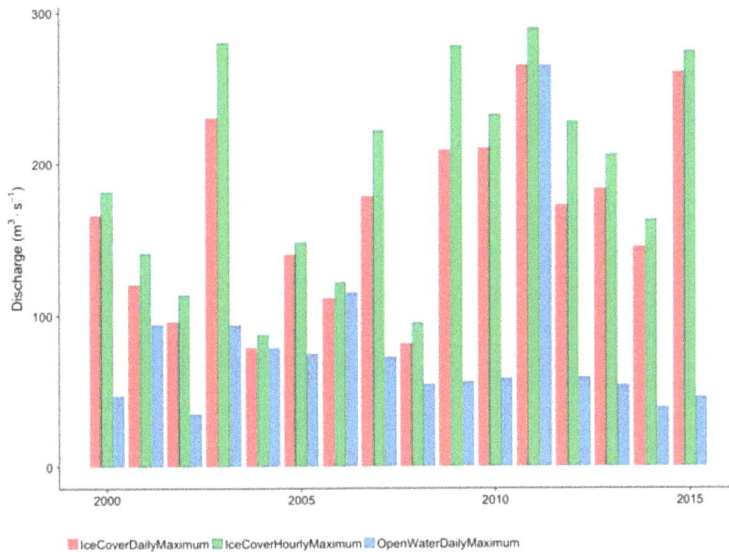

Figure 4. Comparison of winter 1-day maximum, computed for open water conditions (ice-corrected data with a daily time resolution, OpenWaterDailyMaximum), 1-day maximum from hourly raw data (IceCoverHourlyMaximum), and the 1-day maximum for daily raw data (IceCoverDailyMaximum). Note that for 2004, 2006, and 2011, the spring flood occurred in late March; therefore, the daily open water discharge with and without ice is similar and the hourly discharge represents the instantaneous hourly peak.

Based on the data in Figure 4, we see that the ice signal is also present in the daily averaged data, albeit somewhat reduced, but is still marked enough for analysis on a daily time scale to give an indication of the effects of ice on the discharge conditions. In cases in which an early spring flood affects the maximum peak during the winter season, the period of analysis could be shortened to eliminate this from the analysis. Other ice-induced peaks that occur before the start of the spring flood will still be detected. This can be seen by summarizing the number of events larger than 5 h for the winters from 2000/2001 to 2015/2016 (Table 4), noting that winters 2004, 2006, and 2011 all show several peaks.

Table 4. The number of ice-induced events larger than 5 h. The year number indicates the start of the winter (2000—winter 2000/2001).

Year	2000	2001	2002	2003	2004	2005	2006	2007	2008	2009	2010	2011	2012	2013	2014	2015
Duration	22	7	9	8	7	7	7	6	6	3	5	20	6	13	18	19

To compare changes in effects before and after hydropower regulation, indexes were computed for the pre-regulation and post-regulation period on daily data and the results were compared. All measurements are taken from the Syrstad gauge, but it is worth noting that the measurement site pre-regulation was originally a kilometer downstream from the current gauge and was then moved, since the old measurement site was affected by the construction of the Bjørset dam and intake. The series is later spliced by NVE and adjusted to the Syrstad site.

Figure 5 shows the 1-day peak discharge before and after the regulation. There is a significant difference ($p < 0.05$) between peak sizes before and after the regulation. From the figure, we can see that ice had an effect on the flow also before the regulation, and that this has increased after the regulation due to loss of ice cover and a more dynamic ice regime. The effect of ice on the natural flow regime regarding timing of occurrence here is similar to what is seen in panel B of Figure 3.

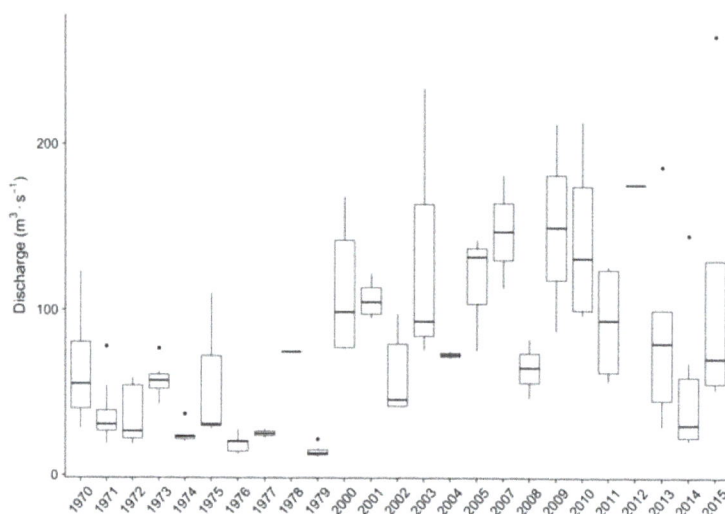

Figure 5. Ice-induced peak discharges before regulation (winters 1970/1971–1979/1980) and after regulation (winters 2000/2001–2015/2016). The bold line is the median, the box shows the interquartile range(IQR), the whiskers 1.5 · IQR, and the circles show outliers.

In Figure 6, a similar comparison is made between the size of daily rises and falls in ice-induced peaks identified in the hydrograph. The difference between drops and rises are also significantly different ($p < 0.05$), with larger drops and rises in discharge after the regulation. Regarding the duration of the peaks, there is no significant difference before and after the regulation. We get more and larger peaks, but the duration of each event is not very different in the regulated and unregulated case. The plot in Figure 7 shows the winter mean index commonly used in the application of IHA in Norway for 10 years before regulation and 16 years after regulation. The plot shows the difference between uncorrected data and ice-corrected data on an important parameter in cold climate impact assessment, and it also illustrates the effect hydropower regulation typically has on the flow regime through increased winter discharge.

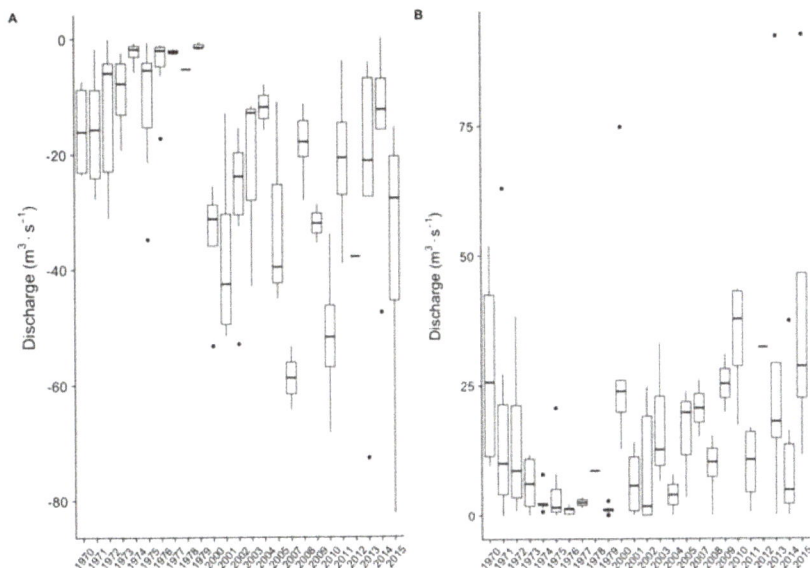

Figure 6. Drops (panel (**A**)) and rises (panel (**B**)) for the pre- and post-regulation periods.

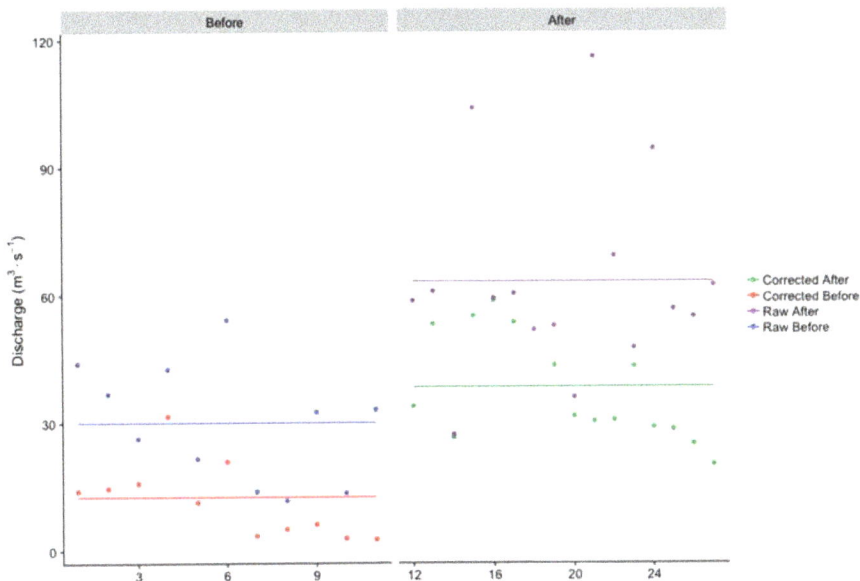

Figure 7. Average winter flow before regulation (year 1–10) for corrected and raw data (red and blue, respectively) and after regulation (year 12–26) for corrected and raw data (green and purple, respectively). The whole lines show the mean of each category of data.

4. Discussion

For many evaluations of flow regimes in rivers during winter, open water data is used even if it is well known that formation and break-up of ice can have a large impact on the flow regime. This paper presents a set of indices that can be used to describe the effect ice has on flow in winter in rivers, information that may be important in environmental analysis during the cold season. The indices are mainly developed for assessing effects of flow regulations in winter, which is known to influence ice formation and ice production [35]. The work builds on the work by Richter, Baumgartner, Powell, and Braun [7] and follows the principles of the Index of Hydrological Alteration. The method presented will add functions to the first cold regions indices, the Cold Region Hydrological Indicators of Change, developed by Peters, Monk, and Baird [27]. The indices presented here adds measures handling duration and frequency of ice-induced peaks and the rise and fall rates of such peaks. These are measures that are particularly relevant for smaller streams with dynamic ice regimes. The evaluation of the differences between recorded raw data and ice-corrected data as shown also highlights the importance of using the right winter data for the analysis and that winter data, without a clear status regarding if they are ice-influenced or not should, should be applied with care.

The derivation of the indexes and the analysis carried out here utilizes the official quality-controlled and ice-corrected data available for many Norwegian gauges operated by Norwegian Water Resources and Energy Directorate (NVE). The ice correction is carried out by NVE to provide discharge series that shows the true inflow from the catchment during winter. The analysis approach is an alternative to the method used by Peters, Monk, and Baird [27], who utilized meta data available in the Canadian discharge database to identify ice-influenced flow periods. Since such information is not available for Norway; the approach used here is to combine the corrected data with the raw data showing the effect of ice formation. The analysis is therefore dependent on the accuracy in the ice corrections, and uncertainties in the corrections will lead to uncertainties in the computed indices. For the study site at Syrstad, the ice corrections should be precise, since both releases from upstream hydropower stations and the production discharge from a downstream station are available. In case such data is missing, a crude approximation can be made from temperature to identify periods in which ice may have an impact on the flow regime. As seen from the results presented here, this will mostly be an approximation, and little quantitative data can be reliably derived from this approach. An alternative approach in the case corrections or meta data on ice is missing—that is, applying a hydrological model to derive the natural inflow to the catchment as a basis for evaluating the magnitude of ice impacts. In this case, the simulated inflow at the gauge could be used as a replacement for the ice-corrected flow data used in this study. Independent of the method used, data for winter length and, particularly, ice cover stability may be difficult to assess without specific ice data for the sites.

The results from the analysis capture the ice dynamics of the flow regime as described in recent literature [13–15], and the evaluation of differences between regulated and natural streams also confirms previous studies on extended ice formation period in rivers regulated from hydropower due to releases of warmer reservoir water into the downstream rivers. The results from this analysis show that, for the rivers studied, data with an hourly time resolution captures more of the dynamics of the ice-flow regime regarding the number of peaks, peak size, and the rate of change than what is seen in the data with a daily resolution. When data are subsequently aggregated into daily time resolution, the signal from short duration ice events are reduced or even removed from the data. On the other hand, since ice-corrected data with an hourly time resolution is not available, one should be careful not to attribute natural instantaneous peaks in the hourly data to ice driven events. For some sites, finer time resolution than one day could be important information related to environmental effects, e.g., Stickler et al. [36] and Alfredsen et al. [37] showed sub-daily responses in Atlantic salmon during several ice formation and release events in the river Orkla. The loss of detail in the daily ice data is, in many ways, similar to the loss of accuracy experienced when analyzing hydro-peaking flow data, e.g., Bevelhimer et al. [38] found that a sub-daily time step is necessary to capture the dynamics of a peaking flow regime. As discussed, it is a drawback that the assessment of hourly peaking relies on daily ice-corrected data. Further, it can also be a problem that hourly data mainly are available for the recent decades, and such data are typically not available for the time periods that are often needed to define the pre-regulation flow conditions. However, as shown in Figures 4 and 7, there is still an ice signal in the daily data, and it is possible to draw information on both timing and the effects ice has on the flow regime.

In a natural river, the winter season often can be divided into a freeze-up period, a stable mid-winter period, and a breakup period, of which each has distinct processes. The typical freeze-up effects are shown in Figure 3B, and the same figure also shows effects of releases of ice on the flow measurements. In a regulated river with releases of production water from reservoirs, the mid-winter, stable ice period is often reduced or completely removed and replaced with a more dynamic ice formation [21], as is seen in Figure 3A. The method described here will catch the temporal distribution of ice effects, and as seen in Table 3, events are timed and identified through the winter period.

A source of uncertainty associated with station-based analysis of hydrologic variability, as presented here, is if the method properly catches the spatial variability of the hydrologic regime in the river. This is even more important for assessment of ice effects than for the open water hydrological analysis, since the spatial variability of ice is larger than the variability of the flow alone, and is also driven by local hydraulic and morphological conditions and the energy balance of the river section. We can assume that observed, ice-induced low flows in a river probably mean that there are extensive ice formations and, likely also, ice-induced high flow at another location in the same river reach. This can be remedied by having a network of discharge gauges along the river, but, at least for Norway, there are few rivers that have the necessary spatial detail in their measurement networks. An option to overcome this is to model ice formation along the river reach and use this to develop a spatially distributed hydrological index. Timalsina, Charmasson, and Alfredsen [11] showed the possibility of simulating ice development along the Orkla river using the MIKE-ICE model, and the findings could have been used to improve the hydrological assessment. The drawbacks of this approach is the significant data needed and the potential problems with setting up the model in smaller and steeper rivers with an even more dynamic ice regime. Turcotte, Morse, Dube, and Anctil [16] presented an energy balance model that can be used to describe ice processes and quantify ice formation in steep rivers, which could be used as a tool to establish the ice regime and thereby better understand the spatial variability of the hydrological indices. This approach is an interesting way to better describe the ice regime of a river, but it also requires detailed data on climate and morphology with a spatial resolution that may not be available for many sites. A third option would be to use river classification or river mapping system on meso-scale, which is applied in many environmental studies, e.g., [39,40]. This is a rapid method of classifying river sections according to simple morphological and hydraulic parameters, and it could be combined with knowledge on ice processes to map probable sites where ice will have an effect on the flow regime, and then this information could be used to assess the spatial distribution of the ice effects measured at the gauge site. Typically, the classification systems' map riffles, runs, and pools, which will have different ice formation processes, and this could then be related to the type of river found at the gauge site.

Figure 8 shows a section of the river Orkla mapped into river types. The Syrstad gauge is indicated on the map in a river type (shallow glide), which can have large frazil ice accumulations in winter, which explains the ice effects on the flow, as seen, e.g., in Figure 2. By comparing river types, the extent of possible sites with similar ice regimes can be found, and, thereby, the spatial distribution of similar ice effects, as observed at the gauge site, can be evaluated. This method can be improved by more field data to confirm the ice formation processes. The linkage between river types and ice formation can be further developed, e.g., by utilizing a statistical model as described by Lind, Alfredsen, Kuglerova, and Nilsson [12] to link simple parameters on river type to ice types and ice processes, and if more detail is needed this could further be developed into a spatial model using a strategy similar to the method described by Lindenschmidt and Chun [41]. This could lead to a method that provides some modelling capability at the same time as it preserves the desired simplicity of the hydrological index method. A similar strategy could be to explore the possibility of linking the hydrological ice indexes to the conceptual ice model developed by Turcotte and Morse [42]. This method links ice cover formation with ice formation processes using simple parameters like channel type, channel size, and winter intensity that can easily be found for any river. This model could also be combined with data from the river classification shown in Figure 8 to give an overview of both the spatial and temporal aspect of ice formation. An important factor to consider when these models are applied to rivers like Orkla at Syrstad is the effect of the hydropower regulation, and a system like the one in Orkla with several reservoir and power plants will have a large effect on the flow regime and the spatial distribution of ice. This must be considered when existing tools developed for natural conditions are used in regulated rivers.

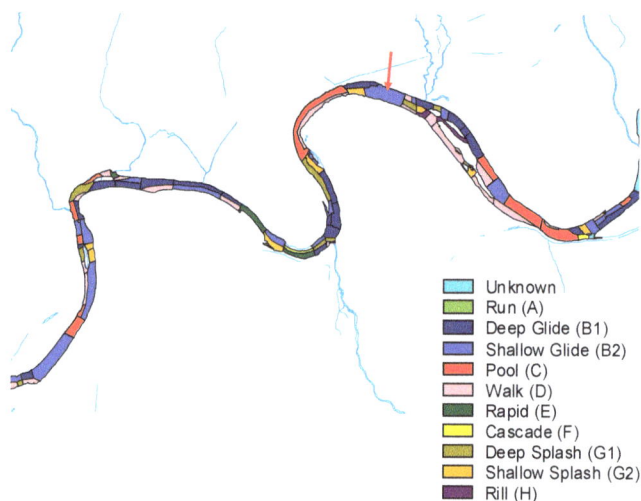

Figure 8. A section of the Orkla river mapped by the classification system described by [39]. The arrow indicates the position of the Syrstad gauge located in a shallow glide, which shows frazil ice accumulation over winter.

The indexes presented here will be directly useful as a complement to the existing tools when measuring the before and after situation in rivers with a regulated discharge. In addition, in the recent measures for hydro-morphological effects made for the water framework directive in Norway [9], ice effects are categorized on a scale from natural to severely modified based on percentage changes in ice cover and/or anchor ice formation. This is data that can be difficult to find directly from measurements, so the derivation of data needed for this assessment must in many cases be based on proxy information. The indices of ice-induced flow effects could be used to evaluate the categories in this classification. The frequency, duration, and temporal distribution of ice-induced peaks could be related to reduced stability of the ice regime, and they are direct effects of the formation of frazil and anchor ice in the river section.

The use of hydrological indexes presented here is mainly focused on river regulations and a situation in which we compare the flow situation before and after the regulation to determine changes that can be critical to the environment and as a basis for finding and evaluating mitigation measures. From this application, it follows that the method also can be used to identify ice effects on a hydrograph and as a measure of the impact of ice on discharge and water levels also in natural rivers, e.g., as a tool for assessing effects of vegetation as shown by Lind and Nilsson [23]. The method is also applicable to other types of studies in which measures of changes in the hydrology are useful. Bin Ashraf et al. [43] use the index of hydrologic alteration to evaluate climate change effects on flow regimes in rivers in Finland. If ice is a case in climate change studies, the combination of changes in temperature and the current indexes of ice-induced flow changes could be used to assess future ice effects. Linnansaari et al. [44] showed that ice had an effect on how Atlantic salmon utilized an artificial habitat in a regulated river, and unpublished data from the same river shows significant ice-driven modification to the water level and thereby the function of the habitat in the same river. The method presented here could be used to evaluate ice effects in such reaches by combining local measurements with a base flow assessment from an upstream unregulated gauge.

5. Conclusions

In combination with the standard indexes of hydrological alteration that give a measure of the true inflow to the catchment, the indexes presented here measure ice modifications to the flow and thereby to relevant habitat characteristics in the river that are directly influenced by the retention, release, and water level increases caused by ice. The proposed ice indexes presented here should, therefore, together with the indexes presented by Peters, Monk, and Baird [27], provide a useful supplement to the open water indexes when impacts of ice on the environment need to be addressed. The results shown here also illustrate the importance of consistency in data used for winter analysis, and the possible effects encountered by utilizing data without ice correction or winter data without any indication if ice has an effect on the discharge or not.

The spatial heterogeneity of ice formation in particularly small rivers may not be properly covered by the gauge-based method presented here. It is, therefore, necessary to be careful if data is extrapolated from the site to larger sections of river, and preferably this should be done with some underlying knowledge of how ice formation processes vary along the river system.

Acknowledgments: A first version of this article was presented at the CRIPE 2017 conference, and the author wishes to acknowledge the comments and input from the discussion at the conference. The author also thanks Dr. Tor Haakon Bakken for providing preliminary data on the hydro morphological indexes developed for the implementation of the water framework directive in Norway. I also wish to thank two anonymous reviewers for their comments to the manuscript.

Conflicts of Interest: The author declares no conflict of interest.

References

1. Arthington, A.H.; Bunn, S.E.; Poff, N.L.; Naiman, R.J. The challenge of providing environmental flow rules to sustain river ecosystems. *Ecol. Appl.* **2006**, *16*, 1311–1318. [CrossRef]
2. Poff, N.L.; Allan, J.D.; Bain, M.B.; Karr, J.R.; Prestegaard, K.L.; Richter, B.D.; Sparks, R.E.; Stromberg, J.C. The natural flow regime. *BioScience* **1997**, *47*, 769–784. [CrossRef]
3. Poff, N.L.; Richter, B.D.; Arthington, A.H.; Bunn, S.E.; Naiman, R.J.; Kendy, E.; Acreman, M.; Apse, C.; Bledsoe, B.P.; Freeman, M.C.; et al. The ecological limits of hydrological alteration (ELOHA): A framework for developing regional flow standards. *Freshw. Biol.* **2010**, *55*, 147–170. [CrossRef]
4. Acreman, M.C.; Dunbar, M.J. Defining environmental flow requirements—A review. *Hydrol. Earth Syst. Sci.* **2004**, *8*, 861–876. [CrossRef]
5. Tharme, R.E. A global perspective on environmental flow assessment; emerging trends in the development and application of environmental flow methodologies for rivers. *River Res. Appl.* **2003**, *19*, 397–441. [CrossRef]
6. Poff, N.L.; Zimmerman, J.K.H. Ecological responses to altered flow regimes: A literature review to inform the science and management of environmental flows. *Freshw. Biol.* **2010**, *55*, 194–205. [CrossRef]
7. Richter, B.D.; Baumgartner, J.V.; Powell, J.; Braun, D.P. A method for assessing hydrologic alteration within ecosystems. *Conserv. Biol.* **1996**, *10*, 1163–1174. [CrossRef]
8. Forseth, T.; Harby, A. *Handbook for Environmental Design in Regulated Salmon Rivers*; NINA: Trondheim, Norway, 2014; Volume 53, p. 90.
9. Bakken, T.H.; Forseth, T.; Harby, A. *Assessment of Hydro-Morphological State and Reference Conditions in Regulated Rivers*; SINTEF Energy: Trondheim, Norway, 2017.
10. Olden, J.D.; Poff, N.L. Redundancy and the choice of hydrologic indices forcharacterizing streamflow regimes. *River Res. Appl.* **2003**, *19*, 101–121. [CrossRef]

11. Timalsina, N.P.; Charmasson, J.; Alfredsen, K.T. Simulation of the ice regime in a norwegian regulated river. *Cold Reg. Sci. Technol.* **2013**, *94*, 61–73. [CrossRef]
12. Lind, L.; Alfredsen, K.; Kuglerova, L.; Nilsson, C. Hydrological and thermal controls of ice formation in 25 boreal stream reaches. *J. Hydrol.* **2016**, *540*, 797–811. [CrossRef]
13. Nafziger, J.; She, Y.; Hicks, F.; Cunjak, R.A. Anchor ice formation and release in small regulated and unregulated streams. *Cold Reg. Sci. Technol.* **2017**, *141*, 66–77. [CrossRef]
14. Stickler, M.; Alfredsen, K.T.; Linnansaari, T.; Fjeldstad, H.-P. The influence of dynamic ice formation on hydraulic heterogeneity in steep streams. *River Res. Appl.* **2010**, *26*, 1187–1197. [CrossRef]
15. Turcotte, B.; Morse, B. Ice processes in a steep river basin. *Cold Reg. Sci. Technol.* **2011**, *67*, 146–156. [CrossRef]
16. Turcotte, B.; Morse, B.; Dube, M.; Anctil, F. Quantifying steep channel freezeup processes. *Cold Reg. Sci. Technol.* **2013**, *94*, 21–36. [CrossRef]
17. Turcotte, B.; Alfredsen, K.; Beltaos, S.; Burrell, B.C. Ice-related floods and flood delineation along streams and small rivers. In Proceedings of the 19th Workshop on the Hydraulics of Ice Covered Rivers, Whitehorse, YT, Canada, 10–12 July 2017; CGU HS Committee on River Ice Processes and the Environment: Whitehorse, YT, Canada, 2017; p. 26.
18. Gridley, N.C.; Prowse, T.D. Physical effects of river ice. In *Environmental Aspects of River Ice*; Prowse, T.D., Gridley, N.C., Eds.; National Hydrology Research Institute: Saskatoon, SK, Canada, 1993; Volume 1, pp. 3–75.
19. Hedger, R.D.; Næsje, T.F.; Fiske, P.; Ugedal, O.; Finstad, A.; Thorstad, E.B. Ice-dependent winter survival of juvenile atlantic salmon. *Ecol. Evol.* **2013**, *3*, 523–535. [CrossRef] [PubMed]
20. Brown, R.S.; Hubert, W.A.; Daly, S.F. A primer on winter, ice and fish: What fisheries biologists should know about winter ice processes and stream-dwelling fish. *Fisheries* **2011**, *36*, 8–26. [CrossRef]
21. Huusko, A.R.I.; Greenberg, L.; Stickler, M.; Linnansaari, T.; Nykänen, M.; Vehanen, T.; Koljonen, S.; Louhi, P.; Alfredsen, K. Life in the ice lane: The winter ecology of stream salmonids. *River Res. Appl.* **2007**, *23*, 469–491. [CrossRef]
22. Heggenes, J.; Alfredsen, K.; Bustos, A.A.; Huusko, A.; Stickler, M. Be cool: A review of hydro-physical changes and fish responsesin winter in hydropower-regulated northern streams. *Environ. Biol. Fishes* **2017**, *21*, 1–21. [CrossRef]
23. Lind, L.; Nilsson, C. Vegetation patterns in small boreal streams relate to ice and winter floods. *J. Ecol.* **2015**, *103*, 431–440. [CrossRef]
24. Lind, L.; Nilsson, C.; Polvi, L.E.; Weber, C. The role of ice dynamics in shaping vegetation in flowing waters. *Biol. Rev.* **2014**, *89*, 791–804. [CrossRef] [PubMed]
25. Lesack, L.F.W.; Marsh, P. River-to-lake connectivities, water renewal, and aquatic habitat diversity in the mackenzie river delta. *Water Resour. Res.* **2010**, *46*, 16. [CrossRef]
26. Monk, W.A.; Peters, D.L.; Baird, D.J. Assessment of ecologically relevant hydrological variablesinfluencing a cold-region river and its delta: The athabascariver and the peace—Athabasca delta, northwestern Canada. *Hydrol. Process.* **2012**, *26*, 1827–1839. [CrossRef]
27. Peters, D.L.; Monk, W.A.; Baird, D.J. Cold-regions hydrological indicators of change (CHIC) for ecological flow needs assessment. *Hydrol. Sci. J.* **2014**, *59*, 502–516. [CrossRef]
28. Gebre, S.B.; Alfredsen, K.T. Investigation of river ice regimes in some norwegian water courses. In Proceedings of the 16th Workshop on the Hydraulics of Ice Covered Rivers, Winnipeg, MB, Canada, 18–22 September 2011; CGU HS Committee on River Ice Processes and the Environment: Winnipeg, MB, Canada, 2011; pp. 1–20.
29. Alfredsen, K. Incorporating ice effects in environmental flow indices. In Proceedings of the 19th Workshop on the Hydraulics of Ice Covered Rivers, Whitehorse, YT, Canada, 10–12 July 2017; CGU HS Committee on River Ice Processes and the Environment: Whitehorse, YT, Canada, 2017.
30. Prowse, T. *River Ice Ecology*; National Water Research Institute, Environment Canada: Burlington, ON, Canada, 2000.
31. Casas-Mulet, R.; Alfredsen, K.; Hamududu, B.; Timalsina, N.P. The effects of hydropeaking on hyporheic interactions based on field experiments. *Hydrol. Process.* **2014**. [CrossRef]
32. Gebre, S.; Alfredsen, K. Contemporary trends and future changes in freshwater ice conditions: Inference from temperature indices. *Hydrol. Res.* **2014**, *45*, 455–478. [CrossRef]
33. Finstad, A.G.; Forseth, T.; Faenstad, T.F.; Ugedal, O. The importance of ice cover for energy turnover in juvenile atlantic salmon. *J. Anim. Ecol.* **2004**, *73*, 959–966. [CrossRef]

34. R-Development Core Team. *R: A Language and Environment for Statistical Computing*; R Foundation for Statistical Computing: Vienna, Austria, 2011.

35. Gebre, S.; Alfredsen, K.; Lia, L.; Stickler, M.; Tesaker, E. Ice effects on hydropower systems—A review. *J. Cold Reg. Eng.* **2013**, *27*, 196–222. [CrossRef]

36. Stickler, M.; Alfredsen, K.; Scruton, D.A.; Pennell, C.; Harby, A.; Økland, F. Mid-winter activity and movement of atlantic salmon parr during ice formation events in a norwegian regulated river. *Hydrobiologia* **2007**, *582*, 81–89. [CrossRef]

37. Alfredsen, K.; Stickler, M.; Scruton, D.A.; Pennel, C.J.; Kelley, C.; Halleraker, J.H.; Harby, A.; Fjeldstad, H.-P.; Heggenes, J.; Økland, F. A telemetry study of winter behaviour of juvenile atlantic salmon (salmo salar) in a river with a dynamic ice regime. In Proceedings of the 5th International Symposium on Ecohydraulics, Madrid, Spain, 12–17 September 2004; de Jalon Lastra, D., Martinez, P., Eds.; International Association of Hydraulic Research: Madrid, Spain, 2004; Volume 1, pp. 133–139.

38. Bevelhimer, M.S.; McManamay, R.A.; O'Connor, B. Characterising sub-daily flow regimes: Implications of hydrologic resolution on ecohydrology. *River Res. Appl.* **2014**. [CrossRef]

39. Borsanyi, P.; Alfredsen, K.; Harby, A.; Ugedal, O.; Kraxner, C. A meso-scale habitat classification method for production modelling of atlantic salmon in norway. *Hydroecol. Appl.* **2004**, *14*, 119–138. [CrossRef]

40. Parasiewicz, P. The mesohabsim model revisited. *River Res. Appl.* **2007**, *23*, 893–903. [CrossRef]

41. Lindenschmidt, K.-E.; Chun, K.P. Geospatial modelling to determine the behaviour of ice cover formation during freeze-up of the dauphin river in manitoba. *Hydrol. Res.* **2014**, *45*, 645–659. [CrossRef]

42. Turcotte, B.; Morse, B. A global river ice classification model. *J. Hydrol.* **2013**, *507*, 134–148. [CrossRef]

43. Ashraf, F.B.; Haghighi, A.T.; Marttila, H.; Kløve, B. Assessing impacts of climate change and river regulation on flow regimes in cold climate: A study of a pristine and a regulated river in the sub-arctic setting of northern europe. *J. Hydrol.* **2016**, *542*, 410–422. [CrossRef]

44. Linnansaari, T.; Alfredsen, K.; Stickler, M.; Arnekleiv, J.V.; Harby, A.; Cunjak, R.A. Does ice matter? Site fidelity and movements by atlantic salmon (*salmo salar* L.) parr during winter in a substrate enhanced river reach. *River Res. Appl.* **2009**, *25*, 773–787. [CrossRef]

Article

RIVICE—A Non-Proprietary, Open-Source, One-Dimensional River-Ice Model

Karl-Erich Lindenschmidt

Global Institute for Water Security, University of Saskatchewan, Saskatoon, SK S7N 3H5, Canada;
karl-erich.lindenschmidt@usask.ca; Tel.: +1-306-966-6174

Academic Editor: Y. Jun Xu
Received: 7 March 2017; Accepted: 26 April 2017; Published: 2 May 2017

Abstract: Currently, no river ice models are available that are free and open source software (FOSS), which can be a hindrance to advancement in the field of modelling river ice processes. This paper introduces a non-proprietary (conditional), open-source option to the scientific and engineering community, the River Ice Model (RIVICE). RIVICE is a one-dimensional, fully-dynamic wave model that mimics key river ice processes such as ice generation, ice transport, ice cover progression (shoving, submergence and juxtapositioning) and ice jam formation, details of which are highlighted in the text. Three ice jam events at Fort McMurray, Alberta, along the Athabasca River, are used as case studies to illustrate the steps of model setup, model calibration and results interpretation. A local sensitivity analysis reveals the varying effects of parameter and boundary conditions on backwater flood levels as a function of the location of ice jam lodgment along the river reach and the location along the ice jam cover. Some limitations of the model and suggestions for future research and model development conclude the paper.

Keywords: Athabasca River; Fort McMurray; ice jam flooding; river ice modelling; RIVICE

1. Introduction

River ice can be a major influence on the fluvial hydraulics and geomorphology of northern, high-latitude rivers. An important feature of river ice is that its presence on rivers raises water levels higher than those occurring during open-water conditions, given the same water discharge. The increased roughness and wetted perimeter length introduced by the ice contributes to the increased staging. Rubble or slush ice also has the potential to jam and dam up a river section, causing even more backwater staging and river banks and levees to overflow and flood the surrounding floodplain. From an ecological perspective, many floodplain and wetland ecosystems require such periodic flooding for their replenishment of moisture, sediment and nutrients, especially in perched ponds and lakes of inland deltas such as, for example in Canada, the Peace-Athabasca Delta in Alberta [1], the Slave River Delta in the Northwest Territories [2,3] and the Saskatchewan River Delta along the Saskatchewan/Manitoba boundary [4,5]. However, ice jams and the flooding they induce can pose threats to, and wreak damage on, many communities located along these rivers. To mention only a few cases in Canada, examples include the Town of Peace River along the Peace River in Alberta [6,7], Fort McMurray on the Athabasca River in Alberta (this study), Winnipeg and Selkirk on the Red River in Manitoba [8,9], St. Raymond on the St. Anne River in Quebec [10] and many communities along the Saint John River in New Brunswick [11]. Floods account for the greatest number of hydrological/meteorological natural disaster events in Canada [12] and, for most Canadian rivers, the annual peak water levels are due to ice jams [13]. Ice jam floods have also been reported on other continents, for instance along the Tornionjoki River in Finland and Sweden [14], the Oder River in Germany and Poland [15] and many rivers in Russia [16]. A review of successful river ice modeling cases is provided in [17]. Other important features of ice jam flooding are its unpredictability and the rapidity of inundation, often making it more dangerous than open-water flood events.

River ice modelling has been a useful tool in understanding key river ice processes that lead to ice jamming and subsequent flooding. The models help us to determine backwater levels, leading to a more comprehensive assessment of the flood hazard and risk posed to communities [6,7,18,19]. They assist in the design of flood protection measures such as crest elevations for river dikes [20], and in the mitigation or exacerbation of the effects of hydro-power dam construction and flow regulation on the severity and frequency of such events [21]. Also, much effort has been dedicated to attempting to forecast ice jams and ice jam flood events [22], with some success, and this remains a field with much research potential. Many of the processes leading to ice jam formation are probabilistic in nature (e.g., ice volume and location of an ice jam); hence, inserting such models in a stochastic framework, such as Monte-Carlo analysis, allows the random nature of these events to be captured and described in terms of frequency distributions. The model must be complex enough to capture all of the main processes leading to ice cover and jam formation, but not be so computationally demanding to discount the model's applicability in a stochastic simulation environment. The River Ice Model (RIVICE) has proven to be very robust, capturing many of the essential river ice processes in a computationally efficient structure to yield high performing results. Examples include good simulation performances for flood water level profiles along the Peace River [7] and dike crest elevation design along the Dauphin River [20]. The model is a non-proprietary, open-source model currently housed at and distributed from the Global Institute for Water Security at the University of Saskatchewan, with permission from Environment Canada. User registration to access the software can be done at http://giws.usask.ca/rivice/. The author is not aware of any other river ice models that are free and open source software (FOSS), which may be a hindrance to the advancement of the field. Hence, an important aim of this paper is to promote a non-commercial option for scientists and engineers who wish to carry out river ice modelling.

In 1988, Environment Canada, along with several consulting firms, government agencies and a consortium of hydro-power companies, initiated the development of the River Ice Model with the aim of providing a non-proprietary numerical model to simulate river ice processes dynamically given a set of hydraulic, meteorological, fluvial geomorphological conditions and the thermal and ice regimes of the river water. These factors steer the simulations to capture key ice processes such as ice generation, ice transport, ice cover formation and progression, hanging dam formation and ice cover melting and ablation. The development of the model was completed by Maurice Sydor (maurice.sydor@videotron.ca), then at Environment Canada, and Rick Carson from KGS Group in Winnipeg. Beta-testing was carried out by the author, then with Manitoba Water Stewardship for ice jam flooding on the Red River [8,23] and a flood risk mitigation project on the Dauphin River [20,24], both rivers in Manitoba, Canada. The first version of the model was released in January 2013. The model has since been applied successfully to winter flow testing of the Qu'Appelle River in Saskatchewan [25,26], ice-jam flood risk assessment and mapping of the Town of Peace River in Alberta [6,7], predicting the impact of climate change on ice-jam frequency at Fort McMurray on the Athabasca River in Alberta [27], determining areas of increased vulnerability to ice cover breakup from hydro-peaking along the South Saskatchewan River in Saskatchewan [28] and investigating ice jam characteristics in the Slave River Delta in the Northwest Territories [29].

2. River Hydraulic and Ice Processes Modelled in RIVICE

To convey an impression of an ice jam, a panoramic view of a jam is shown in Figure 1. The river and ice jam are small enough to allow the jam to be viewed in its entirety from one vantage point on the ground. The figure shows the toe of the jam lodged against a downstream intact ice cover. Bridge piers often slow down and arrest the flow of ice runs to create a rubble ice accumulation. The head of the jam shows the elevated water level surface due to the backwater effect imposed by the jam's damming of water.

Figure 1. A panoramic view of an ice jam in its entirety.

For an open water case, RIVICE solves the full dynamic wave of the Saint Venant equations in one dimension (hydraulic variation in the longitudinal direction; lateral and vertical variations are averaged), solved using an implicit finite difference scheme. To close the momentum and continuity equations of the Saint Venant formulations, Manning's equation is applied:

$$v = \frac{1}{n_b} r_H^{2/3} s^{1/2} \tag{1}$$

which relates the mean flow velocity v at a cross-section to the river bed slope s, the cross-sectional hydraulic radius r_H (= cross-sectional flow area/wetted perimeter) and the Manning's roughness coefficient n_b. RIVICE can also simulate the flow dynamics for an ice-covered river by increasing the wetted perimeter to include the width of the underside of the cover. For a solid ice cover, the roughness coefficient now becomes a composite value n_c of both the bed and ice roughness coefficients, n_b and n_i respectively:

$$n_c = \left(\frac{n_b^{3/2} + n_i^{3/2}}{2} \right)^{2/3} \tag{2}$$

n_b is assumed constant for all in-stream hydraulic conditions of a river reach. For ice covers formed by ice rubble or slush the roughness n can be set as a function of ice thickness h through a relationship established by Nezhikhovskiy [30]:

$$n = 0.0302 * \ln(h) + 0.0445 \tag{3}$$

The equation is extended in RIVICE to allow the user to define the magnitude of the roughness coefficient at an ice thickness of 8 m (n_{8m}) [31]:

$$n = \frac{n_{8m}}{0.105} (0.0302 * \ln(h) + 0.0445) \tag{4}$$

The following summary of the ice processes modelled in RIVICE refers to Figure 2.

In the initial stages of river freeze-up, border ice often forms along the river banks, particularly in areas where flow conditions are more quiescent allowing a layer of skim ice and a subsequent thermally thickened bottom layer to form in the slower moving water. The border ice can extend towards the middle of the river from both banks, and can eventually meet mid-river, if flow conditions permit. In its simplest form, the user can indicate a maximum width and the simulation time for the border ice to attain that width in RIVICE. A dependency on cumulative degree days of freezing [32] or a more complex dependency on flow velocities and heat flux from the water to the atmosphere [33] are also options offered in RIVICE. The thickness of the border ice is defined by the user. If the water level drops or rises a certain input displacement, as defined by the user, the border ice breaks off and flows downstream to accumulate at the next ice cover front. A more detailed description of border ice formulation is provided in [34]. However, an increase in complexity does not necessarily yield more accurate results, due to the inherent uncertainty in the structure and parameters of border ice formulations. Nonetheless, the equations do provide an orientation to the magnitude of border ice volumes, which can be a large source of rubble ice for downstream ice cover formation. This is particularly important in wider rivers with low gradients of velocity currents along the banks and large fluctuations of the water level, where the border ice can potentially break off at a hinge crack along the bank and float as rubble ice downstream. Large fluctuations can occur through hydro-peaking of the flow by upstream dam operations, for example, or through the backwater staging from a downstream ice jam, particularly in low sloping river reaches.

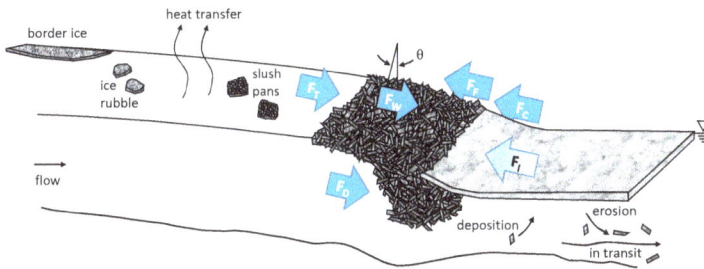

Figure 2. Depiction of the river ice processes modelled in the River Ice Model (RIVICE). Forces on the ice cover include F_T (thrust), F_D (drag), F_W (weight in the sloping direction), F_F (friction), F_C (cohesion) and F_I (internal resistance).

The border ice covers can extend from both banks until they meet within the channel middle to form a bridging where the flow of ice formed upstream can lodge to initiate further ice cover formation. Lodgments may also occur at river constrictions, such as between bridge piers, at islands and at reaches that are narrower than the rest of the river. Other than through border ice bridging, the location of lodgments cannot be predicted by RIVICE but must be set by the user, indicated by the cross-section number. Efforts have been made, with some success through geospatial modelling, to correlate geomorphological features of a river system (e.g., sinuosity, channel width, slope) to lodgment locations [35,36] which may provide an initial orientation to where lodgments are most likely to occur.

Ice progression from a lodgment requires an upstream ice source to provide ice that can accumulate at the lodgment. One source, rubble ice broken off from border ice, was already discussed above. A volume of rubble ice per time step can also be explicitly provided as input, at the upstream boundary to represent any ice volume that has floated from upstream of the modelled river section. This may include fragments from upstream broken up ice covers or breached ice jams. Frazil in the form of slush pans is the third source of ice and is generated along the open water stretch upstream of the ice cover front, when the water temperature T_w is at $0\,^\circ$C and the overlying air temperature T_a is below freezing. The heat transfer q simply becomes:

$$q = H(T_w - T_a) \tag{5}$$

with H, the heat transfer coefficient, typically ranging between 15 and 25 W/m^2/$^\circ$C, depending on how conducive the site conditions are to heat transfer. Many factors can affect the water-to-air transfer of heat, including wind speed, the degree of sheltering of the river from the wind (high sloping banks with trees provide more sheltering) and the amount of longwave radiation (cloudy conditions) that can be an important heat source that slows down the production of frazil ice. Only the open-water areas contribute to the heat loss; heat loss through moving ice is considered negligible.

The suspended frazil ice conglomerates to form frazil flocs, which become buoyant enough to float to the water surface. Here they further conglomerate to form slush pans that float downstream and add to the existing ice accumulation at the downstream lodgment. The main ice properties of the rubble ice and slush pans parameterised in RIVICE are their thickness ST and porosity PS.

Once the ice floes reach the lodgment or ice accumulation, the ice cover progression can occur in three ways—shoving, submergence and juxtaposition—depending on flow velocity and ice-cover forcing criteria:

Shoving: During the ice simulations, a balance of the external forces exerted on the ice cover is compared to the resistive forces within the cover. External forces include the thrust of the flowing water on the ice cover front F_T, the drag force exerted by the flow on the ice cover underside F_D, the vertical component of the weight of the ice cover along the direction of the river's slope F_W, the friction of the ice along the river banks F_F, and the cohesive forces F_C along the banks when freezing of the ice to the banks occurs. Friction is parameterised by $K1TAN$ which quantifies the amount the longitudinal compressive forces in the cover shed laterally to the river banks. The resistive force within the ice cover F_I is parameterised by the redistribution of the longitudinal forces vertically within the ice cover as it thickens $K2$. When the summation of the external forces exceed the resistive forces within the ice cover $(F_T + F_D + F_W - F_F - F_C > F_I)$, the ice shoves downstream and thickens the cover. This "telescoping" of the ice cover continues until the ice is thick enough for the internal resistive force to exceed the summation of the external forces. Properties of the ice cover include its porosity PC, thickness of the ice front FT, and the thickness of the ice cover downstream of the lodgment h.

A quantitative description of the forces is provided elsewhere [24,31,37]. The cohesive and friction forces have been extended in the model to include further distribution of the force components to additional "banks" at cross-sections with islands or bridge piers:

$$F_C = 2c \cdot h \cdot L \cdot SHED \tag{6}$$

$$F_F = 2f \cdot h \cdot L \cdot K1TAN \cdot SHED \tag{7}$$

where c is the area of cohesion at the ice-bank interface, f is the compressive stress within the ice cover, h is the average ice thickness, L is the distance between cross-sections and $SHED$ is a shed factor indicating the number of bank pairs to which loads can be shed to increase the resistance to the external forces. $SHED$ is set to 1 for no additional banks other than the left and right bank of the cross-section, 2 if an island/pier (four banks) is situated in the river, 3 for two islands/piers (six banks), and so on.

Submergence: If the densimetric Froude number F_r at the ice cover front is high enough, the rubble ice and slush pans submerge under the ice front and travel downstream in transit under the ice cover. F_r is a non-dimensional ratio of the inertial forces of the flow (represented by flow velocity v) to the gravitational forces (represented by the square-root of the gravitational acceleration g and the water depth at the ice cover front D). F_r can also be formulated as a function of the densities of water and ice, ρ and ρ_i respectively, the porosity of the ice cover PC and the ratio of the ice thickness h to the water depth D [38]:

$$F_r = \frac{v}{\sqrt{gD}} > \sqrt{2\frac{(\rho - \rho_i)}{\rho}(1 - PC)\frac{h}{D}\left(1 - \frac{h}{D}\right)} \tag{8}$$

The ice can then deposit on the ice underside if the mean flow velocity under the ice drops below a threshold value v_{dep}. Deposited ice can also erode and be transported downstream again if the flow velocity exceeds a higher threshold value v_{erode}.

Juxtaposition: If the Froude number is low enough and the ice cover's internal resistance exceeds the summation of the external forces applied to it, then the rubble ice and slush pans will stack up against each other along the water surface and protract the ice accumulation in the upstream direction.

Subroutines are also embedded in the RIVICE source code to simulate water quality. Variables that can be simulated include salinity, water temperature, dissolved oxygen, biochemical oxygen demand, organic and inorganic components of nitrogen and phosphorus, phytoplankton, zooplankton, fecal coliforms and conservative and decaying lignins. Reaeration, substance decay rates, phytoplankton respiration and nutrient-limited growth, zooplankton grazing, excretion and respiration, are some of the parameters that steer the water-quality simulations. Further model development and testing is required for the coupling of the water quality components to the river ice processes

3. Study Site

A map of the site, Athabasca River at Fort McMurray in Alberta, Canada, is provided in Figure 3. Fort McMurray is particularly vulnerable to ice jam flooding since the Athabasca River's geomorphological setting at that location makes this river section conducive to ice jamming, for the following reasons:

1. There is a substantial change in river slope (from approximately 0.0010 to 0.0003 m/m) and widening of the river cross-section (from approximately 300 to 700 m) immediately at the town;
2. There are many rapids along the upstream stretch to the Town of Athabasca (approximately 300 km) along which much ice rubble and water accumulates which, once released, can discharge a surge of water and ice into the Fort McMurray area;
3. The confluence of Clearwater River lies at Fort McMurray and can be a source of additional ice and water, exacerbating an already hazardous ice flood situation along the Athabasca River (see Figure 4); water can also back up in Clearwater River from a jam on the Athabasca River to flood Fort McMurray's downtown area;
4. Many obstacles, both natural (e.g., islands and sand bars) and anthropogenic (e.g., bridges) in origin, can impede the flow of ice, increasing the area's susceptibility to ice jam formation;
5. The Athabasca River's flow is in a northerly direction, an orientation favourable to ice jam formation, since the ice cover in the upstream, southerly areas generally breaks up sooner due to warmer weather conditions (increased ice cover ablation and runoff from melting snow), transporting ice to downstream, northerly areas where the temperatures are cooler and the ice cover remains intact and competent longer to resist the flow of ice.

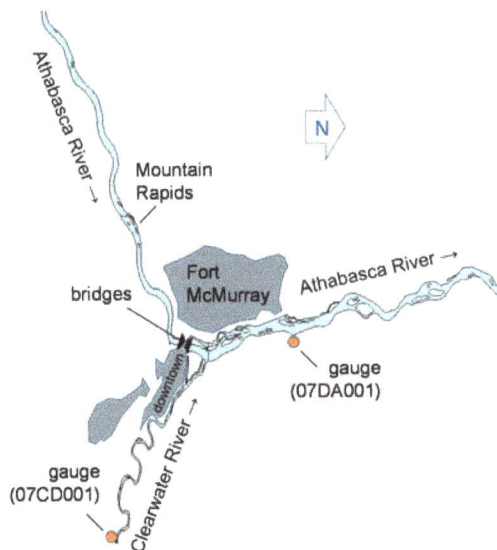

Figure 3. Fort McMurray on the Athabasca River at the Clearwater River confluence.

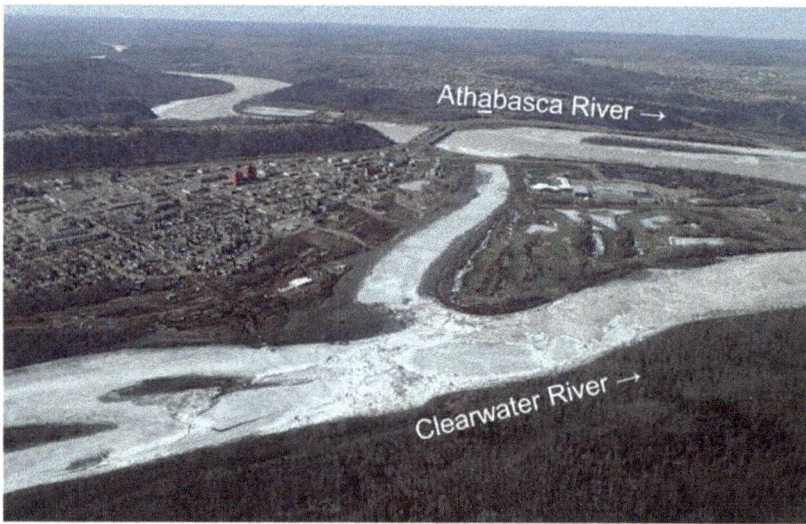

Figure 4. An aerial photo taken 8 April 2015 showing the confluence of the Athabasca and Clearwater rivers in close proximity to downtown Fort McMurray, when there was an ice jam several kilometres along the Athabasca River and an intact ice cover on the Clearwater River (*courtesy of Gov't of Alberta*).

Beltaos tested his model assuming a rectangular channel using approximations of the ice jam parameters typical for breakup jams at Fort McMurray [39]. Other modelling exercises have been carried out along the Athabasca River but with a different focus (ice jam releases) in a different reach along the Athabasca River (between the Town of Athabasca and Fort McMurray, a stretch that extends upstream from the one studied here) [40,41]. The upstream stretch is much steeper and generally geomorphologically different than the one presented in this paper.

4. Model Setup

Three ice jam events were investigated in each of the years 1977, 1978 and 1979, to lay out the steps for setting up the RIVICE model. The configurations of each jam, in regard to the river hydraulic and ice regimes and the extent of the subsequent flooding, are quite different, allowing the performance of the RIVICE model to be tested.

Bathymetry: Cross-sections of the river are required to capture the cross-sectional flow areas along the river, an important pre-requisite for accurate simulations of flows, current velocities and water level elevations. An example of a cross-section is shown in Figure 5, indicating the difference in width and flow area between the steeper stretch immediately upstream of Fort McMurray and the less steep stretch immediately downstream of the town. The cross-sections consist of surveyed elevation points along transects of a river bed.

Lateral inflows: Lateral inflows, which include tributary inflows and overland runoff, are simulated as a discharge per unit length along a river stretch. In our test case, the inflow from Clearwater River is incorporated as its discharge is divided up along the 100 m length, which corresponds to the approximate width of its mouth.

Boundary conditions: Typically, water flow rate serves as an upstream boundary condition whereas a water level elevation designates the downstream boundary condition, although rating curves can also define the boundaries. These values were derived from the records of the Water Survey of Canada gauges "Athabasca River below McMurray" (# 07DA001) and "Clearwater River at Draper" (# 07CD001).

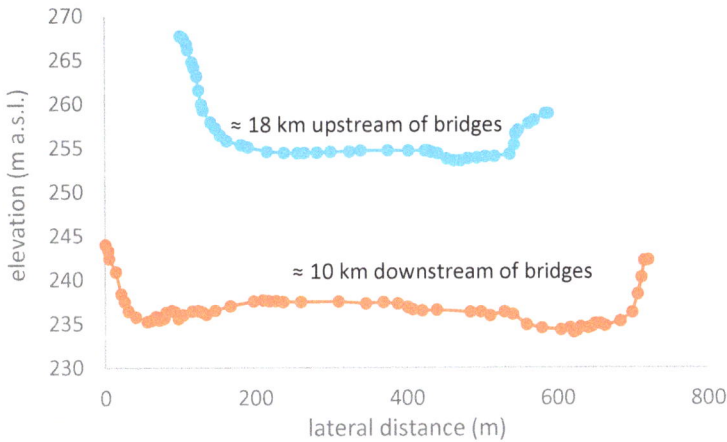

Figure 5. Examples of cross-sections used for model setup.

Surveyed water level profiles: During the ice jam events, water levels were surveyed along the banks of the river to provide a longitudinal profile of the flood water level elevation [42]. In difficult to access areas near the jam, aerial photographs were taken of flood water along the banks with reference features that could then be surveyed after the event had passed. A similar method was carried out by the author along the Dauphin River in Manitoba using a road surface adjacent to the river as a reference elevation for the encroached ice jam flood water, as shown in Figure 6. A road surface profile was surveyed before the flood event to provide backwater level elevations during the flight.

Figure 6. Encroached floodwater frozen to road surface, as indicated by the arrow. The road surface profile was surveyed before the flood event so that the backwater elevation could be determined at the time of the flight.

Parameters: Important parameters used for model calibration can be grouped according to the characteristics or function of a component of the hydraulic/ice regime (Table 1). Parameter values and ranges were extracted from the literature [42,43] and then calibrated for each event to fit the simulations to the water level observations.

Table 1. Parameters and boundary conditions for model calibration.

Parameter	Description	Units	1977	1978	1979
rubble ice:					
PS	porosity of rubble ice	–	0.3	0.3	0.3
ST	thickness of rubble pans	m	2.5	1.2	1.5
ice cover:					
V_{ice}	volume of incoming ice	$m^3/\Delta t$	2000	2000	2000
PC	porosity of ice cover	–	0.5	0.5	0.5
FT	thickness of ice cover front	m	2.5	1.0	1.5
ice jam lodgment:					
h	thickness of ice downstream of jam	m	1.0	0.75	1.0
x	cross-section number of lodgement [1]	–	550	491	783
ice transport:					
$v_{deposit}$	threshold of deposition velocity	m/s	1.2	1.2	1.2
v_{erode}	threshold of erosion velocity	m/s	1.8	1.8	1.8
hydraulic roughness:					
n_{8m}	ice roughness	$s/m^{1/3}$	0.14	0.16	0.11
n_{bed}	river bed roughness	$s/m^{1/3}$	0.025	0.025	0.025
strength properties:					
K1TAN	lateral: longitudinal stresses	–	0.18	0.18	0.18
K2	longitudinal: vertical stresses	–	8.4	8.4	8.4
boundary conditions:					
Q	upstream discharge	m^3/s	1000	1500	1366
W	downstream water level	m a.s.l.	235.0	235.0	235.5

Note: [1] corresponding chainage = 27,500, 24,500 and 39,150 m for 1977, 1978 and 1979, respectively.

5. Results and Discussion

5.1. Calibration and Validation

Figure 7 shows the longitudinal simulated profiles of the ice cover (black) during the ice jam events of each year studied; 1977, 1978 and 1979. Included, for reference, is the thalweg (orange) and the open-water profile (blue line) attained with the same upstream and tributary discharge and downstream water level boundary conditions. The open-water profile reveals how much staging occurs by the ice cover and jam. A Manning's roughness coefficient n_b of 0.025 was calibrated to fit the simulated open-water level elevation at the gauge with the water level elevation recorded by the gauge (blue dot).

Surveyed water level elevations attained at the peak of each ice jam flood event (pink diamonds) were included to guide the calibration of the remaining river ice parameters and assess model performance. The optimum values are provided in Table 1. Generally, there is good agreement between the ice jam water profile and the surveyed water elevation points. The 1978 ice jam event was difficult to calibrate and yielded larger deviations between simulated and observed water level profiles. This is due to the fact that the discharge upstream of the jam (1850 m^3/s, sensu [12]) was far greater than that recorded at the gauge downstream of the jam (544 m^3/s). This is a large discrepancy in flows. To fulfil continuity, the extra upstream water may have been diverted around the jam, which has not been considered in the model. A similar diversion was reported during the 1977 event by Doyle [43], who stated that "flood wave [washed] over ice on right side of island downstream of bridges but ice sheet remains intact while sheet on left side of island is broken up and all flow is directed around left side of island". Increasing the discharge to fit the simulated ice jam water profile to the surveyed water level points required the ice cover downstream of the jam toe to lift slightly in order to allow for additional flow under the jam toe and maintain computational stability. Using an upstream boundary discharge of 1500 m^3/s, a "Pareto" balance was attained between slightly underestimated upstream water level profiles and a slightly overestimated ice cover elevation downstream of the ice jam. There was some discrepancy between the upstream discharge and downstream gauge flow reading for the other studied events, but not of the same magnitude as the 1978 ice jam. Andres and Doyle [42] estimated a discharge range of 1135–1600 m^3/s upstream of the 1977 ice jam, with 934 m^3/s recorded at the gauge. For 1979, the upstream flow was estimated to range between 1300 and 1850 m^3/s when the gauge reading showed 1480 m^3/s.

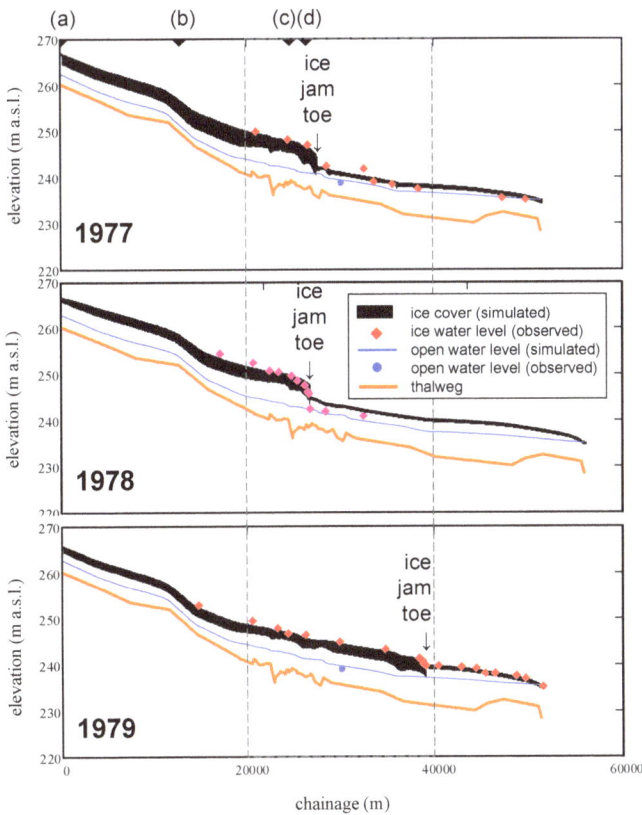

Figure 7. Ice and corresponding backwater level profiles of the 1977 (**top**), 1978 (**middle**) and 1979 (**bottom**) ice jam events. Open water profiles of the same discharge included to indicate the amount of staging. At the top, indicated locations are: (a) Cascade Rapids, (b) Mountain Rapids, (c) McEwan bridges and (d) Clearwater River confluence.

The 1978 ice jam event is a special case because its toe was situated at the bridge piers, which provided additional support and strength to the jam to maintain its high volume of rubble ice and high backwater staging levels. Some of the extra strength required to support such a large jam can be incorporated in the model using an increased *SHED* factor of 5, to account for the bridge piers in the water, which shed much of the longitudinal compressive stresses in the ice cover laterally to the banks and piers. However, not all of them may have been shed, and some of the longitudinal stresses of a jam of that size were potentially shed vertically onto the banks and, perhaps, on points along the jam where the ice grounded on the river bed or a sand bar. This shedding of forces has not been included in the model and is a topic to be tackled in future model development.

5.2. Local Sensitivity Analysis

Each jam lodgment was also situated at different locations along the river, each location having slight differences in their fluvial geomorphology. The 1977 jam occurred the furthest upstream in the steepest section of the studied reach and the 1979 ice jam was located furthest downstream in the least steep slope of the reach. These differences in the jams' geomorphological settings will have different impacts on the river ice processes and parameters. Such effects can be quantified using a sensitivity analysis to show how simulated backwater levels react to slight changes in the parameter and boundary condition settings.

Figure 8 provides the local sensitivities of each parameter to the output variable backwater level elevation, both 5 km and 15 km upstream of the ice jam lodgment of each studied event. Positive sensitivity means that an increase/decrease in the parameter yields an increase/decrease in the output variable. When a parameter induces a negative sensitivity to the output variable, the variable increases or decreases opposite to the value change in the parameter.

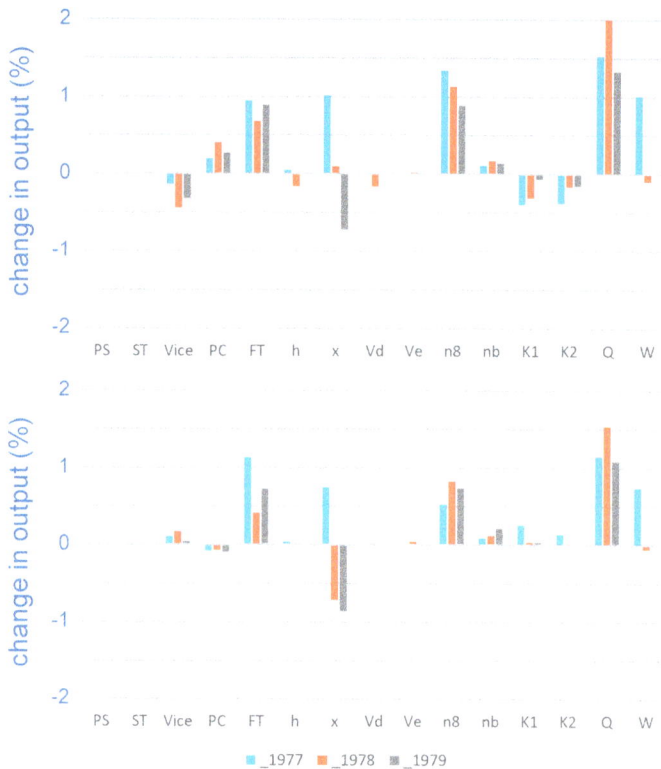

Figure 8. Sensitivities of model parameters and boundary conditions on backwater levels 5 km (**top panel**) and 15 km (**bottom panel**) from the ice jam lodgments for the 1977, 1978 and 1979 ice jam flood events.

Rubble ice: Ice jam formation is not sensitive to the parameters related to the ice rubble characteristics, *PS* and *ST*, regardless of the location along the jam. Any rubble or slush ice simply adds to the accumulation of the ice volume in the jam, regardless of the dimensions of the ice blocks or pans.

Ice jam cover: Backwater levels 15 km upstream of a lodgment are positively sensitive to the inflowing ice volume V_{ice} and negatively sensitive to the ice cover porosity *PC*. Hence, an increase in the compaction of the ice in the jam due to more ice of lesser porosity will lead to increased water level staging far upstream of the jam. However, in close proximity to the jam lodgment (5 km upstream from the jam toe), the sensitivities reverse due to the added weight of the upstream ice abutted to the lodgment. Staging all along the ice jam is positively sensitive to the ice front thickness *FT* since more thrust force against the cover causes more shoving against the jam ice cover and subsequent backwater staging.

Lodgment: An indication as to the thickness of the intact ice cover *h* before the ice jam events is provided in [42]. However, variations in the thickness of the intact ice cover downstream of the ice jam have little effect on the staging upstream of the jam. Geomorphology is also an important influence on floodwater levels, as reflected in the variation of sensitivities of the ice jam lodgment location *x*. This highlights the importance of accurately representing the bathymetry in the model setup.

Ice transport: The ice transport parameters, v_d and v_e, have little effect on the ice jam morphology or water levels since the main mechanism of ice cover thickening is the retraction of the ice cover by shoving.

Hydraulic roughness: Backwater level rises are very sensitive to the ice jam underside roughness parameter n_8, more so than river bed roughness n_b, particularly in close proximity to the ice jam lodgment. The point closer to the lodgment (5 km upstream of the ice jam toe) induces a negative sensitivity of the parameters *K1TAN* and *K2* on backwater levels, indicating the importance of the frictional and resistive forces in reducing further thickening and stabilizing the ice jam. Increases in these parameters will increase friction of the ice rubble against the banks and amongst each other within the ice accumulation. This becomes less prevalent at the point further upstream of the lodgment (15 km upstream of the ice jam toe) where sensitivities lessen and, particularly for the 1977 event, even become slightly positive.

Boundary conditions: Flow *Q* is one of the most sensitive factors influencing water level elevation and becomes more sensitive as we move further upstream along the jam. More flow will induce greater thrust and drag forces on the ice cover. Further upstream, the river slope is steeper, which increases the weight of the ice on the cover to promote shoving. The downstream water level boundary *W* only has an influence on the 1977 water levels.

6. Discussion on the Broader Applicability of RIVICE

The applicability of RIVICE has been shown here using one case study, that of ice jam flooding along the Athabasca River at Fort McMurray at latitude 56.7° N. The model has also been successfully applied to a more northerly river, the Slave River near Fort Resolution in the Northwest Territories of Canada at latitude 61.3° N and more southerly regions such as the Red River at Winnipeg and Selkirk in Manitoba at the approximately latitude 50° N. The mean winter flows for the months between December and February of the Athabasca River at Fort McMurray is 174 m^3/s but the model has also been successfully applied to rivers with mean January to February flows as high as 1123 m^3/s, such as at the Town of Peace River on the Peace River and as low as 4 m^3/s for the upper Qu'Appelle River in Saskatchewan. Although these examples all stem from Canadian rivers, more recently RIVICE has also been applied to the Oder River, the lower reach of which constitutes the border between Germany and Poland. Although these examples all stem from regions with continental climates, currently RIVICE is being implemented for the Exploits River in Newfoundland, a region which has a more maritime climate.

Due to its relatively short computational times and the availability of the source code, RIVICE can be embedded in a Monte-Carlo framework to allow stochastic hydraulic modelling of ice jam flood events. Such an approach is well suited to ice jam modelling because ice jamming is a stochastic process and can best be described in probabilistic, not deterministic, terms. The stochastic approach allows the author to currently engage in research to develop ice jam flood forecasting systems in which RIVICE serves as an integral part in providing ensembles of backwater level predictions to determine probabilities of flood level exceedances. Such forecasting systems are being developed for the Athabasca River at Fort McMurray and the Exploits River at Badger under the project "Near real-time Ice-Related Flood Hazard Assessment (RIFHA)" funded by the Canadian Space Agency and managed by C-Core, a remote sensing company in St. John's, Newfoundland. Under the Global Water Futures research program at the University of Saskatchewan (http://gwf.usask.ca/), funded in part by a $77.8-million grant from the Canada First Excellence Research Fund, ice jam flood forecasting systems are in planning to be developed for the Porcupine River at Old Crow in the Yukon Territories and the Red and Assiniboine rivers at Winnipeg, Manitoba.

7. Conclusions

RIVICE captures many of the major river ice processes within the channel that can lead to ice jamming and subsequent flooding. Flood waters that overtop the river banks and divert around the jam are not accounted for; however, developments are underway to couple RIVICE to a two-dimensional storage cell floodplain model that will allow water to be exchanged between the river channel and the floodplain. Future studies should also concentrate on extending the model to incorporate the additional forces that support the ice jam vertically when it is grounded at the river banks or on shallower points of the river bed such as at sandbars.

The local sensitivity analysis shows varying importance of the parameters and boundary conditions to the ice jamming. Both the location of the ice jam lodgment and the position along the jam ice cover show differences in the sensitivities to backwater staging. An example is the varying sensitivities to staging along different locations of the jam, imputed by the parameters associated with the frictional and internal resistive forces and properties of the ice jam accumulation (e.g., porosity).

One limitation of RIVICE is that it can only form ice covers through the juxtapositioning of frazil-generated or rubble-transported ice. Hence, it is only suited for rivers where mean flow velocity, at freeze-up or ice jamming, exceeds 0.4 m/s, a common threshold between thermally- and mechanically-driven ice formation. Additionally, a thermal module still needs to be integrated into the model to allow for thermal thickening of the ice cover once the cover has been established along the river. This is a topic of future research and model development.

Acknowledgments: The author thanks Faye Hicks for providing the cross-sectional data of the Athabasca River's bathymetry. Also special thanks to Bernard Trevor and Nadia Kovachis-Watson from the Government of Alberta for sharing the photo in Figure 4.

Conflicts of Interest: The authors declare no conflict of interest.

References

1. Prowse, T.D.; Conly, F.M.; Church, M.; English, M.C. A review of hydroecological results of the northern river basins study, Canada. Part 1. Peace and Slave rivers. *River Res. Appl.* **2002**, *18*, 429–446. [CrossRef]
2. Chu, T.; Das, A.; Lindenschmidt, K.-E. Monitoring the variation in ice-cover characteristics of the Slave River, Canada using RADARSAT-2 data. *Remote Sens.* **2015**, *7*, 13664–13691. [CrossRef]
3. Chu, T.; Lindenschmidt, K.-E. Integration of space-borne and air-borne data in monitoring river ice processes in the Slave River, Canada. *Remote Sens. Environ.* **2016**, *181*, 65–81. [CrossRef]
4. Sagin, J.; Sizo, A.; Wheater, H.; Jardine, T.D.; Lindenschmidt, K.-E. A water coverage extraction tool for optical satellite data: Application in a large river delta. *Int. J. Remote Sens.* **2015**, *36*, 764–781. [CrossRef]
5. MacKinnon, B.; Sagin, J.; Baulch, H.; Lindenschmidt, K.-E.; Jardine, T. Influence of hydrological connectivity on winter limnology in floodplain lakes of the Saskatchewan River Delta, SK. *Can. J. Fish. Aquat. Sci.* **2016**, *73*, 140–152. [CrossRef]

6. Lindenschmidt, K.-E.; Das, A.; Rokaya, P.; Chun, K.P.; Chu, T. Ice jam flood hazard assessment and mapping of the Peace River at the Town of Peace River. Presented at the 18th Workshop on the Hydraulics of Ice Covered Rivers, Quebec City, QC, Canada, 18–21 August 2015.

7. Lindenschmidt, K.-E.; Das, A.; Rokaya, P.; Chu, T. Ice jam flood risk assessment and mapping. *Hydrol. Process.* **2016**, *30*, 3754–3769. [CrossRef]

8. Lindenschmidt, K.-E.; Sydor, M.; Carson, R.W.; Harrison, R. Ice jam modelling of the Lower Red River. *J. Water Resour. Prot.* **2012**, *4*, 1–11. [CrossRef]

9. Wazney, L.; Clark, S.P. The 2009 flood event in the Red River Basin: Causes, assessment, and damages. *Can. Water Resour. J.* **2016**, *41*, 56–64. [CrossRef]

10. Turcotte, B.; Morse, B. Ice-induced flooding mitigation at St. Raymond, QC, Canada. Presented at the 18th Workshop on the Hydraulics of Ice Covered Rivers, Quebec City, QC, Canada, 18–21 August 2015.

11. Beltaos, S.; Burrell, B.C. Extreme ice jam floods along the Saint John River, New Brunswick, Canada. In *Extremes of the Extremes: Extraordinary Floods*; IAHS Publication: Reykjavik, Iceland, 2002; pp. 9–14.

12. Thistlethwaite, J.; Feltmate, B. Weather hardened flood insurance. *Can. Underwrit.* **2013**, *80*, 42–44.

13. *Causes of Flooding*; Environment and Climate Change Canada: Ottawa, ON, Canada, 2013. Available online: https://ec.gc.ca/eau-water/default.asp?lang=En&n=E7EF8E56-1 (accessed on 27 April 2017).

14. Ahopelto, L.; Huokuna, M.; Aaltonen, J.; Koskela, J.J. Flood frequencies in places prone to ice jams, case city of Tornio. Presented at the 18th Workshop on the Hydraulics of Ice Covered Rivers, Quebec City, QC, Canada, 18–21 August 2015.

15. Kögel, M.; Das, A.; Marszelewski, W.; Carstensen, D.; Lindenschmidt, K.-E. Developing an ice-jam flood forecasting system for the Oder River; proof-of-concept (in German). *Wasserwirtschaft* **2017**, *5*. [CrossRef]

16. Agafonova, S.A.; Frolova, N.L.; Krylenko, I.N.; Sazonov, A.A.; Golovlyov, P.P. Dangerous ice phenomena on the lowland rivers of European Russia. *Nat. Hazards* **2016**. [CrossRef]

17. Shen, H.T. Mathematical modeling of river ice processes. *Cold Reg. Sci. Technol.* **2010**, *62*, 3–13. [CrossRef]

18. Burrell, B.; Huokuna, M.; Beltaos, S.; Kovachis, N.; Turcotte, B.; Jasek, M. Flood hazard and risk delineation of ice-related floods: Present status and outlook. In Proceedings of the 18th CGU-HS CRIPE Workshop on the Hydraulics of Ice Covered Rivers, Quebec City, QC, Canada, 18–21 August 2015.

19. Xia, J.; Falconer, R.A.; Lin, B.; Tan, G. Numerical assessment of flood hazard risk to people and vehicles in flash floods. *Environ. Model. Softw.* **2011**, *26*, 987–998. [CrossRef]

20. Lindenschmidt, K.-E.; Sydor, M.; Carson, R. Modelling ice cover formation of a lake-river system with exceptionally high flows (Lake St. Martin and Dauphin River, Manitoba). *Cold Reg. Sci. Technol.* **2012**, *82*, 36–48. [CrossRef]

21. Jasek, M.; Pryse-Phillips, A. Influence of the proposed Site C hydroelectric project on the ice regime of the Peace River. *Can. J. Civ. Eng.* **2015**, *42*, 645–655. [CrossRef]

22. Beltaos, S.; Tang, P.; Rowsell, R. Ice jam modelling and field data collection for flood forecasting in the Saint John River, Canada. *Hydrol. Process.* **2012**, *26*, 2535–2545. [CrossRef]

23. Lindenschmidt, K.-E.; Sydor, M.; Carson, R. Ice jam modelling of the Red River in Winnipeg. Presented at the 16th CRIPE Workshop on the Hydraulics of Ice Covered Rivers, Winnipeg, MB, Canada, 18–22 September 2011; pp. 274–290.

24. Lindenschmidt, K.-E.; Sydor, M.; van der Sanden, J.; Blais, E.; Carson, R.W. Monitoring and modeling ice cover formation on highly flooded and hydraulically altered lake-river systems. Presented at the 17th CRIPE Workshop on the Hydraulics of Ice Covered Rivers, Edmonton, AB, Canada, 21–24 July 2013; pp. 180–201.

25. Lindenschmidt, K.-E.; Sereda, J. The impact of macrophytes on winter flows along the Upper Qu'Appelle River. *Can. Water Resour. J.* **2014**, *39*, 342–355. [CrossRef]

26. Lindenschmidt, K.-E.; Carstensen, D. The upper Qu'Appelle water supply project in Saskatchewan, Canada–Upland Canal ice study. *Österreichische Wasser-und Abfallwirtschaft* **2015**, *67*, 230–239. [CrossRef]

27. Rokaya, P.; Morales-Marin, L.A.; Wheater, H.; Lindenschmidt, K.-E. Hydro-climatic variability and implications for ice-jam flooding in the Athabasca River Basin in western Canada. *Cold Reg. Sci. Technol.* **2017**, submitted.

28. Liu, N.; Kells, J.; Lindenschmidt, K.-E. Mapping river ice cover breakup induced by hydropower operations. Presented at the CRIPE 18th Workshop on the Hydraulics of Ice Covered Rivers, Quebec City, QC, Canada, 18–21 August 2015.

29. Zhang, F.; Mosaffa, M.; Lindenschmidt, K.-E. Using remote sensing data to parameterize ice jam modeling for a northern inland delta. *Water* **2017**, *9*, 306. [CrossRef]

30. Nezhikhovskiy, R.A. Coefficients of roughness of bottom surface on slush-ice cover. In *Soviet Hydrology*; American Geophysical Union: Washington, DC, USA, 1964; pp. 127–150.

31. RIVICE—User's Manual. Environment Canada. Available online: http://giws.usask.ca/rivice/Manual/RIVICE_Manual_2013-01-11.pdf (accessed on 27 April 2017).

32. Newbury, R. The Nelson River: A Study of Subarctic River Processes. Ph.D. Theiss, John Hopkins University, Baltimore, MD, USA, 1968.

33. Matousek, V. Regularity of the freezing-up of the water surface and heat exchange between the water body and water surface. In Proceedings of the IAHR Ice Symposium, Hamburg, Germany, 27–31 August 1984; pp. 187–201.

34. Haresign, M.; Toews, J.S.; Clark, S. Comparative testing of border ice growth prediction models. Presented at the 16th Workshop on River Ice, Winnipeg, MB, Canada, 18–22 September 2011.

35. Lindenschmidt, K.-E.; Chun, K.P. Geospatial modelling to determine the behaviour of ice cover formation during river freeze-up. *Hydrol. Res.* **2014**, *45*, 645–659.

36. Lindenschmidt, K.-E.; Das, A. A geospatial model to determine patterns of ice cover breakup and dislodgement behaviour along the Slave River. *Can. J. Civ. Eng.* **2015**, *42*, 675–685. [CrossRef]

37. Lindenschmidt, K.-E. *Winter Flow Testing of the Upper Qu'Appelle River*; Lambert Academic Publishing: Saarbrucken, Germany, 2014.

38. Pariset, E.; Hausser, R. Formation and evolution of ice covers on rivers. *Trans. Eng. Inst. Can.* **1961**, *5*, 41–49.

39. Beltaos, S. Numerical computation of river ice jams. *Can. J. Civ. Eng.* **1993**, *20*, 88–99. [CrossRef]

40. She, Y.; Hicks, F. Modeling ice jam release waves with consideration for ice effects. *Cold Reg. Sci. Technol.* **2006**, *45*, 137–147. [CrossRef]

41. She, Y.; Andrishak, R.; Hicks, F.; Morse, B.; Stander, E.; Krath, C.; Keller, D.; Abarca, N.; Nolin, S.; Tanekou, F.N.; et al. Athabasca River ice jam formation and release events in 2006 and 2007. *Cold Reg. Sci. Technol.* **2009**, *55*, 249–261. [CrossRef]

42. Andres, D.D.; Doyle, P.F. Analysis of breakup and ice jams on the Athabasca River at Fort McMurray, Alberta. *Can. J. Civ. Eng.* **1984**, *11*, 444–458. [CrossRef]

43. Doyle, P.F. *1977 Breakup and Subsequent Ice Jam at Fort McMurray*; Alberta Research Council, Transportation and Surface Water Engineering Division: Edmonton, AB, Canada, 1977.

water

MDPI

Article

Using Remote Sensing Data to Parameterize Ice Jam Modeling for a Northern Inland Delta

Fan Zhang *, Mahtab Mosaffa, Thuan Chu and Karl-Erich Lindenschmidt

Global Institute for Water Security, Saskatoon, SK S7N 3H5, Canada; mahtab.mosaffa@gmail.com (M.M.); thuan.chu@usask.ca (T.C.); karl-erich.lindenschmidt@usask.ca (K.-E.L.)
* Correspondence: fan.zhang@usask.ca; Tel.: +1-306-881-6508

Academic Editor: Hongjie Xie
Received: 27 January 2017; Accepted: 23 April 2017; Published: 27 April 2017

Abstract: The Slave River is a northern river in Canada, with ice being an important component of its flow regime for at least half of the year. During the spring breakup period, ice jams and ice-jam flooding can occur in the Slave River Delta, which is of benefit for the replenishment of moisture and sediment required to maintain the ecological integrity of the delta. To better understand the ice jam processes that lead to flooding, as well as the replenishment of the delta, the one-dimensional hydraulic river ice model RIVICE was implemented to simulate and explore ice jam formation in the Slave River Delta. Incoming ice volume, a crucial input parameter for RIVICE, was determined by the novel approach of using MODIS space-born remote sensing imagery. Space-borne and air-borne remote sensing data were used to parameterize the upstream ice volume available for ice jamming. Gauged data was used to complement modeling calibration and validation. HEC-RAS, another one-dimensional hydrodynamic model, was used to determine ice volumes required for equilibrium jams and the upper limit of ice volume that a jam can sustain, as well as being used as a threshold for the volumes estimated by the dynamic ice jam simulations using RIVICE. Parameter sensitivity analysis shows that morphological and hydraulic properties have great impacts on the ice jam length and water depth in the Slave River Delta.

Keywords: river ice; modeling; RIVICE; MODIS; the Slave River Delta

1. Introduction

In cold regions, the ice regime of rivers can be divided into three main phases during the winter: river freeze-up, continuous solid ice cover, and ice cover breakup. With ice cover breakup, rubble ice in rivers can pose threats to infrastructure adjacent to the river via jamming and subsequent flooding. However, some ecosystems, particularly inland deltas (e.g., the Slave River Delta), rely on ice jam flooding for replenishment of moisture and sediment. One of the main contributions of flow in the Slave River is from the Peace River. Ice jam development can be separated into three phases: formation, extension, and release. Studying ice jam processes can help extend our understanding of flooding and drying behaviors in river deltas such as the Slave River Delta. Ice jams have rough undersides, which increase flow resistance leading to backwater effects. Within the ice jam, cohesion is a force due to additional freezing which can uphold the structural integrity of the ice jam. Beltaos applied the Rising Limb Analysis Method (RLAM) to analyze hydrodynamic forces in the lower Mackenzie River [1]. He found that pre-breakup conditions, such as Cumulative Melting Degree Days, as well as the rate of flow increases, are important factors in ice jam formation and release. He substantiated these findings with a hydrodynamic model for simulating ice jam waveforms.

River ice jamming and subsequent flooding can significantly impact local aquatic ecosystems [2]. On the other hand, the reduced frequency of ice-jam flooding can lead to drying trends and subsequent shifts in flora and fauna composition and succession. Beltaos stated the effects of flow regulation by the dam construction on the river and climate change factors with respect to the drying trends in the Peace-Athabasca Delta [3]. Based on calculations under different flow conditions, he concluded that regulation is the dominant contributor to the lower frequency and severity of ice jam flooding. Prowse and Conly pointed out that ice-jam-induced backwater flowing into the Peace-Athabasca Delta [4] is a major source of that delta's water supply. Although the Slave River Delta is approximated 400 km downstream of the Peace-Athabasca Delta, and its flows are augmented by the Lake Athabasca and the Athabasca River systems, flow regulation may still affect the Slave River Delta [5].

Inland deltas are particularly susceptible to ice jamming and subsequent flooding. Examples include flooding along the Peace-Athabasca Delta [6], and the lower Red River in Manitoba [7]. To understand the formation and development of ice jams and the flood hazard they induce, the river ice hydraulic model RIVICE was implemented and calibrated to determine the conditions required for ice jamming. RIVICE has proven to be an effective tool for predicting ice jam formation and its impacts on flooding [8].

Due to safety issues, it is difficult to carry out field measurements on ice jams. To obtain the information necessary for better understanding river ice processes, researchers have taken advantage of satellite remote sensing technology, such as MODIS and RADARSAT-2, to monitor and characterize river ice behavior [9]. RADARSAT-2, which is a commercial Earth observation radar satellite, was launched in December 2007 to monitor the environment for natural resource management. It provides high-quality data of the earth's surface features with a finer resolution (2 m/8 m) but at longer revisit intervals (24 days) than MODIS [10]. Due to the longer wavelength emitted and received by the sensor, an advantage of RADARSAT-2 is that image acquisition is not hindered by weather conditions. Lindenschmidt et al. used RADARSAT-2 to measure ice thickness along the Red River, and determined the relationship between snow depth, ice thickness, and backscattering signal [11]. Due to the high penetration and resolution of RADARSAT-2 signals, it is possible to analyze ice types based on RADARSAT-2 backscatter. Chu and Lindenschmidt classified river ice types using RADARSAT-2 imagery [12]. In another study of the Slave River, different breakup patterns were tracked along the river, using RADARSAT-2 satellite imagery [13]. These researchers used space-borne remote sensing technology (RADARSAT-2 and MODIS) to determine ice extent and ice types. However, it is difficult to acquire ice cover data along the Slave River in consecutive days by RADARSAT-2. Hence, MODIS can be utilized to monitor daily ice changes instead of RADARSAT-2. MODIS (MODerate-resolution Imaging Spectroradiometer) regularly scans the surface of the earth, albeit with a coarse resolution, permitting the monitoring of daily changes in river ice, which is essential for ice cover breakup research [14]. MODIS has been used to facilitate various river-ice-related research, including the determination of the extent of ice and prediction of ice-related hazards along the Susquehanna River [15,16], and ice breakup monitoring on the Mackenzie River [17]. Chaouch et al. designed an automated algorithm to monitor river ice using MODIS data [15]. In addition to river ice data, researchers have also obtained climatic, snow cover and surface temperature data from MODIS products [18–20]. One drawback of MODIS is that it is an optical sensor dependent on cloud cover and daylight conditions.

Studies focused on monitoring and modeling ice jams in the Slave River Delta are sparse. Our research goal was to obtain more understanding of river ice formation and deformation processes in the Slave River Delta by focusing on the breakup period during which ice jam flooding is most prevalent.

2. Study Site and Methods

2.1. Study Site

The Slave River flows approximately 434 km northward from the Peace River and Peace-Athabasca Delta in Alberta to Great Slave Lake. This study focuses mainly on the delta of the river, the Slave River Delta, which is located at the mouth of the river at Great Slave Lake. The study area of this research extends from the Jean River to Great Slave Lake, approximately 25 km along the most downstream part of the main channel of the river (Figure 1). The drop in elevation at this part of the river is very low, with a gradient of 0.00003 m/m. The Resdelta Channel is the main channel, from which four other tributary channels, the Middle, Steamboat, and Nagle channels, and the Jean River, branch through the Slave River Delta. The mean annual flow of the Slave River is 3400 m³/s [5]. Regulation of flow by the W.A.C Bennett Dam in the upper reaches of the Peace River moderates the seasonal difference in the flow, typically resulting in increasingly low natural winter flows and decreasingly high natural summer flows. The ice-cover season in the Slave River Delta extends from November to May. Before 1980, the frequency of high-magnitude discharge was once every two years; since 1980, this frequency has been once every five years [21]. This decreasing trend in flood frequency past several decades shows the drier conditions in the Slave River delta.

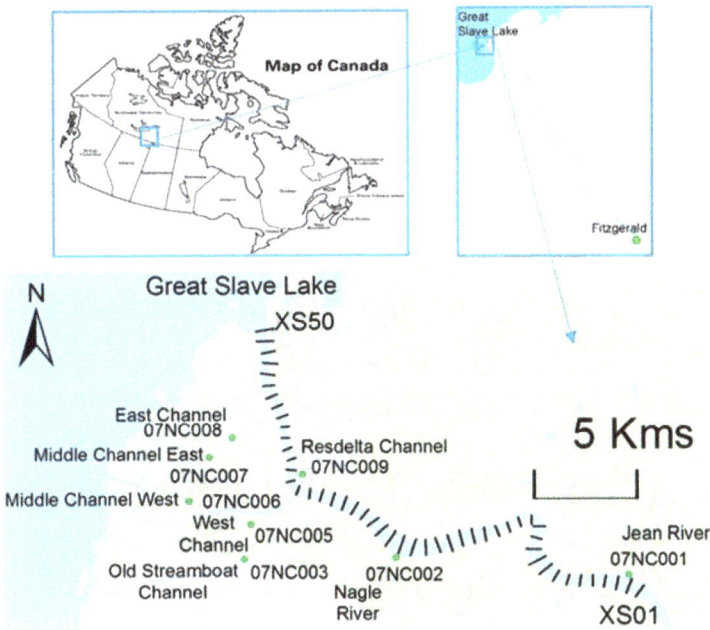

Figure 1. The Slave River Delta with important channels. (The dark lines in the bottom panel represent cross-sections spaced 500 m apart. Widths of cross-sections vary between 500 and 1200 m).

Geomorphological factors of the river system, such as the sinuosity, channel width, and slope, have a great influence on the river ice regime and, due to its high flows and low gradient, make the Slave River Delta more conducive to ice jam flooding than the rest of the river [22].

2.2. Methods

2.2.1. RIVICE Model

The hydraulic model, RIVICE, was used to simulate ice jamming in the Slave River Delta. RIVICE is a one-dimensional model based on the continuity and full dynamic wave equations of flow and momentum. Although there are some limitations in one-dimensional models, they are widely used and have proven to be reasonably accurate. RIVICE can be used to simulate ice processes such as frazil ice generation, ice lodgment, ice cover progression through juxtaposition, ice deposition, hanging dam formation, and telescoping leading to ice thickening [23].

The model setup was as follows: Fifty cross-sections of the Slave River spaced 500 m apart (Figure 1) were extracted from a bathymetric model, different parts of which were surveyed between 2013 and 2015 and used to set up the RIVICE model. Cross-sections were also available for the complete reach of the Slave River, approximately 10 km apart, as part of an extensive water surveying campaign carried out in 1980. This campaign included gauging water levels and flows along the main channel and side-channels of the delta. These data served as a basis for calibration and validation of open-water and ice-covered conditions of the flow regime. The gauging stations included the following: Jean River (07NC001), Nagle River (07NC002), and Resdelta Channel (07NC009) (Figure 1). Only water level and flow data from 1980 are available at these stations in the Delta.

MODIS data were used to determine ice volumes approximated by:

$$\text{Ice Volume} = \text{Average width of ice cover} \times \text{Ice cover length} \times \text{Average ice thickness} \quad (1)$$

All three factors on the right hand of Equation (1) can cause errors in ice volume estimation, which is an input parameter in the RIVICE model. Overestimated ice volume may lead to higher backwater levels. To improve the accuracy of the ice thickness calculation and enhance the reliability of the RIVICE model, aerial photos were used to validate the MODIS ice cover estimation (shown in the Results section).

When simulating an ice jam, both the upstream discharge boundary and incoming ice volume data are necessary. For other years, only the discharge data from the Fitzgerald station (07NB001) (Figure 1) can be drawn upon for the ice jam modeling. This station, located in Alberta, has provided continuous daily flow data for the past 68 years [24].

Before running the model, boundary conditions need to be set, particularly the downstream water levels and the upstream discharge. After running RIVICE, simulated water level profiles of the river at different time steps are generated and compared to the gauge data. The fit can be fine-tuned by adjusting the parameter settings. For example, Manning's coefficient of bed roughness n_{bed} (Table 1) is adjusted until the open water simulations are reasonably in line with recorded water levels. The n_{8m} (Table 1) coefficient can be adjusted to calibrate the roughness along the ice cover underside and match simulated ice-cover conditions with back-water staging observations.

Table 1. Parameters used in RIVICE model.

Parameters	Description	Units
Hydraulic roughness		
n_{bed}	River bed roughness	$s/m^{1/3}$
n_{8m}	Ice roughness	$s/m^{1/3}$
Boundary conditions		
Q	Upstream discharge	m^3/s
W	Downstream water level	m a.s.l. [1]
Ice cover characteristics		
v_d	Ice deposit velocity	m/s
v_e	Ice erosion velocity	m/s
V_{ice}	Inflowing ice volume	m^3
FT	Thickness of ice cover front	m
PC	Porosity of ice cover	—
X	Ice bridge location (from XS1 in Figure 1)	m
Strength properties		
K1TAN	Lateral: longitudinal stresses	—
K2	Longitudinal: vertical stresses	—
Slush ice characteristics		
PS	Porosity of slush	—
ST	Thickness of slush pans	m

Note: [1] For water level, m a.s.l. means "meters above sea level".

Parameters that define the condition and state of the ice regime were also included, such as water temperature, ice cover strength parameters, ice transport parameters and ice cover and flow characteristics. Initial values from the literature and other studies were used to set up the model, but were then adjusted and "fine-tuned" during the calibration process. The model run parameter settings from the Monte-Carlo analysis were chosen for the calibration, as this yielded the best fit between the simulation and observed data.

To quantify which factor influences the formation of ice jamming the most, a local parameter sensitivity analysis was carried out to calculate the sensitivity of different parameters on ice jam flooding. Local sensitivity (e) is defined as in [25]:

$$e = \frac{\Delta O}{\Delta P} \times \frac{P}{O} = \frac{(O_0 - O_i)}{(P - 1.1P)} \times \frac{P}{O_0} = \frac{O_0 - O_i}{-0.1 O_0} \tag{2}$$

in which P refers to the parameter, such as n_{bed}, n_{8m}, etc. (Table 1); O_0 is the initial calibrated output variable; and O_i is the output from a change in a parameter value. Sensitive parameters have higher impacts on the output variable and generate higher ΔO values. Some parameters such as inflowing ice volume (v_{ice}), downstream water level, and discharge (Q) are determined by measurements including remote sensing and direct gauge data. Thus, errors from remote sensing data processing may lead to the inaccuracy of the local sensitivity calculation. To improve the accuracy of the analysis, more than one remote sensing technology was utilized in this research.

2.2.2. Remote Sensing Dataset

RADARSAT-2 uses microwaves (radar) for ranging and distance measuring. The waves are transmitted obliquely to the normal surface of the earth and the sensor then receives the backscatter signals from features on the surface of the earth which reveal different properties and characteristics of the surface features [26]. After processing these signals, a two-dimensional image can be constructed from which further analyses of the surface or sub-surface features can be made, which, in our case, is the ice cover of the river. Radar data were used to help determining the breakup time of the Slave River, and more importantly, to be compared with MODIS image, improving our understanding of the ice phenology.

MODIS is an optical remote sensing satellite sensor. In this study, MOD09GQ (MODIS Surface Reflectance) was applied to assess river ice progression. Since MODIS satellites provide products of daily scans of the earth's surface, it is possible to acquire a continuous dataset of fine temporal resolution, allowing better tracking of ice cover breakup behavior and potentially identifying factors that cause ice cover retraction [22]. In the MODIS images of the river, brighter pixels represent ice whereas darker ones correspond to open water. The contrast between the light and dark pixels permits the determination of the length of the ice-covered river channel from which ice volume can be calculated. MODIS datasets from 2000 to 2015 were downloaded from the NASA website [27]. Through pixel analyses of the MODIS images, ice cover information can be attained, which are crucial input data for the RIVICE model. ArcGIS [28] was used to pre-process the raw MODIS imagery data and obtain the ice-covered channel length. In ArcMap, the centerline and cross-sections were marked on the MODIS image from which the length of the ice-covered portion was measured. Using these, together with the average ice cover width and estimated ice thickness, ice volume was determined.

2.2.3. Ice Volume Calculation by Using MODIS

Various methods for ice volume evaluation have been used, including modeling, direct measurements and the combination of these two [19,29]. We present a novel technique in which remote sensing datasets are applied to calculating the extent of the ice cover and, using an empirical equation to estimate ice thicknesses (h) as a function of both Cumulative Freezing Degree Days (CFDD) and the average channel width in the Slave River Delta, derive an ice volume for downstream ice jamming. After comparing many equations for ice thickness (details about two other equations can be found in [30]), Stefan's equation showed the highest consistency with gauged data. Although it is more accurate and reliable to measure the ice thickness using direct measurement or remote sensing technology, the Stefan equation can still reasonably estimate thermal ice growth, even if it does not consider snow depth, which can affect the accuracy of ice thickness calculations. However, while snow depth data were not available for the Slave River Delta, the Stefan equation still provided reasonable ice thickness estimates. RADARSAT-2 will be utilized for ice thickness estimation in future research to compliment the ice thickness calculation. Plots of the CFDD for all winters from 2000 to 2015 and the winter of the year 1980 is shown in Figure 2. The relationship between ice thickness and CFDD is $h = c\sqrt{CFDD}$, where c is a constant. Thawing effects on ice volume can be accounted for by determining the cumulative average air temperature degrees surpassing $-5\,^{\circ}\text{C}$. Thus, thawing effects were used to attenuate CFDD in spring.

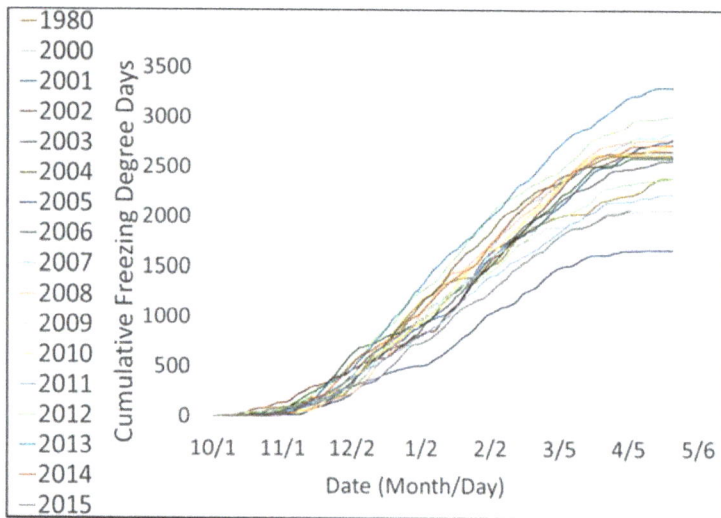

Figure 2. Plots of cumulative freezing degree days for winters (1 October–19 April) between the years 1980 and 2015 at Fort Smith, Northwest Territories.

MOD09GQ band2 (wavelength: 841–876 nm, resolution: 250 m) datasets were used to establish time series to estimate the ice cover situation of the Slave River [31]. Different pixel values in the MODIS images represent different backscatters to which threshold values are applied to distinguish water from ice. By creating the histogram, the distribution of pixel values for each image can be determined. Clear water provides little backscatter signal to the satellite since it absorbs longer wavelengths and the reflected shorter wavelengths tend to scatter in the air. Hence, longer wavelengths appear more sensitive to the difference between water and ice. Since MOD09GQ band2 is longer in wavelength than band1, it has been chosen for our ice phenology research. A heat balance calculation based on upstream river temperature may help determine how much ice is melting and moving respectively. This calculation requires more extensive field sampling and can be addressed in future work. Due to this characteristic of clear water, it appears dark on MODIS imagery. On the other hand, ice provides more reflectance and appears lighter than clear water on the MODIS imagery [32]. This difference leads to the significantly different pixel values in the MODIS imagery. Pixel values, with only one feature variable (a pixel is either ice or water), are normally distributed. Theoretically, a mixture of two normal distributions forms a bimodal distribution. Figure 3 shows an example of the MODIS pixel-value distribution from 17 November 2002. On that day, both ice and water appear in the river, so that two distinct peaks (local maxima) representing ice and water can be found in the histogram, which is consistent with the bimodal distribution theory. There is a big difference between the pixel values of the ice and water, and a value separating the two establishes the threshold. On cloudy days, backscatter values are difficult or even impossible to attain. Hence ice volumes for those days have to be interpolated between clear-day images [12]. The algorithm Harmonic Analysis of Time Series (HANTS) was used to replace cloud-contaminated pixels with Fourier series values from MODIS images [33]. After interpolation, pixel values change so that another threshold must be applied to the interpolated images. Applying the threshold to the MODIS image pixel-value distributions, percentages of water and ice can be measured as well as ice volumes. During the ice breakup period, the diminishing ice of the upstream reach is considered to be inflowing ice at the downstream ice jam. We do make allowance for the recession of the ice front by tracking the location of the ice front at a daily time resolution using the MODIS imagery. Ice motion velocities in the year 2000–2015 were approximately 48–100 km/day. Based on the ice front movement, inflowing ice can be tracked. One point of uncertainty is that not all of the ice downstream of the ice front may be moving; rather, some could be melting while stationary, due to warm inflow river water. However, it is very difficult or perhaps impossible to distinguish how much ice has moved from upstream to the delta, as opposed to just melting before the next image was acquired. A shorter time interval between image acquisitions would be necessary.

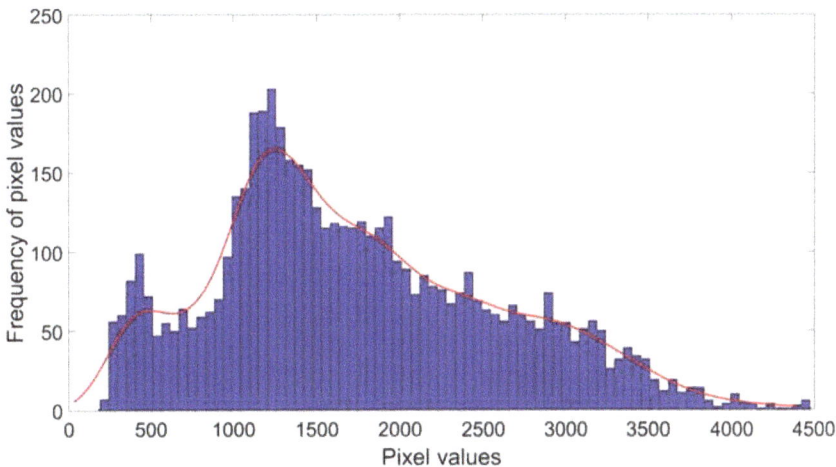

Figure 3. The histogram of pixel values of MODIS imagery.

Ice volume calculation processes are shown in Figure 4 Using the Geographic Information System (GIS) [28], the water surface area of the Slave River Delta was measured, which is 12.8 km^2. Percentages of the ice cover on the Slave River Delta are calculated from MODIS datasets. An example of calculating ice percentages is shown in Figure 4.

2.2.4. Data Preparation for Model

Essential inputs to the RIVICE control files include upstream discharge boundaries, downstream water level boundaries, incoming ice volume, the location of ice lodgments, and cross-sectional data. The water level data can be obtained from gauge stations (Figure 5). Flow data of Resdelta Channel (07NC009) (Figure 1) and Fitzgerald station are shown in Figure 6 and were used as the upstream discharge boundary. The downstream water levels were acquired at Great Slave Lake, which is used as the boundary condition for modelling according to the daily water level data at Yellowknife Bay station (07SB001).

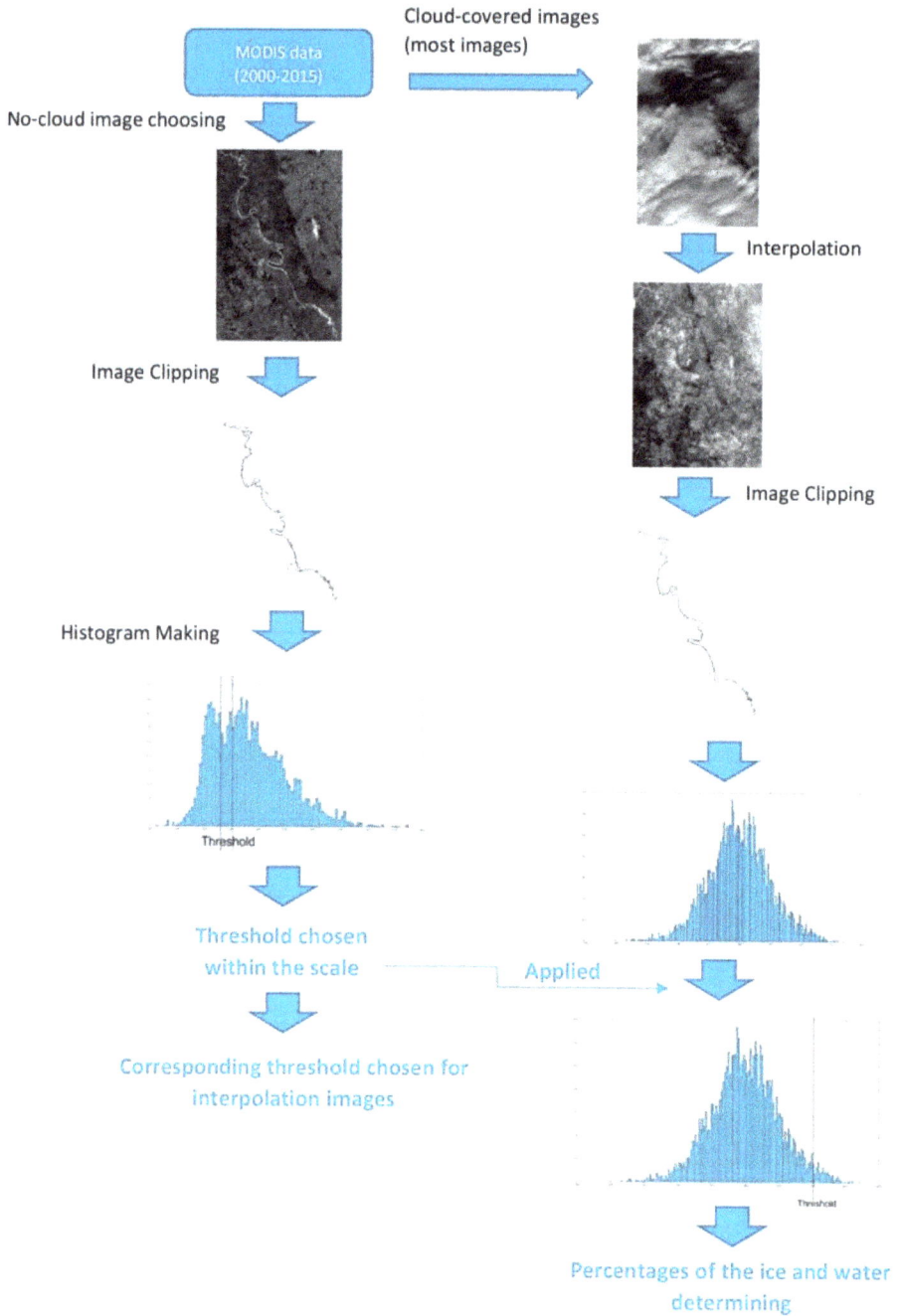

Figure 4. Steps taken in determining ice cover percentages.

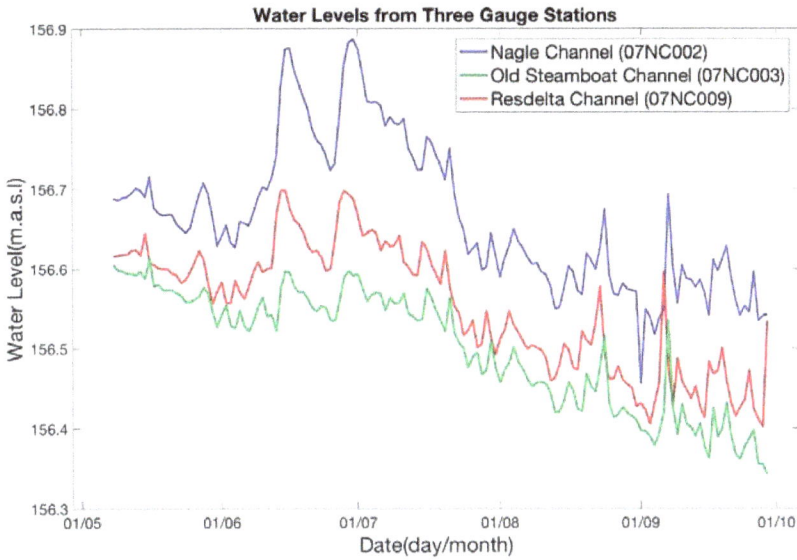

Figure 5. Water level data for open water case of the three gauge stations.

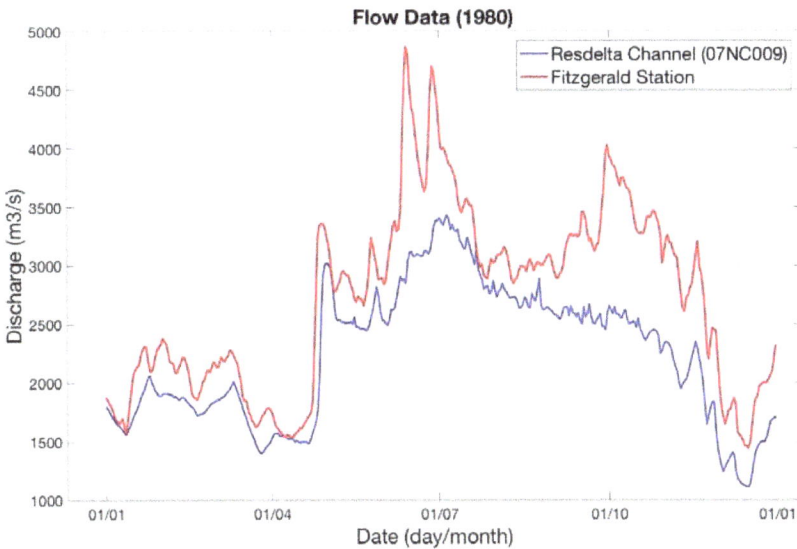

Figure 6. Measured flow data at Resdelta Channel and Fitzgerald Station.

3. Results

3.1. Ice Volume

The CFDD was determined for each winter from mean daily air temperature records between 1971 and 2015. In the Stefan equation, the constant c depends on many factors such as meteorological conditions, geomorphological parameters, and water quality. It is derived empirically as the slope of a linear relationship between measured ice thicknesses and $\sqrt{\text{CFDD}}$ (Figure 7). Based on the field survey and relationship between ice thicknesses and CFDD at the Slave River Delta and its tributaries [13], the constant c was estimated to be 0.021. The blue trendline shows the relationship between the ice thickness and $\sqrt{\text{CFDD}}$ of the Slave River near Fort Smith. The constant c equals 0.018. Hence, 0.02 has been chosen as the constant, and the trendline with a slope of 0.02 m/m is shown by the orange line in Figure 7.

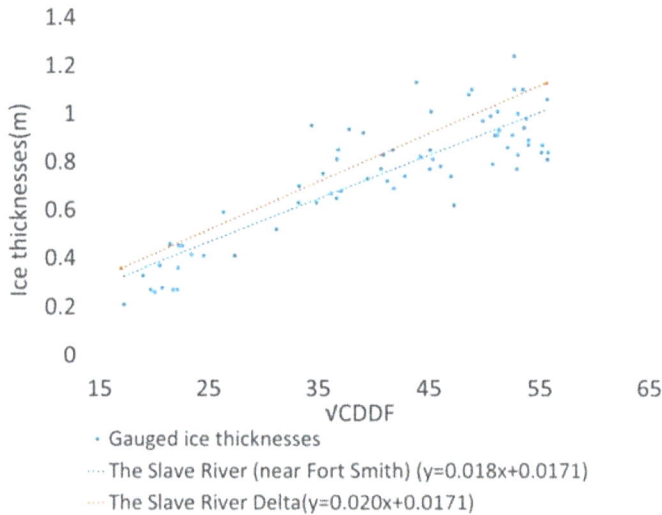

Figure 7. Relationship between the cumulative freezing degree days and ice thicknesses.

After determining ice percentage, ice thickness, and the ice-covered water surface area, ice volumes were calculated.

It is difficult to determine the ice cover extent from MODIS images during cloud-covered periods, particularly for the breakup events of 2007 and 2010. The ice cover thresholds for these two years cannot be determined. Thus, ice volumes cannot be calculated for these two events. Aerial photos acquired along the Slave River on 29 April 2015 were used to validate the ice volume estimation result of MODIS (Figure 8). The RADARSAT-2 image acquired on 30 April was compared with MODIS data for the same day (Figure 9). The noticeable difference between images from these two satellites is partly due to the different acquisition times, other reasons can be found in [12].

In Figure 10, three ice volume clouds (scatters showing the relationship between ice volumes and water levels) are shown. The cluster of blue dots show the modeling results of HEC-RAS, and the red ones and green ones show RIVICE modeling results with and without thawing effects. Thawing comes into effect for days on which average temperatures are above −5 °C. Clearly, the ice volumes are lower when thawing is incorporated in the calculation. The RIVICE clusters yield lower backwater elevations than HEC-RAS, which is to be expected, because not every jam that is formed is expected to be a fully-evolved equilibrium jam as premised by HEC-RAS. In HEC-RAS, stresses acting on an ice jam are calculated based on the ice jam force balance equation, while the deposition and erosion of ice covers are not taken into account [23,34]. In RIVICE, ice jam formation and development were considered in a dynamic balance process.

The incoming ice volume of 1.1 million m^3 will lead to backwater staging effects causing ice jam flooding, which replenishes moisture in the delta during ice-cover breakup. When the incoming ice volume is higher than 1.5 million m^3, the backwater effects are substantial.

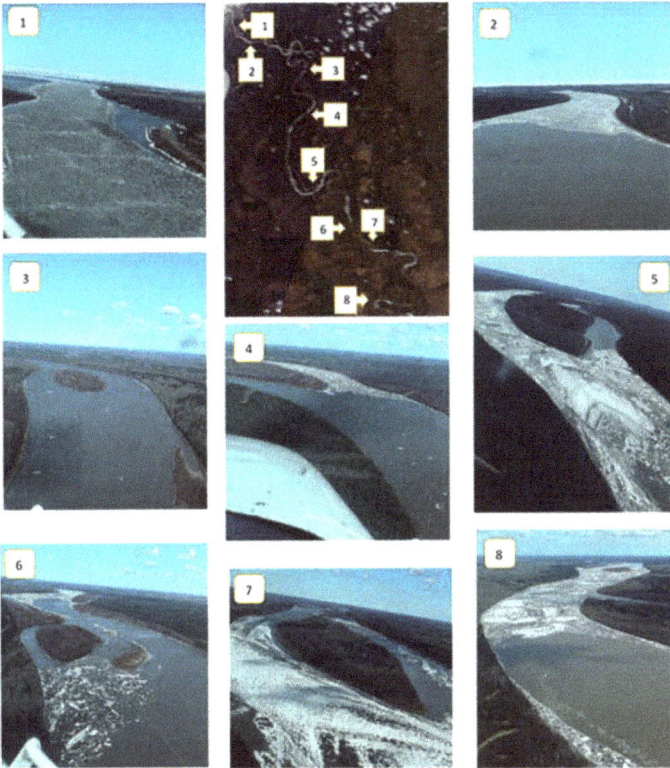

Figure 8. Examples of aerial photographs used to validate the MODIS ice cover estimation (the number shows the location of the aerial photos. Data were acquired on 29 April 2015).

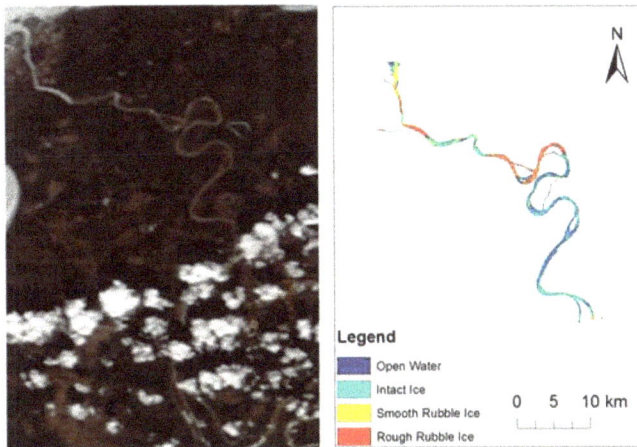

Figure 9. Comparison of RADARSAT-2 (right panel) and MODIS (left panel) image (data were acquired on 30 April 2015).

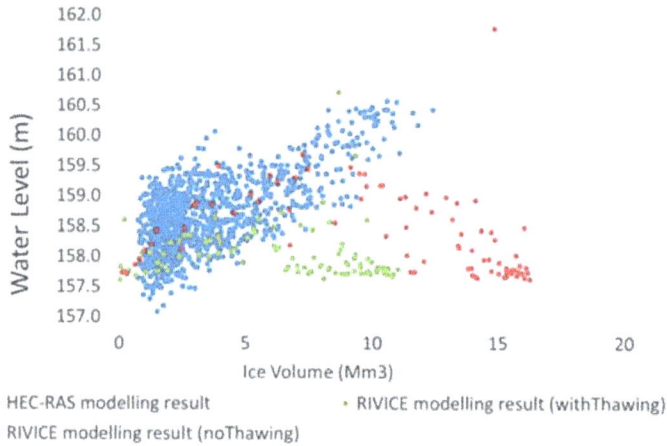

Figure 10. Ice volume calculation results from HEC-RAS and RIVICE.

3.2. Model Calibration

Water levels from 1980 recorded during the open-water and ice-covered period were used to calibrate the RIVICE model.

Open water conditions: In 1980, during the open-water period, the daily flow reached a maximum of 4870 m^3/s, on 13 June 1980 (Figure 6), when the water level elevation of Great Slave Lake was 156.43 m [24]. These values were used as the boundary conditions in RIVICE. The minimum daily flow during the open water period in 1980 was 2660 m^3/s, which occurred on 21 May 1980. Comparisons of gauged and simulated data are provided in Figure 11, showing good agreement between the two.

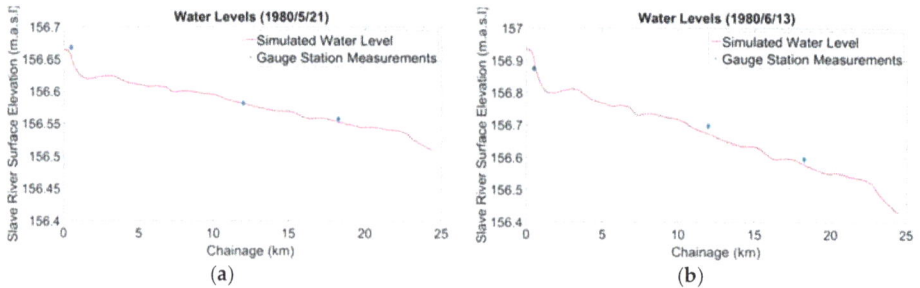

Figure 11. Calibration for RIVICE simulation under open water conditions: (**a**) water level profile under minimum daily flow; (**b**) water level profile under maximum daily flow.

The change of water levels at the Jean River gauge (07NC001) occurred approximately one day earlier than at the Nagle Channel gauge (07NC002) and two days earlier than at the Resdelta River gauge (07NC009) (Figure 1).

Ice-covered conditions: RIVICE was also implemented to simulate ice-covered conditions of the main channel in the Slave River Delta. The water level of the Resdelta River gauge (07NC009) (Figure 1) on 31 January 1980, was used to calibrate the model.

Figure 12 shows good agreement between the modeling result and the gauge recording of Resdelta (07NC009).

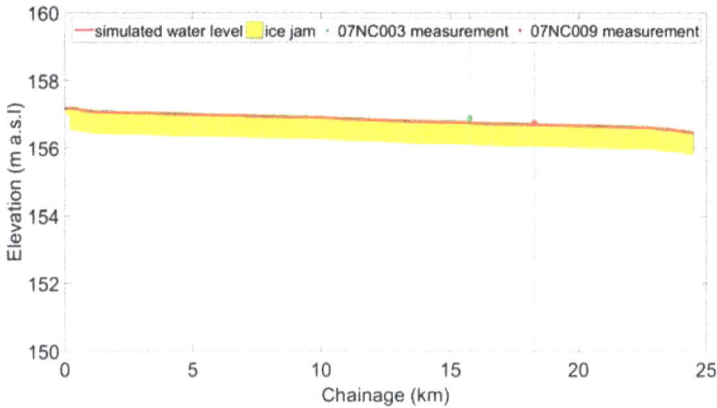

Figure 12. Calibrated ice-covered water profile for the Slave River Delta.

Ice-jammed conditions: Simulating ice jams is more complicated than simulating open water and ice-covered conditions since more factors need to be used in RIVICE calculations to simulate ice jam formation and progression. Water level data at the Great Slave Lake (156.557 m) and the Resdelta Channel (156.884 m) on 27 April 1980 were used to set up and calibrate the model, respectively. After defining the incoming ice volume on ice jam, the water level profile can be determined (Figure 13). Gauged water levels and HEC-RAS modeling results are applied to calibrate the ice-jam modeling results. HEC-RAS is a one-dimensional hydrodynamic model which is mainly used for simulating steady and unsteady flow routing based on an implicit four-point finite difference scheme. The HEC-RAS model simulates only equilibrium ice jams, i.e., in steady state, not dynamically, as is the case in RIVICE. Hence, the HEC-RAS result can be seen as the maximum backwater level that can be determined if the ice jam formation is not limited by the supply of incoming rubble ice.

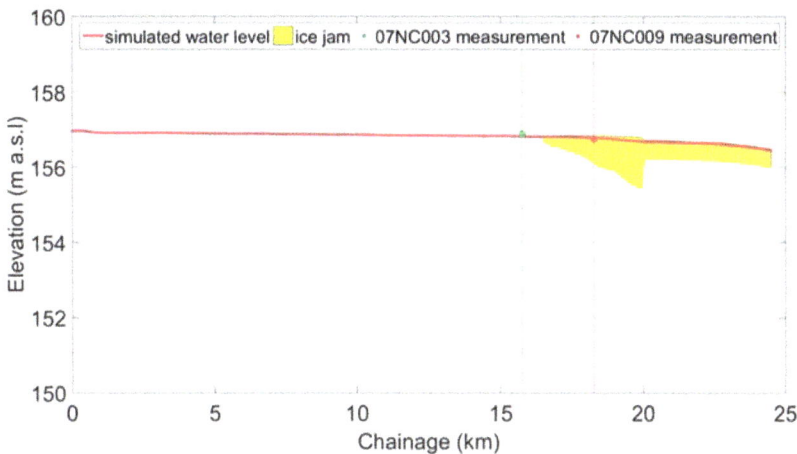

Figure 13. Simulated ice-jammed water profile for the Slave River Delta.

Some of the gauges in the delta were installed during the winter, prior to ice-cover breakup. Ice jams regularly occur along this reach during the final breakup of the Slave River. The error between simulated values and observed ones is shown in Figure 14, which is acceptable.

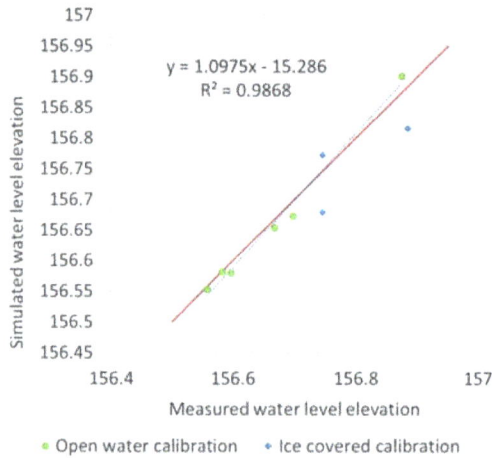

y = 1.0975x - 15.286
R² = 0.9868

Figure 14. Error between simulated values and observed data.

3.3. Ice Jam Flooding

Figure 15 shows the floodwater extensions, calculated with HEC-RAS, within the floodplain for the ponding cover water surface at elevations of 156, 157, and 158 m. For instance, when the backwater level exceeds 156 m, the area to the left-hand side of the red line will be flooded. To determine the backwater effects of ice jamming, which causes flooding, the RIVICE model was run three times under the following conditions: open water, ice-covered, and ice-jammed conditions. These three scenarios were simulated using the same boundary downstream water level elevation (156.45 m), but different upstream water flows. Associated with the HEC-RAS simulated flood extension under different backwater levels, the RIVICE modeling results can help determine the maximum possible flooding extent in the floodplain under different situations.

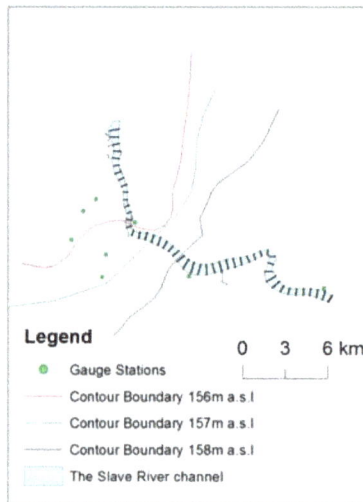

Figure 15. Floodplain under different simulated water levels (HEC-RAS result); Floodplain under different simulated water levels (HEC-RAS result).

3.4. Local Sensitivity Analysis

Applying equation (2) to the local sensitivity analysis, e-values are positively proportional to parameter sensitivity. Given that, parameter sensitivity can be evaluated based on e-values.

Hydraulic properties, including upstream discharge (Q), downstream water level (W), and the riverbed roughness coefficient (n_{bed}), have the greatest impact on the backwater levels. The upstream discharge (Q) has the greatest negative impact on the ice jam length.

Ice transporting properties (v_{dep}, v_e, Table 1) have little impact on the ice jamming since the main process of ice jam formation is ice thickening as forces shove and retract the ice cover. Surprisingly, ice volume (V_{ice}) has some impact on the ice jamming length but not on backwater staging (Figure 16).

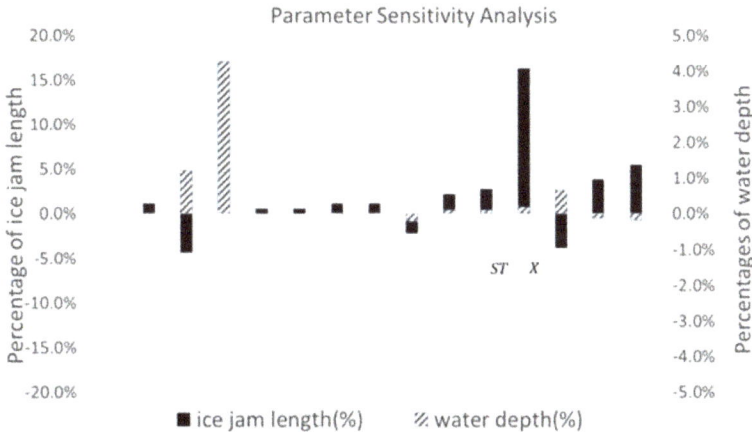

Figure 16. Local parameter sensitivity analysis result.

Physical properties of the ice (FT, PC, PS, ST) have some influence on the ice jam morphology. Increasing porosity will increase compaction between ice blocks when shoved, leading to a reduction in ice jam length. An increase in ice front thickness means more surface area is available for water shoving on the ice front to retract the ice jam cover, which corresponds to the shortening of the ice jam.

The location of the lodgment X has the largest positive impact on the ice jam length, potentially due to varying cross-section flow dimensions along the river. Moving the location further downstream leads to a decrease in backwater staging and shortening of the ice jam cover.

Mechanical properties, K1 and K2, have a significant impact on the ice jam length and partly on the water level, which is consistent with the research in [35]. Ice cover internal resistance will increase with an increase in K1, which provides more resistance to shoving. Therefore, more ice can stay in place and the ice jam length is extended. Greater internal resistance corresponds to less thickening of the ice cover, allowing more flow underneath the ice cover and decreasing backwater effects.

4. Discussion

Both numerical modeling and remote sensing images are useful tools for studying river ice processes. These two methods were combined successfully to characterize ice jamming within the Slave River Delta. The excellent agreement between modeling results and measurement readings endorses the reliability of the river ice model RIVICE. Under open water conditions, this agreement verified the hydraulic properties set for the Slave River Delta. Under ice-covered conditions, with the limitation of sparse gauged data (07NC009), model calibration cannot be firmly supported by the field measurements. HEC-RAS estimated a greater ice volume than RIVICE, but lower than when no thawing effects are considered, which is reasonable.

The local sensitivity analysis indicates that ice cover morphological properties (X, n_{8m}) and hydraulic properties (Q, W, and n_{bed}) have the highest influence on ice jam shape and backwater staging for this low-sloping river section. This is because increasing upstream discharge will raise water levels and increase flow velocities, leading to the additional drag and thrust forcing on the ice cover. The additional forces will lead to the retraction of the ice cover at the front, corresponding to a shortening of the ice jam. Because discharge has the largest effect on ice jam formations, we conclude that discharge variation has the greatest impact on ice jam evolution and subsequent flooding in the delta. Various research shows that discharge variation of the river is sensitive to climate change [36–38]. Coupling a climate model with the hydraulic model could help in understanding and predicting the ice jam behavior within the context of climate change, which will be the scope of our future research. The significant impact of location on the lodgment (X) indicates the morphological factor in ice jam formation and deformation is worth studying further, which concurs with the result in [39]. On the other hand, ice transport characteristics, such as ice deposition and erosion, have little impact on jamming for this reach. In future research, global sensitivity analysis will be carried out for the Slave River Delta.

Remote sensing imagery provides a source of information about ice cover extent and ice breakup sequences, which is helpful for determining the ice volumes required to form and shape ice jams. MODIS imagery facilitated researchers' quantifying of the ice volume, which will stimulate future investigation into the usage of remote sensing in the ice jam modeling area. However, MODIS is restricted to daytime operation. Cloud has a great adverse influence on MODIS information acquisition. It is difficult to determine ice cover extent from MODIS images during cloud-covered periods. Hence MODIS data should be supplemented with RADARSAT-2 imagery, which is not influenced by cloud cover or restricted to daytime acquisition. This approach will allow us to determine thresholds during cloudy days to improve ice volume calculations, and determine ice volumes for specific days. The downside of RADARSAT-2 is the lower revisit time. However, the combination of RADARSAT-2 and MODIS may enhance the accuracy of ice volume calculation, which is a topic for further research.

Author Contributions: Fan Zhang ran the RIVICE simulations and extracted ice volumes from the satellite imagery. Mahtab Mosaffa ran the HEC-RAS simulations and determined flood extents in Figure 15. Thuan Chu devised the method of determining threshold values between ice and water distribution from MODIS imagery. Karl-Erich Lindenschmidt conceptualized this research, helped setup the RIVICE model and assisted in the manuscript write-up.

Conflicts of Interest: The authors declare no conflict of interest.

References

1. Beltaos, S. Hydrodynamic and climatic drivers of ice breakup in the lower Mackenzie River. *Cold Reg. Sci. Technol.* **2013**, *95*, 39–52. [CrossRef]
2. Helland, I.P.; Finstad, A.G.; Forseth, T.; Hesthagen, T.; Ugedal, O. Ice-cover effects on competitive interactions between two fish species. *J. Anim. Ecol.* **2011**, *80*, 539–547. [CrossRef] [PubMed]
3. Beltaos, S. Comparing the impacts of regulation and climate on ice-jam flooding of the Peace-Athabasca Delta. *Cold Reg. Sci. Technol.* **2014**, *108*, 49–58. [CrossRef]
4. Prowse, T.D.; Conly, F.M. Effects of climatic variability and flow regulation on ice-jam flooding of a northern delta. *Hydrol. Process.* **1998**, *12*, 1589–1610. [CrossRef]
5. Public Works and Government Services of Canada. *Slave River—Water and Suspended Sediment Quality in the Transboundary Reach of the Slave River, Northwest Territories*; Report Summary; Aboriginal Affairs and Northern Development Canada: Slave River, AB, Canada, 2012.
6. Beltaos, S.; Prowse, T.D.; Carter, T. Ice regime of the lower Peace River and ice-jam flooding of the Peace-Athabasca Delta. *Hydrol. Process.* **2006**, *20*, 4009–4029. [CrossRef]

7. Lindenschmidt, K.E.; Sydor, M.; Carson, R.W.; Harrison, R. Ice jam modelling of the Lower Red River. *J. Water Resour. Prot.* **2012**, *4*, 16739. [CrossRef]
8. Lindenschmidt, K.E.; Das, A.; Rokaya, P.; Chu, T.A. Ice-jam flood risk assessment and mapping. *Hydrol. Process.* **2016**, *30*, 3754–3769. [CrossRef]
9. Chu, T.; Das, A.; Lindenschmidt, K.E. Monitoring the variation in ice-cover characteristics of the Slave River, Canada using RADARSAT-2 data—A case study. *Remote Sens.* **2015**, *7*, 13664–13691. [CrossRef]
10. Government RADARSAT Data Services. *A New Satellite, A New Vision*; Canadian Spave Agency: Sanit-Hubert, QC, Canada, 2010.
11. Lindenschmidt, K.-E.; Syrenne, G.; Harrison, R. Measuring Ice Thicknesses along the Red River in Canada Using RADARSAT-2 Satellite Imagery. *J. Water Resour. Prot.* **2010**, *2*, 923–933. [CrossRef]
12. Chu, T.; Lindenschmidt, K.E. Integration of space-borne and air-borne data in monitoring river ice processes in the Slave River, Canada. *Remote Sens. Environ.* **2016**, *181*, 65–81. [CrossRef]
13. Das, A.; Sagin, J.; Van der Sanden, J.; Evans, E.; McKay, K.; Lindenschmidt, K.E. Monitoring the freeze-up and ice cover progression of the Slave River. *Can. J. Civ. Eng.* **2015**, *42*, 609–621. [CrossRef]
14. Vermote, E.F.; Kotchenowa, S.Y.; Bay, J.P. *MODIS Surface Reflectance User's Guide*; NASA: Greenbelt, MD, USA, 2015; p. 35.
15. Chaouch, N.; Temimi, M.; Romanov, P.; Cabrera, R.; McKillop, G.; Khanbilvardi, R. An automated algorithm for river ice monitoring over the Susquehanna River using the MODIS data. *Hydrol. Process.* **2014**, *28*, 62–73. [CrossRef]
16. Kraatz, S.; Khanbilvardi, R.; Romanov, P. River ice monitoring with MODIS: Application over Lower Susquehanna River. *Cold Reg. Sci. Technol.* **2016**, *131*, 116–128. [CrossRef]
17. Muhammad, P.; Duguay, C.; Kang, K.K. Monitoring ice break-up on the Mackenzie River using MODIS data. *Cryosphere* **2016**, *10*, 569–584. [CrossRef]
18. Shi, L.; Wang, H.; Zhang, W.; Shao, Q.; Yang, F.; Ma, Z.; Wang, Y. Spatial response patterns of subtropical forests to a heavy ice storm: A case study in Poyang Lake Basin, southern China. *Nat. Hazards* **2013**, *69*, 2179–2196. [CrossRef]
19. Maskey, S.; Uhlenbrook, S.; Ojha, S. An analysis of snow cover changes in the Himalayan region using MODIS snow products and in-situ temperature data. *Clim. Chang.* **2011**, *108*, 391–400. [CrossRef]
20. Geldsetzer, T.; van der Sanden, J.; Brisco, B. Monitoring lake ice during spring melt using RADARSAT-2 SAR. *Can. J. Remote Sens.* **2010**, *36*, S391–S400. [CrossRef]
21. Brock, B.E.; Wolfe, B.B.; Edwards, T.W.D. Spatial and temporal perspectives on spring break-up flooding in the Slave River Delta, NWT. *Hydrol. Process.* **2008**, *22*, 4058–4072. [CrossRef]
22. Lindenschmidt, K.-E.; Das, A. A geospatial model to determine patterns of ice cover breakup along the Slave River1. *Can. J. Civ. Eng.* **2015**, *42*, 675–685. [CrossRef]
23. Environment Canada RIVICE Steering Committee. *RIVICE Model-User's Manual*; Environment Canada: Winnipeg, MB, Canada, 2013.
24. Environment Canada, Historical Hydrometric Data. Available online: https://wateroffice.ec.gc.ca/mainmenu/historical_data_index_e.html (accessed on 25 April 2017).
25. Lindenschmidt, K.E.; Rode, M. Linking hydrology to erosion modelling in a river basin decision support and management system. In Integrated Water Resources Management. 2001; pp. 243–248. Available online: http://iahs.info/uploads/dms/iahs_272_243.pdf (accessed on 27 April 2017).
26. Canada Center for Remote Sensing. *Fundamentals of Remote Sensing*; CCRS: Ottawa, ON, Canada, 2009.
27. NASA Level-1 and Atmosphere Archive & Distribution System (LAADS) Distributed Active Archive Center (DAAC), MOD09GQ. Available online: https://ladsweb.nascom.nasa.gov/ (accessed on 27 April 2017).
28. *ArcGIS Desktop*; Version:10.3.1.4959; Environmental Systems Research Institute (ESRI): Redlands, CA, USA, 2015.
29. Su, H.; Wang, Y.P. Using MODIS data to estimate sea ice thickness in the Bohai Sea (China) in the 2009–2010 winter. *J. Geophys. Res. Oceans* **2012**, *117*, C10018. [CrossRef]
30. Carstensen, D. *Eis im Wasserbau—Theorie, Erscheinungen, Bemessungsgrößen*; Selbstverl: Dresden, Germany, 2008.
31. Vermote, E.; Nixon, D. *MOD09 (Surface Reflectance) User's Guide*; NASA: Greenbelt, MD, USA, 2011.
32. Khorram, S. *Remote Sensing*; Springer: Berkeley, CA, USA, 2012; p. 134.
33. Roerink, G.J.; Menenti, M.; Verhoef, W. Reconstructing cloudfree NDVI composites using Fourier analysis of time series. *Int. J. Remote Sens.* **2000**, *21*, 1911–1917. [CrossRef]

34. Brunner, G.W.; United States Army Corps of Engineers; Institute for Water Resources (U.S.); Hydrologic Engineering Center (U.S.). *HEC-RAS River Analysis System-Hydraulic Reference Manual*; US Army Corps of Engineers, Institute for Water Resources, Hydrologic Engineering Center: Davis, CA, USA, 2016.

35. Lindenschmidt, K.E.; Sydor, M.; Carson, R.W. Modelling ice cover formation of a lake-river system with exceptionally high flows (Lake St. Martin and Dauphin River, Manitoba). *Cold Reg. Sci. Technol.* **2012**, *82*, 36–48. [CrossRef]

36. Pietroniro, A.; Leconte, R.; Toth, B.; Peters, D.L.; Kouwen, N.; Conly, F.M.; Prowse, T. Modelling climate change impacts in the Peace and Athabasca catchment and delta: III—Integrated model assessment. *Hydrol. Process.* **2006**, *20*, 4231–4245. [CrossRef]

37. Li, Z.; Huang, G.H.; Wang, X.Q.; Han, J.C.; Fan, Y.R. Impacts of future climate change on river discharge based on hydrological inference: A case study of the Grand River Watershed in Ontario, Canada. *Sci. Total Environ.* **2016**, *548*, 198–210. [CrossRef] [PubMed]

38. Tao, B.; Tian, H.Q.; Ren, W.; Yang, J.; Yang, Q.C.; He, R.Y.; Cai, W.J.; Lohrenz, S. Increasing Mississippi river discharge throughout the 21st century influenced by changes in climate, land use, and atmospheric CO_2. *Geophys. Res. Lett.* **2014**, *41*, 4978–4986. [CrossRef]

39. Lindenschmidt, K.E.; Chun, K.P. Evaluating the impact of fluvial geomorphology on river ice cover formation based on a global sensitivity analysis of a river ice model. *Can. J. Civ. Eng.* **2013**, *40*, 623–632. [CrossRef]

water MDPI

Article

Updated Smoothed Particle Hydrodynamics for Simulating Bending and Compression Failure Progress of Ice

Ningbo Zhang [1] , Xing Zheng [1,* and Qingwei Ma [1,2]

[1]   College of Shipbuilding Engineering, Harbin Engineering University, Harbin 150001, China;
      zhangningbo@hrbeu.edu.cn (N.Z.); q.ma@city.ac.uk (Q.M)
[2]   School of Mathematics, Computer Science & Engineering, University of London, London EC1V 0HB, UK
*   Correspondence: zhengxing@hrbeu.edu.cn

Received: 8 September 2017; Accepted: 7 November 2017; Published: 12 November 2017

Abstract: In this paper, an updated Smoothed Particle Hydrodynamics (SPH) method based on the Simplified Finite Difference Interpolation scheme (SPH_SFDI) is presented to simulate the failure process of ice. The Drucker–Prager model is embedded into the SPH code to simulate the four point bending and uniaxial compression failure of ice. The cohesion softening elastic–plastic model is also used in the SPH_SFDI framework. To validate the proposed modeling approach, the numerical results of SPH_SFDI are compared with the standard SPH and the experimental data. The good agreement demonstrated that the proposed SPH_SFDI method including the elastic–plastic cohesion softening Drucker–Prager failure model can provide a useful numerical tool for simulating failure progress of the ice in practical field. It is also shown that the SPH_SFDI can significantly improve the capability and accuracy for simulating ice bending and compression failures as compared with the original SPH scheme.

Keywords: ice failure; SPH_SFDI; Drucker–Prager model; bending; uniaxial compression; cohesion softening elastic–plastic model

---

1. Introduction

With the increasing activities in Arctic regions, the method for the accurate calculation of the corresponding ice loads on structures are crucial to the design of marine structures operating in ice [1]. To simulate ice–water or ice–ship interactions effectively, it is necessary to have a reasonable study and understanding of the ice failure progress. The bending failure is the common failure behavior of the ice and is important for ships in ice–ship interactions because of the inclined contact interfaces with the ice [2,3]. In addition, under the compression of a ship or structure, compressive failure (crushing) is also easy to occur during the failure process of the ice. Thus, it is of high importance to study the bending and compression failure progress of the ice.

In the past few years, many full-scale tests and model tests of the ice failure have been investigated [4–8]. However, the experimental data are highly dispersed because of different experimental equipment, different test methods, and different measurements of the ice specimens. In addition, some simplified empirical models are also used to study the ice failure and the interaction of ice and structure [9–11]. However, these simplified empirical models only focus on some main aspects of ice failure and are lack of the study of dynamics and some changing details during the ice failure progress. Thus, it is very important to develop a reliable numerical ice model to simulate the sea ice failure in the bending and compression process, especially the current studies on the behavior of sea ice failure are not adequate; although some obtained numerical approaches were proposed to simulate sea ice failure, these studies mainly focus on the ice–structure interaction rather than on the failure properties of sea ice itself.

The existing numerical methods for simulating ice destruction mainly include the finite element method (FEM) [12], the discrete element method (DEM) [13] or their coupled forms [14,15]. To some extent, these methods can achieve good simulation results for ice failure. However, due to the different discretization schemes of FEM and DEM, the numerical continuity is not easily guaranteed on the solid boundary in coupling method of FEM and DEM. For FEM, it has some common drawbacks relevant to simulating ice failure, which are mesh tangling, and using erosion to predict failure patterns [16]. Deb and Pramanik [17] pointed out that DEM needs to make extensive calibration work to identify the parameters for both deformability and strength.

In recent years, SPH method is emerging as a potential tool for simulating the large deformation and failure behavior of solids. Because of its Lagrangian behaviors, cracks may initiate and propagate immediately and naturally after the yielding of SPH particles. Therefore, SPH can simulate the large deformations, failure behaviors effectively and accurately. Thus, it can be easily used to solve solid failure problems, including the fracture, crushing and fragmentation with the application of solid state constitutive relation.

SPH was originally introduced by Lucy [18] and Gingold and Monaghan [19] to solve astrophysical problems. In recent years, SPH method has been successfully applied to a wide range of problems, which include fluid flows [20], geophysical flows [21], water wave dynamics [22–24] and wave–structure interaction problems [25,26]. Currently, there are two different approaches in the SPH formulation: the "weakly compressible" (WCSPH) [20,21] and the "truly incompressible" (ISPH) [27–29], the first of which is employed in this paper. Libersky and Petschek [30] applied SPH to solid mechanics firstly. Benz and Asphaug [31] extended their work to simulation of the fracture process in brittle solids. Then Randles and Libersky [32] used SPH successfully to study dynamic response of solid material with large deformations. Bui et al. [33] applied SPH to model large deformation or post-failure of soil. Deb and Pramanik [17] and Douillet-Grellier et al. [34] simulated the brittle fracturing process of rock by SPH, respectively. Zhang et al. [35] tried to use the SPH method based on the failure model [17] to study the fracture of ice. This paper draws on the elastic–perfectly plastic constitutive equation in Bui et al. [33] and combines the cohesion softening law in Whyatt and Board [36] and Drucker–Prager yield criterion to reflect the plasticity and brittleness of ice during the failure progress. Recently, Das [16] used SPH mode in LS-DYNA to simulate ice beam in four-point bending. In his study, the Von Mises yield criterion is embedded into the SPH to identify the failure of ice. Besides, in his approach once failure is reached, the deviatoric stress components are scaled directly to zero without the cohesion softening and stress correction used in this paper.

The main contribution of the paper lies in the following two aspects. On the one hand, the cohesion softening elasto-plastic constitutive model integrated with the Drucker–Prager yield criterion with plastic flow rules has been prospectively implemented to simulate the plastic failure of ice in the SPH framework. As far as we know, similar failure investigations have been widely used in soil and rock mechanics but almost not known in the ice field. The validation of numerical results shows that this approach can accurately simulate the failure behavior of ice in the practical field. On the other hand, the standard SPH algorithms lack some kinds of high accuracy due to the formulation of its first-order derivative. To improve this situation the Simplified Finite Difference Interpolation (SFDI) method proposed by Ma [37] is used to improve the shear stress and strain rate formulations thus more reasonable failure path and pattern during the ice failure process can be achieved. The enhanced performance of SPH_SFDI method (compared with the standard SPH method) in predicting a more precise force and stress of ice field is also demonstrated by the robust comparisons between the numerical results and experimental data. It needs to be pointed out that the SPH_SFDI method is applied only to 2D test cases in this paper.

2. Governing Equations

The governing equations in the SPH method include the mass conservation equation and momentum conservation equation, which written in the Lagrangian form, are given as:

$$\frac{D\rho}{Dt} = -\frac{1}{\rho}\frac{\partial v^\alpha}{\partial x^\alpha} \tag{1}$$

$$\frac{Dv^\alpha}{Dt} = \frac{1}{\rho}\frac{\partial \sigma^{\alpha\beta}}{\partial x^\beta} + g^\alpha \tag{2}$$

where α and β are the Cartesian components in x, y and z directions; v is the particle velocity; ρ is the ice density; $\sigma^{\alpha\beta}$ is the tress tensor of ice particles; g is the gravitational acceleration; and D/Dt is the particle derivative following motion. The ice constitutive relation need to be applied into the system to solve the governing equations (Equations (1) and (2)). In this paper, the stress tensor can be divided into two parts, which is same with [32] and includes the hydrostatic pressure and deviatoric shear stress:

$$\sigma^{\alpha\beta} = \frac{1}{3}\sigma^{\gamma\gamma}\delta^{\alpha\beta} + s^{\alpha\beta} \tag{3}$$

in which $\delta^{\alpha\beta}$ = Kronecker delta and satisfies the following conditions: $\delta^{\alpha\beta} = 1$ if $\alpha = \beta$ or $\delta^{\alpha\beta} = 0$ when $\alpha \neq \beta$.

2.1. Ice Elasto-Plastic Constitutive Model

To simulate the ice failure process, an elasto-plastic constitutive model [33] is applied into SPH in this paper. The components of the strain rate $\dot{\varepsilon}^{\alpha\beta}$ are given by:

$$\dot{\varepsilon}^{\alpha\beta} = \frac{1}{2}\left(\frac{\partial v^\alpha}{\partial x^\beta} + \frac{\partial v^\beta}{\partial x^\alpha}\right) \tag{4}$$

For an elasto-plastic material, the strain rate $\dot{\varepsilon}^{\alpha\beta}$ can be divided into the elastic strain rate tensor $\dot{\varepsilon}_e^{\alpha\beta}$ and the plastic strain rate tensor $\dot{\varepsilon}_p^{\alpha\beta}$. The elastic strain rate tensor $\dot{\varepsilon}_e^{\alpha\beta}$ follows the generalized Hooke's law:

$$\dot{\varepsilon}_e^{\alpha\beta} = \frac{\dot{s}^{\alpha\beta}}{2G} + \frac{1-2v}{3E}\dot{\sigma}^{\gamma\gamma}\delta^{\alpha\beta} \tag{5}$$

which $\dot{s}^{\alpha\beta}$ = the deviatoric shear stress rate tensor; G and E are the shear modulus and Young's modulus, respectively; and v is Poisson's ratio. The plastic strain rate tensor $\dot{\varepsilon}_p^{\alpha\beta}$ is obtained according to the flow rule:

$$\dot{\varepsilon}_p^{\alpha\beta} = \dot{\lambda}\frac{\partial Q}{\partial \sigma^{\alpha\beta}} \tag{6}$$

where $\dot{\lambda}$ is the plastic multiplier rate, and Q is the plastic potential function which determines the development direction of plastic strain. The plastic multiplier $\dot{\lambda}$ is computed through the consistency condition, which is given by:

$$dF = \frac{\partial F}{\partial \sigma^{\alpha\beta}}d\sigma^{\alpha\beta} = 0 \tag{7}$$

According to Equations (5) and (6), the total strain rate tensor can be expressed as:

$$\dot{\varepsilon}^{\alpha\beta} = \frac{\dot{s}^{\alpha\beta}}{2G} + \frac{1-2v}{3E}\dot{\sigma}^{\gamma\gamma}\delta^{\alpha\beta} + \dot{\lambda}\frac{\partial Q}{\partial \sigma^{\alpha\beta}} \tag{8}$$

According to Equations (3) and (8), the general stress–strain equation of the elastic–plastic ice material can be given by:

$$\dot{\sigma}^{\alpha\beta} = 2G\dot{e}^{\alpha\beta} + K\dot{\varepsilon}^{\gamma\gamma}\delta^{\alpha\beta} - \dot{\lambda}\left[\left(K - \frac{2G}{3}\right)\frac{\partial Q}{\partial\sigma^{mn}}\delta^{mn}\delta^{\alpha\beta} + 2G\frac{\partial Q}{\partial\sigma^{\alpha\beta}}\right] \tag{9}$$

in which α and β are free indexes, m and n are dummy indexes which denote the Cartesian components x, y with the Einstein convention applied to repeated indices; $e^{\alpha\beta} = \dot{\varepsilon}^{\alpha\beta} - \frac{1}{3}\dot{\varepsilon}^{\gamma\gamma}\delta^{\alpha\beta}$ is the deviatoric shear strain rate tensor; $K = E/(3(1-2v))$ is the elastic bulk modulus; and $G = E/(2(1+v))$ is the shear modulus.

The plastic multiplier rate $\dot{\lambda}$ of an elasto-plastic material can be calculated by substituting Equation (9) into Equation (7) as follows:

$$\dot{\lambda} = \frac{2G\dot{e}^{\alpha\beta}\frac{\partial F}{\partial\sigma^{\alpha\beta}} + \left(K - \frac{2G}{3}\right)\dot{\varepsilon}^{\gamma\gamma}\frac{\partial F}{\partial\sigma^{\alpha\beta}}\delta^{\alpha\beta}}{2G\frac{\partial F}{\partial\sigma^{mn}}\frac{\partial Q}{\partial\sigma^{mn}} + \left(K - \frac{2G}{3}\right)\frac{\partial F}{\partial\sigma^{mn}}\delta^{mn}\frac{\partial Q}{\partial\sigma^{mn}}\delta^{mn}} \tag{10}$$

3. Failure Model in the SPH Framework

3.1. Drucker–Prager Model

The Drucker–Prager yield criterion (Figure 1) has been widely used in soil and rock mechanics but is almost unknown in the ice field. In this paper, the Drucker–Prager yield criterion with flow rules has been prospectively used to determine the plastic regime of the ice. The validation of numerical results in Section 5 shows that the Drucker–Prager yield criterion can be used to identify the occurrence of the plastic deformation of ice particles in SPH.

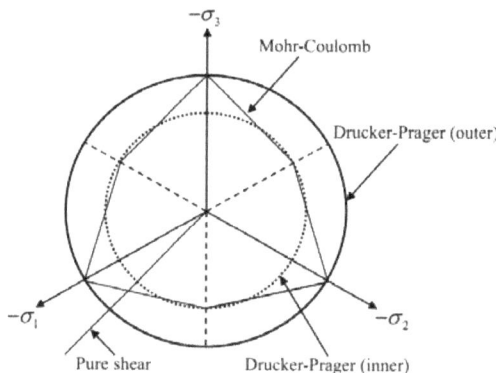

Figure 1. The yield surface π-plane section of Drucker–Prager.

In this study, the Drucker–Prager yield criterion can be expressed as following:

$$F(\sigma^{\alpha\beta}, c) = \sqrt{J_2(s^{\alpha\beta})} + \alpha_\phi I_1(\sigma^{\alpha\beta}) - \xi c = 0 \tag{11}$$

in which c is the ice cohesion, $J_2(s^{\alpha\beta})$ is the second invariant of the stress tensor, and $I_1(\sigma^{\alpha\beta})$ is one third of the first invariant of the stress tensor. The parameters α_ϕ and ξ are defined as:

$$\alpha_\phi = \frac{6\sin\phi}{\sqrt{3}(3-\sin\phi)}, \quad \xi = \frac{6\cos\phi}{\sqrt{3}(3-\sin\phi)} \tag{12}$$

where ϕ is the friction angle. In addition, the plastic potential function is also used to completely define the relationship between the stress and strain. The flow rules are usually applied into SPH to simulate solid fracture, including the associated flow rule and non-associated flow rule. In the associated flow rules, the plastic potential function has the same form with the yield criterion, namely as:

$$Q(\sigma^{\alpha\beta}, c) = \sqrt{J_2(s^{\alpha\beta})} + \alpha_\phi I_1(\sigma^{\alpha\beta}) - \xi c \tag{13}$$

The non-associated plastic potential function is taken to be:

$$Q(\sigma^{\alpha\beta}, c) = \sqrt{J_2(s^{\alpha\beta})} + \overline{\eta} I_1(\sigma^{\alpha\beta}) \tag{14}$$

where parameter $\overline{\eta}$ is related to dilatancy angle φ, which can be expressed as:

$$\overline{\eta} = \frac{6 \sin \varphi}{\sqrt{3}(3 - \sin \varphi)} \tag{15}$$

Substituting Equation (13) into Equations (9) and (10), the stress–strain relationship with the associated plastic flow rule is given by:

$$\dot{\sigma}^{\alpha\beta} = 2G\dot{e}^{\alpha\beta} + K\dot{\varepsilon}^{\gamma\gamma}\delta^{\alpha\beta} - \dot{\lambda}\left(\eta K\delta^{\alpha\beta} + \frac{G}{\sqrt{J_2}}s^{\alpha\beta}\right) \tag{16}$$

When $F(\sigma^{\alpha\beta}, c) < 0$, it is in pure elasticity condition:

$$\dot{\sigma}^{\alpha\beta} = \begin{cases} 2G\dot{e}^{\alpha\beta} + K\dot{\varepsilon}^{\gamma\gamma}\delta^{\alpha\beta} & if \ F(\sigma^{\alpha\beta}, c) < 0 \\ 2G\dot{e}^{\alpha\beta} + K\dot{\varepsilon}^{\gamma\gamma}\delta^{\alpha\beta} - \dot{\lambda}\left(\eta K\delta^{\alpha\beta} + \frac{G}{\sqrt{J_2}}s^{\alpha\beta}\right) & else \end{cases} \tag{17}$$

where the plastic multiplier rate $\dot{\lambda}$ is calculated for the ice model by:

$$\dot{\lambda} = \frac{\eta K\dot{\varepsilon}^{\gamma\gamma} + \frac{G}{\sqrt{J_2}}s^{\alpha\beta}\dot{\varepsilon}^{\alpha\beta}}{\eta^2 K + G} \tag{18}$$

The stress–strain equation of the ice model with the non-associated flow rule is obtained by taking Equation (14) into Equations (9) and (10) as follows:

$$\dot{\sigma}^{\alpha\beta} = \begin{cases} 2G\dot{e}^{\alpha\beta} + K\dot{\varepsilon}^{\gamma\gamma}\delta^{\alpha\beta} & if \ F(\sigma^{\alpha\beta}, c) < 0 \\ 2G\dot{e}^{\alpha\beta} + K\dot{\varepsilon}^{\gamma\gamma}\delta^{\alpha\beta} - \dot{\lambda}\left(\overline{\eta} K\delta^{\alpha\beta} + \frac{G}{\sqrt{J_2}}s^{\alpha\beta}\right) & else \end{cases} \tag{19}$$

where the plastic multiplier rate $\dot{\lambda}$ can be written as:

$$\dot{\lambda} = \frac{\overline{\eta} K\dot{\varepsilon}^{\gamma\gamma} + \frac{G}{\sqrt{J_2}}s^{\alpha\beta}\dot{\varepsilon}^{\alpha\beta}}{\eta\overline{\eta} K + G} \tag{20}$$

It can be seen from the above description that the main difference between the associative and non-associative models is reflected in the dilatancy angle. In the associated flow rule ice model, the Dilatancy angle is always equal to the friction angle, whereas dilatancy angle is optional in the non-associated flow rule. It should be noted that according to the comparative analysis in the following Section 5, non-associative flow rule yield more stable and precise numerical results than associative flow rule in this paper.

3.2. Numerical Errors in Computational Plasticity

In computation for plastic deformation of elastic–plastic material using the Drucker–Prager yield criterion, the numerical errors are easy to occur, which corresponds to the following condition:

$$- \alpha_\phi I_1^n + k_c < \sqrt{J_2^n} \tag{21}$$

In this study, a stress-rescaling procedure based on Bui et al. [33] is adopted to modify the stress. The stress components are modified according to the following relation:

$$\tilde{\sigma}^{\alpha\beta} = r^n s^{\alpha\beta} + \frac{1}{3} I_1(\sigma^{\alpha\beta}) \delta^{\alpha\beta} \tag{22}$$

This scaling factor r^n at the time step n is defined by:

$$r^n = \frac{-\alpha_\phi I_1^n + \xi c}{\sqrt{J_2^n}} \tag{23}$$

In addition, if the condition $-\alpha_\phi I_1^n + \xi c < 0$ is satisfied at the time step n, the normal stress components need to be adjusted to the new correct values $\tilde{\sigma}^{\alpha\beta}$:

$$\tilde{\sigma}^{\alpha\beta} = \sigma^{\alpha\beta} + \frac{1}{3} \left(I_1^n(\sigma^{\alpha\beta}) - \frac{\xi c}{\alpha_\phi} \right) \delta^{\alpha\beta} \tag{24}$$

When $-\alpha_\phi I_1^n + k_c < \sqrt{J_2^n}$ is satisfied, the stress tensor needs to take the plastic correction, which can be expressed as:

$$\tilde{\sigma}^{\alpha\beta} = \begin{cases} \sigma^{\alpha\beta} + \frac{1}{3} \left(I_1(\sigma^{\alpha\beta}) - \frac{\xi c}{\alpha_\phi} \right) \delta^{\alpha\beta} & if \ (-\alpha_\phi I_1 + \xi c) < 0 \\ \frac{-\alpha_\phi I_1^n + \xi c}{\sqrt{J_2^n}} s^{\alpha\beta} + \frac{1}{3} I_1(\sigma^{\alpha\beta}) \delta^{\alpha\beta} & else \end{cases} \tag{25}$$

3.3. Cohesion Softening

In this paper, the cohesion softening law [36] needs to be used in the Drucker–Prager model to simulate the reduction of the ice strength under external loading numerically. In addition, the cohesion softening model can imply the time dependency of ice failure, which is validated in Section 5.2. The model of cohesion softening is realized by making cohesion c a purely linear function of the accumulated plastic strain (Figure 2), which is similar to:

$$c = c(\bar{\varepsilon}_p) = c_0 + k(\bar{\varepsilon}_p) \tag{26}$$

The specific cohesion softening law in this paper is shown as:

$$c = \begin{cases} c_0 - k\bar{\varepsilon}_p & if \ c > c_R \\ c_R & else \end{cases} \tag{27}$$

k is the specific softening coefficient and c_R is the minimum cohesion. The accumulated plastic strain $\bar{\varepsilon}_p$ can be obtained by the associative softening law as:

$$\dot{\bar{\varepsilon}}_p = -\dot{\lambda} \frac{\partial F}{\partial c} = \dot{\lambda} \xi \tag{28}$$

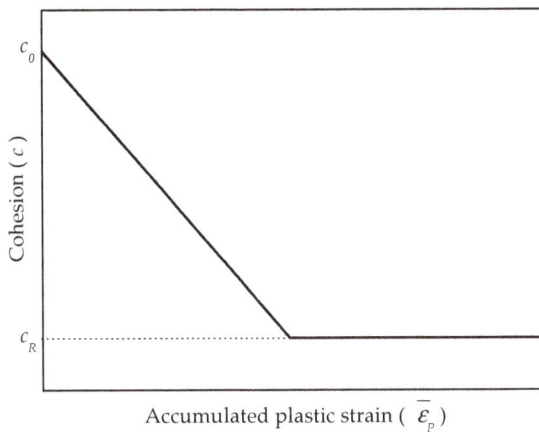

Figure 2. The softening relationship between the cohesion and the accumulated plastic strain.

Because the relationship between cohesion and accumulated plastic strain is a purely mathematical construct, it is difficult to obtain an exact characterization of this relationship. According to Figure 3, different cohesion softening laws can get different results of the cohesion softening and the stress–strain relationships, which can make the simulation of widespread material failure behaviors possible, and can include both brittle and ductile failure. The higher order mathematical equation for cohesion softening law may need to simulate more complex and precise material failure behaviors. More details about the cohesion softening law can be seen in Whyatt and Board [36].

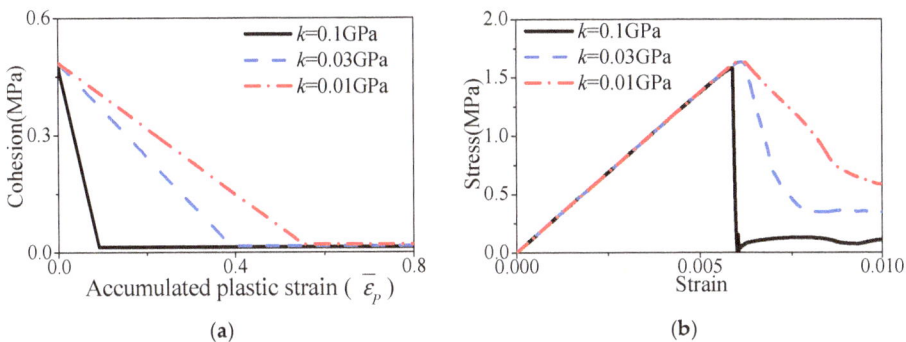

(a)

(b)

Figure 3. The cohesion softening results by the different softening coefficient k in the uniaxial compression test in section 5.2: (**a**) the relationship between the cohesion and accumulated plastic strain; and (**b**) the corresponding stress–strain curves.

4. SPH Formulations and Corrective SPH Method

4.1. The Particle Approximation and Spatial Derivatives of SPH

In SPH method, the computational domain is discretized into a set of particles which carry some variables such as pressure, stress, velocity, density, etc. The smoothing kernels are used to approximate a continuous flow field. The basic principle of SPH expression is that, for any quantity of particle i, whether a scalar or a vector, it can be approximated by the direct summation of the relevant quantities of its neighbor particles j, which is shown as:

$$f(\mathbf{r}_i) = \sum_{j=1}^{N} \frac{m_j}{\rho_j} f(\mathbf{r}_j) W(\mathbf{r}_{ij}) \tag{29}$$

and its gradient can be shown as:

$$\nabla f(\mathbf{r}_i) = \sum_{j=1}^{N} \frac{m_j}{\rho_j} f(\mathbf{r}_j) \nabla_i W(\mathbf{r}_{ij}) \tag{30}$$

where i and j are the referred particle and its neighbor, respectively, and $W(\mathbf{r}_{ij})$ is a kernel function and has different forms. In this paper, the cubic B-spline kernel proposed by Monaghan and Lattanzio [38] was used:

$$W(\mathbf{r}_{ij}, h) = \alpha_d \begin{cases} \frac{2}{3} - q^2 + \frac{1}{2}q^3 & 0 \le q < 1 \\ \frac{1}{6}(2 - q^3) & 1 \le q < 2 \\ 0 & q \ge 2 \end{cases} \tag{31}$$

where $q = r/h$, $\alpha_d = 15/(7\pi h^2)$ for 2D cases, and h is equal to 1.2–1.4dx (dx is the initial particle spacing).

In SPH, the mass conservation Equation (1) can be approximated as follows:

$$\frac{D\rho_i}{Dt} = \sum_{j=1}^{N} m_j \left(v_i^\alpha - v_j^\alpha \right) \frac{\partial W_{ij}}{\partial x_i^\alpha} \tag{32}$$

where ρ_i is the density of particle i with velocity component v_i; and m_j is the mass of particle j which has velocity component v_j. The most widely used SPH approximation of the momentum equation (Equation (2)) is:

$$\frac{dv_i^\alpha}{dt} = \sum_{j=1}^{N} m_j \left(\frac{\sigma_i^{\alpha\beta}}{\rho_i^2} + \frac{\sigma_j^{\alpha\beta}}{\rho_j^2} - \Pi_{ij} \cdot \delta^{\alpha\beta} \right) \frac{\partial W_{ij}}{\partial x_i^\beta} + g^\alpha \tag{33}$$

where Π_{ij} is the artificial viscosity, which was proposed by Monaghan [39].

Finally, the position of particle i in SPH is calculated based on the following equation:

$$\frac{Dx_i^\alpha}{Dt} = v_i^\alpha \tag{34}$$

In addition, the XSPH method [39] is used to solve problems involving the tension. In this method, particle i is defined based on an average velocity, which is shown as:

$$\frac{Dx_i^\alpha}{Dt} = v_i^\alpha + \bar{\varepsilon} \sum_{j=1}^{N} \frac{m_j}{\rho_j} \left(v_j^\alpha - v_i^\alpha \right) W_{ij}, \bar{\varepsilon} \in [0, 1] \tag{35}$$

4.2. Artificial Stress Method

An artificial stress method presented by Monaghan [40] and Gray et al. [41] was used in many papers to remove numerical instability [42] caused by the clumping of SPH particles when SPH is applied to solid mechanics. This method adopts an artificial repulsive force. The artificial repulsive force proposed in Gray et al. [41] is used in this paper and takes the form:

$$\frac{dv_i^\alpha}{dt} = \sum_{j=1}^{N} m_j \left(\frac{\sigma_i^{\alpha\beta}}{\rho_i^2} + \frac{\sigma_j^{\alpha\beta}}{\rho_j^2} - \Pi_{ij} \cdot \delta^{\alpha\beta} + f_{ij}^n \left(R_i^{\alpha\beta} + R_j^{\alpha\beta} \right) \right) \frac{\partial W_{ij}}{\partial x_i^\beta} + g^\alpha \tag{36}$$

where n is the variable exponent based on the smoothing kernel. f_{ij} is defined as

$$f_{ij} = \frac{W_{ij}}{W(\Delta d, h)} \tag{37}$$

where Δd is the initial distance between neighbor particles. h is set to be $1.2\Delta d$ for the cubic B-spline kernel in this paper.

The $R_i^{\alpha\beta}$ and $R_j^{\alpha\beta}$ in Equation (36) is the artificial stress tensor of particles i and j, respectively, with the correction parameter ε (Gray et al. [41]) :

$$R_i^{\alpha\beta} = sc(\overline{R}_i^{\alpha\alpha} + \overline{R}_i^{\beta\beta}) \tag{38}$$

$$R_i^{\alpha\alpha} = c^2 \overline{R}_i^{\alpha\alpha} + s^2 \overline{R}_i^{\beta\beta} \tag{39}$$

$$R_i^{\beta\beta} = s^2 \overline{R}_i^{\alpha\alpha} + c^2 \overline{R}_i^{\beta\beta} \tag{40}$$

$$\overline{R}_i^{\alpha\alpha} = \begin{cases} -\varepsilon \frac{\overline{\sigma}_i^{\alpha\alpha}}{\rho^2} & if\ \overline{\sigma}_i^{\alpha\alpha} > 0 \\ 0 & else \end{cases} \tag{41}$$

The same rule applies for $\overline{R}_i^{\beta\beta}$ with $\alpha\alpha$ replaced by $\beta\beta$.

Where $\overline{\sigma}_i^{\alpha\alpha}$ and $\overline{\sigma}_i^{\beta\beta}$ are the new components of the stress tensor in the rotated frame:

$$\overline{\sigma}_i^{\alpha\alpha} = c^2 \sigma_i^{\alpha\alpha} + 2sc\sigma_i^{\alpha\beta} + s^2 \sigma_i^{\beta\beta} \tag{42}$$

$$\overline{\sigma}_i^{\beta\beta} = s^2 \sigma_i^{\alpha\alpha} + 2sc\sigma_i^{\alpha\beta} + c^2 \sigma_i^{\beta\beta} \tag{43}$$

where $c = \cos\theta_i$ and $s = \sin\theta_i$. θ_i is the angle of roiration for particle i, which statisfies

$$\tan 2\theta_i = \frac{2\sigma_i^{\alpha\beta}}{\sigma_i^{\alpha\alpha} - \sigma_i^{\beta\beta}} \tag{44}$$

More details about the artificial stress can be found in Gray et al. [41]. For the tests discussed in this study, the parameter ε and n are equal to 0.3 and 4, respectively, to solve the tensile instability problems in SPH.

4.3. Boundary Conditions

In this paper, we deal with boundary conditions by two types of particles: solid boundary particles and mirror particles.

The solid boundary is fixed by the particles, which may prevent the real ice particles from penetrating the solid wall (Figure 4). The boundary particles contribute to the velocity and stress gradients for the real ice particles near the boundary. These boundary particles have the same velocity density as the solid wall and their density is set equal to reference density. The stresses of the boundary particles on the solid boundary are calculated by using:

$$\sigma_w^{\alpha\beta} = \frac{\sum_{i=1}^{N} \sigma_i^{\alpha\beta} W_{wi}}{\sum_{i=1}^{N} W_{wi}} \tag{45}$$

where $\sigma_w^{\alpha\beta}$ is the stress of the particle w on a boundary solid boundary; i is its neighboring particle and i can only be the real ice particle; and N is the number of particles in the support domain of wall boundary particle w.

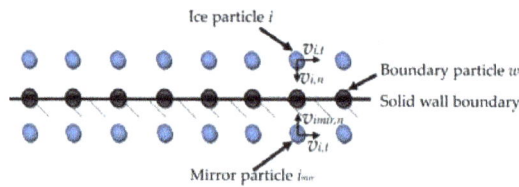

Figure 4. The treatment of the solid boundary.

In addition, the mirror particle (Figure 4) method following Libersky and Petschek [43] is also used to simulate the solid boundary with the free-slip condition. For each real particle i that is close to the wall, a mirror particle i_{mir} is set by a direct reflection of particle i across the boundary. The mirror particle i_{mir} has the same tangential velocity ($v_{i_{mir},t}$) with that of real particle: $v_{i_{mir},t} = v_{i,t}$ to simulate the free-slip boundary condition. The normal velocity ($v_{i_{mir},n}$) of i_{mir} is set opposite to that of real particle $v_{i_{mir},n} = v_{i,n}$ to prevent the real particles from penetrating the boundary as shown in Figure 4. The density and stress tensors of mirror particles are set to be equal to those of real ice particles.

4.4. Corrective SPH Method

The strain rate of the tensor Equation (4) needs to be converted into the discrete form to get the stress rate based on the generalized Hooke's law. In standard SPH, the strain rate is obtained by:

$$\dot{\varepsilon}^{\alpha\beta} = \frac{1}{2}\left[\sum_{j=1}^{N} m_j\left(v_j^\alpha - v_i^\alpha\right)\frac{\partial W_{ij}}{\partial x_i^\beta} + \sum_{j=1}^{N} m_j\left(v_j^\beta - v_i^\beta\right)\frac{\partial W_{ij}}{\partial x_i^\alpha}\right] \tag{46}$$

The standard SPH algorithm is lack of high accuracy due to kernel approximation of its first-order derivative, such as Equation (46). To overcome the shortcomings in first order derivative accuracy of the original SPH, this paper adopts the Simplified Finite Difference Interpolation (SPH_SFDI) method to calculate the strain rate of the ice particles, more details about SFDI method can be found in Ma [37]. According to the results in Zheng et al. [44], SFDI can be a very good option as a high order accuracy. For the purpose of the completion of theory, the formulas of strain rate of the tensor in 2D case can be shown as:

$$\dot{\varepsilon}^{\alpha\beta} = \frac{1}{2}\left(\sum_{j=1,j\neq i}^{N} \frac{n_{i,\alpha}B_{ij,\beta} - n_{i,\beta}B_{ij,\alpha}}{n_{i,x}n_{i,y} - n_{i,\alpha\beta}^2}\left(v_j^\alpha - v_i^\alpha\right) + \sum_{j=1,j\neq i}^{N} \frac{n_{i,\beta}B_{ij,\alpha} - n_{i,\alpha}B_{ij,\beta}}{n_{i,x}n_{i,y} - n_{i,\alpha\beta}^2}\left(v_j^\beta - v_i^\beta\right)\right) \tag{47}$$

where $n_{i,m} = \sum_{j=1,j\neq i}^{N} \frac{(r_j^m - r_i^m)^2}{|r_j - r_i|^2}W\left(r_{ij}\right)$, $n_{i,mk} = \sum_{j=1,j\neq i}^{N} \frac{(r_j^m - r_i^m)(r_j^k - r_i^k)}{|r_j - r_i|^2}W\left(r_{ij}\right)$, $B_{ij,m} = \frac{(r_j^m - r_i^m)}{|r_j - r_i|^2}W\left(r_{ij}\right)$,

in which $\alpha = x, y, \beta = x, y$ and $m = x, k = y$ or $m = y, k = x$, and N is the neighbor particle number of particle i, r_j^m is the component of the position vector in x or y direction. Similarly, the derivative of other variables can also be calculated by this corrective method.

To justify that the SFDI method is more effective than the standard SPH method for the strain rate calculation. Figure 5 shows the comparison of the bending stress in the middle of the ice beam for four-point bending of the ice beam which will be discussed in Section 5.2. In Figure 5, the standard formula is referred to Equation (46) and the SFDI is referred to Equation (47).

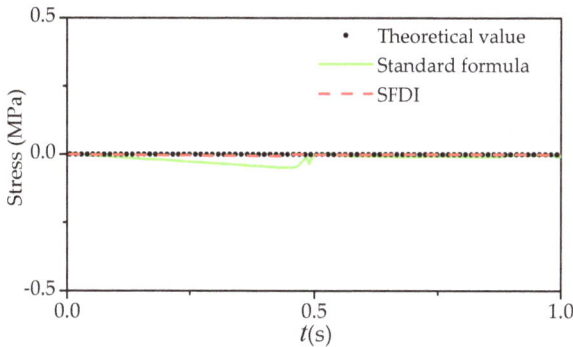

Figure 5. The stress comparison of theoretical value and numerical results: the traditional formula (Equation (46)) and the SFDI scheme (Equation (47)).

According to the comparison of Figure 5, the results from SFDI scheme can get better agreement with the theoretical value than the ones of traditional equation. Especially when the fracture failure start to occur in the ice beam at about $t = 0.45$ s, the stress values by standard formula deviates from theoretical results obviously. The source of discrepancy is expected to be that the accuracy of strain rate in the standard formula is less than the ones in the SFDI scheme.

In SPH_SFDI, the main procedures of numerical implementation of failure model of ice are shown as follows:

(1) Calculate the values of $\dot{\varepsilon}^{\alpha\beta}$ and $\dot{\sigma}^{\alpha\beta}$ from Equations (47), (17) or (19).
(2) Calculate the stress components $\sigma^{\alpha\beta}$ based on the obtained stress rate $\dot{\sigma}^{\alpha\beta}$.
(3) Check the stress state and judge whether the corresponding stress need to be corrected: if $-\alpha_\phi I_1^n + k_c < \sqrt{J_2^n}$, the stress need to be modified by Equation (25).
(4) Implement Cohesion softening model based on Equation (27).

5. Numerical Simulations

In this section, we firstly use the elastic vibration of a cantilever beam to verify the feasibility of SPH_SFDI method in solid mechanics. To test the effectiveness of the SPH_SFDI for simulating the failure progress of ice, two typical tests are included: the ice four-point bending and uniaxial ice compressive test. The enhanced performance of the SPH_SFDI algorithm will be demonstrated through the quantitative comparisons with the standard SPH and experimental data.

5.1. Elastic Vibration of a Cantilever Beam

The elastic vibration of a cantilever beam is used as a benchmark test to verify the reliability of the SPH_SFDI model for the calculation of solid mechanics. The cantilever beam is shown in Figure 6, the dynamic load P is acting at the free end of the cantilever beam. The length $L = 48$ m, the height is $D = 12$ m, the elastic modulus is $E = 3.0 \times 10^7$ N/m^2, the Poisson's ratio is $v = 0.3$, and the mass density is $P = 1$ kg/m^3. External excitation force $P = 1000g(t)$ and $g(t)$ is a function related to time.

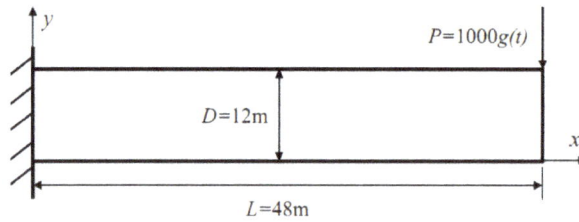

Figure 6. The cantilever beam and the dynamic loads.

A simple harmonic load $g(t) = \sin \omega t$ is considered. ω is the frequency of harmonic load and in this case $\omega = 27$ s^{-1}. Figure 7 shows the comparison of the displacement in y direction of the free end of the cantilever beam (y) between the SPH and SPH_SFDI results with 10,000 particles and the finite element method (FEM) solution from Long [45]. This shows that the displacement time histories computed by the SPH_SFDI method shares a better agreement with the FEM data than the SPH result.

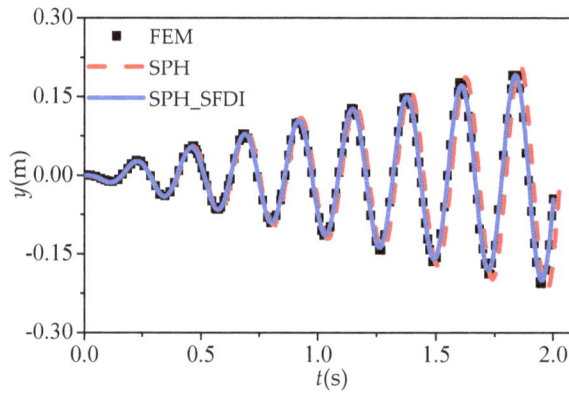

Figure 7. Comparison of SPH results with experimental data: displacement y versus time plot.

To evaluate the enhanced performance of SPH_SFDI method further, the convergence properties of the SPH and SPH_SFDI models are now examined in terms of the displacement y. For this purpose, the time histories of displacement computed by SPH and SPH_SFDI are presented in Figure 8 with the different particle numbers. Figure 9 gives the convergence tests on displacement, in which N is the total particle number and different values using 1600, 3600, 6400 and 10,000 are analyzed here. The relative error Err is defined as the errors between FEM result and SPH, SPH_SFDI results, which are calculated by $Err = |y_0 - y|/y_0$, where y is the computed displacement by SPH and SPH_SFDI from $t = 0.0$ s to $t = 2.0$ s, y_0 is the displacement of FEM from $t = 0.0$ s to $t = 2.0$ s. It is shown in Figure 8 that the error of force decreases as the particle number increases unanimously for both the SPH and SPH_ SFDI approaches. This indicates the convergence of all numerical models. However, the error magnitude of SPH_SFDI is much smaller than that of SPH. Besides, we could also conclude from Figure 9 that the convergence of SPH_ SFDI method is much better than that of the SPH, in that the errors of the former reduce more rapidly following the refinement of spatial resolutions.

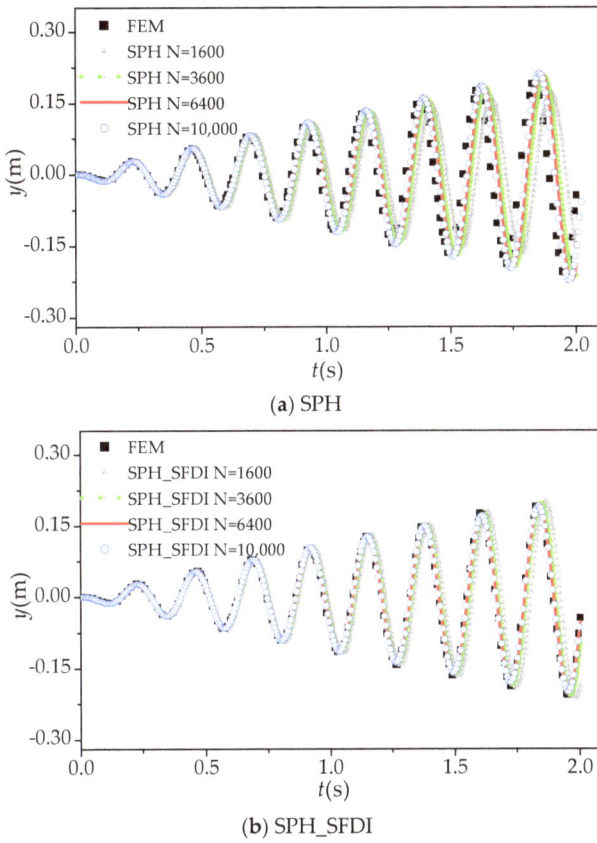

Figure 8. The time histories of displacement *y* obtained by: (**a**) SPH and (**b**) SPH_SFDI with different particle numbers.

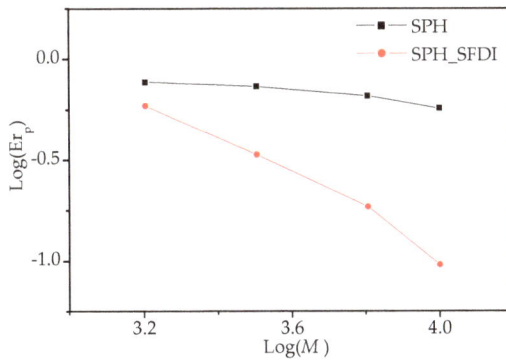

Figure 9. Convergence tests of displacement by different SPH methods.

5.2. Four-Point Pending of Ice Beam

The ice four-point bending experiment was conducted by Kujala et al. [4]. In their work, a loading rig was used to bend the ice beam upward during the experiments. In addition, they used a hydraulic cylinder to push two moving supports to produce a force, which were located 1 m apart in the middle of the beam, so it can bend the ice beam upward. At the same time, two fixed supports, which were 4 m apart, were placed at both ends of the beam to against the ice beam. The detailed resulting measurements can be found in Ehlers and Kujala [46].

In this section, the ice beam, the upper and lower supports are modeled with the particles, which is shown in Figure 10. In total, 2768 particles are used for generating the ice beam. The length and the height of the beam are $L = 4.325$ m and $H = 0.4$ m, respectively. The velocity of two moving upward supports is 0.00275 m/s. The elastic modulus of the ice beam is $E = 4.5$ GPa, the cohesion is $c = 0.58$ Mpa, and the friction angle is $36°$. The dilatancy angle φ in the non-associative plastic rule is one-third of the friction angle ($\varphi = \phi/3$).

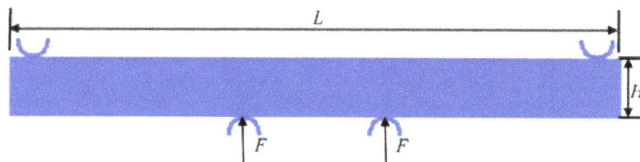

Figure 10. Computational model of the ice beam.

To show the fracture patterns clearly, Figure 11 gives the results obtained by SPH_SFDI using the non-associative flow rule at different time. As shown in Figure 11a, two fracture cracks obviously occur at the upper area by two moving supports and the ice beam breaks into three sections. Then the cracks in the ice beam widen and two sections of ice beams on either side of the two bottom supports sink downward as shown in Figure 12b. As the two supports move up slowly, the cracks in the ice beam widen and the ice beam eventually breaks into three sections as shown in Figure 12c. It need to highlight the point that in our numerical results, due to the complete symmetry of the characteristics of ice beam and the external loading and supporting condition, the fracture location of the ice beam is almost completely symmetric, which is not completely consistent with that in the experimental results. In addition, there are slight crushing failures at the place contacted with the upper two fixed supports, as the two upper supports are fixed and the ice beam has a tendency to move upward.

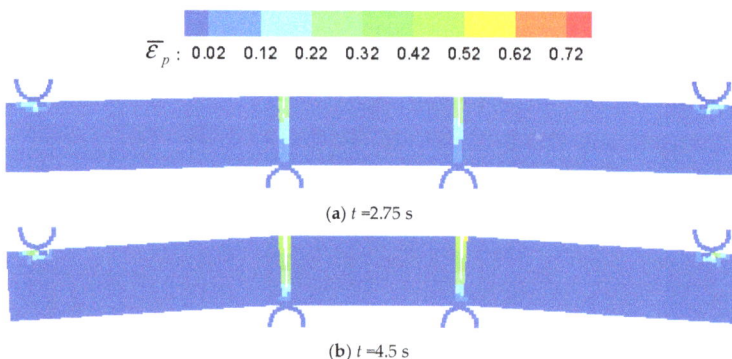

$\overline{\varepsilon}_p:$ 0.02 0.12 0.22 0.32 0.42 0.52 0.62 0.72

(a) $t = 2.75$ s

(b) $t = 4.5$ s

Figure 11. *Cont.*

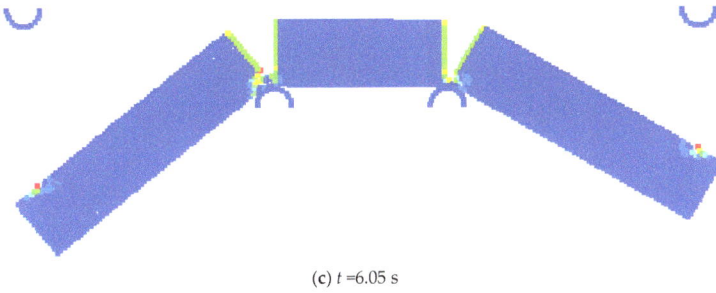

(c) *t* =6.05 s

Figure 11. Predicted fracture patterns of accumulated plastic strain by SPH_SFDI at different time.

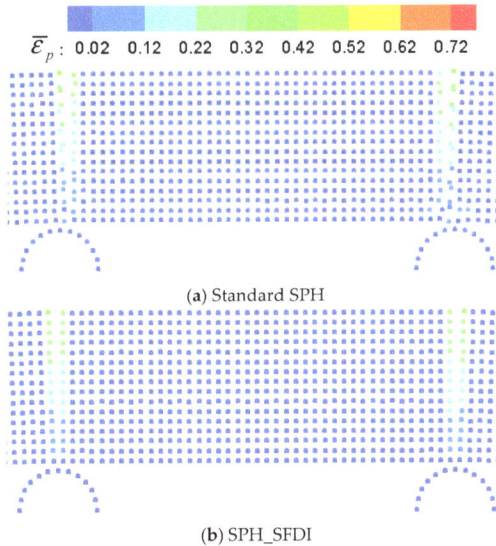

$\overline{\varepsilon}_p$: 0.02 0.12 0.22 0.32 0.42 0.52 0.62 0.72

(a) Standard SPH

(b) SPH_SFDI

Figure 12. Enlarged partial views of accumulated plastic strain of the failure path in the ice beam: (a) standard SPH; and (b) SPH_SFDI.

The snapshots of the failure path predicted by SPH and SPH_SFDI using non-associative flow rule at *t* = 2.75 s are shown in Figure 7. According to the results of Figure 7, the ice beam breaks into three segments which can be obtained both by the standard SPH and SPH_SFDI. The fracture points of the horizontal coordinate on lower two moving supports by standard SPH show the apparent inward deviation compared with the ones of SPH_SFDI. Furthermore, it can be easily observed that the particle distributions for the results of standard SPH are in chaotic, whereas the results of SPH_SFDI are more stable and reliable.

To show the accuracy of numerical solutions of standard SPH and SPH_SFDI, Figure 13 gives the comparison of the force time histories among stand SPH, SPH_SFDI and the experimental data [46]. The relationship between external force and flexural stress according to ITTC [47] is defined as:

$$F = \frac{\sigma B H^2}{6L_0} \tag{48}$$

where B is the width of 3D ice beam. We use the same value of B as that in the experiment of Kujala et al. [4]. L_0 is the distance between the fixed support and the bottom moving support on the same side. σ is flexural stress generated by bending of ice beam.

According to the results in Figure 13, the numerical results of SPH_SFDI have obviously better agreement with the experimental data than the ones of standard SPH results. With the accuracy improvement of the gradient approximation, the force time histories and fractured crack of the ice beam bending can get more accurate and reliable results than the ones of standard SPH.

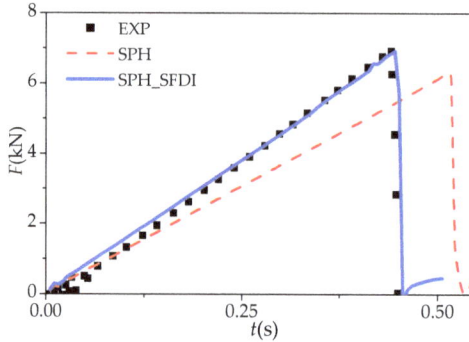

Figure 13. Comparisons of external force among experimental data and different SPH results.

To validate that the numerical model can simulate the failure of ice beam at different times effectively, the failure of ice beam with the same material parameters above under two extra moving velocities of upward supports, such $V_1 = 0.001842$ m/s and $V_2 = 0.003225$ m/s are also considered. Figure 14 gives the comparison of the force versus time curve under different moving velocities of two upward supports among SPH_SFDI and the experimental data [46]. In addition, in Table 1, the results from SPH_SFDI has been compared with experiment tests in terms of the failure force F', failure time t' and the corresponding deflection δ. The good agreement between the numerical results and the experiment data can be obtained clearly in Figure 14 and Table 1, although there exists some little difference. Thus, the presented SPH_SFDI model including the cohesion softening model can imply the time dependent of ice failure and get good simulated results for ice failure with different loading rates.

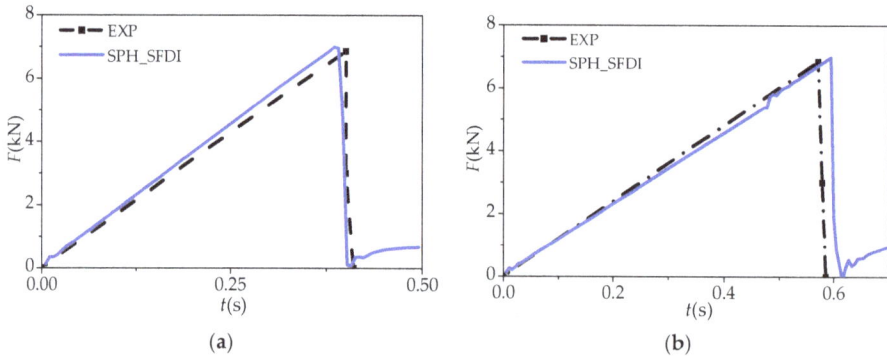

(a)   (b)

Figure 14. Comparison of force versus time plot of the ice beams by numerical results and experimental data with different velocities of two moving upward supports: (**a**) $V_1 = 0.001842$ m/s; and (**b**) $V_2 = 0.003225$ m/s.

Table 1. Comparison of SPH_SFDI and experiment data with different loading velocities.

Approach	$F'(V_1)$	$t'(V_1)$	ffi(V_1)	$F'(V_2)$	$t'(V_2)$	ffi(V_2)
EXP	6.87 kN	0.40 s	1.29 mm	6.87 kN	0.57 s	1.05 mm
SPH_SFDI	6.95 kN	0.39 s	1.24 mm	6.98 kN	0.59 s	1.15 mm

Note: $V_1 = 0.001842$ m/s; $V_2 = 0.003225$ m/s.

To show the effects of different plastic flow rules, Figure 15 gives the comparisons of the force time histories by SPH_SFDI with associative plastic flow (Equation (17)) and non-associative plastic flow (Equation (19)). To show the difference between associative and non-associative plastic flow clearly, Figure 16 gives the snapshots of the cracks in the brittle failure process obtained by SPH_SFDI with associative and non-associative flow rule at $t = 2.75$ s. The force obtained by the associative rule is basically consistent with the experimental data and the failure paths are also consistent with the non-associative flow rule. According to the results of Figure 16, the particles on the cracks, especially near the left bottom support, are slightly disordered by associative flow rule. In comparison, these particles near the same domain are more regular and reliable by non-associative flow rule.

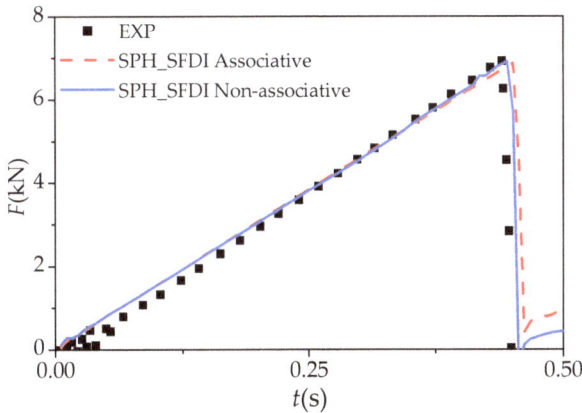

Figure 15. Comparisons of the force time histories among experimental data and SPH_SFDI results with different flow rules.

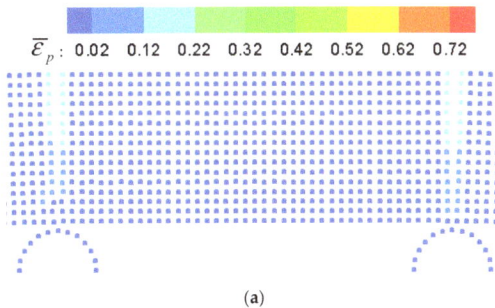

(a)

Figure 16. *Cont.*

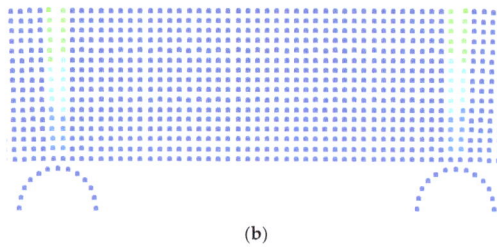

(b)

Figure 16. Snapshot of the failure paths of accumulated plastic strain in SPH_SFDI results with different flow rules: (**a**) associative flow rule; and (**b**) non-associative flow rule.

5.3. Uniaxial Compressive Test of Ice Specimen

In this section, it will justify the efficiency of SPH_SFDI scheme for the ice compressed behavior simulation. Uniaxial compression of ice specimen is one of the most introduced benchmarks in this field. A two-dimensional rectangle ice specimen will be considered. The width (D) and height (H) of the ice specimen are 7 cm and 17.5 cm, respectively. The schematic geometry of this model can be shown in Figure 17. An axial velocity with vertical downward is loaded on the upper platen, which is of the value 0.0034675 m/s. The experiment of the same scale model was conducted by Li et al. [7] and Zhang [48]. Two rigid plates on the top and bottom deal with solid boundary, which can support the cuboid. The top plate could be moved freely in the vertical direction with a certain velocity, which focuses on the compressed ice behavior. The ice specimen has the cohesion $c = 0.45$ Mpa, the friction angle is 22.5°. The dilatancy angle φ in the non-associative plastic rule is set to be one-third of the friction angle ($\varphi = \phi/3$).

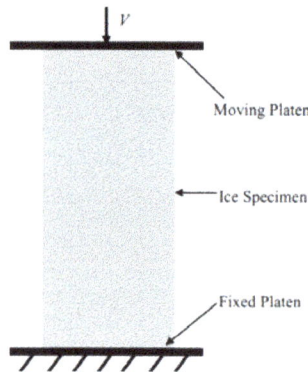

Figure 17. The sketch of the ice specimen of uniaxial compression.

Figure 18a illustrates the comparisons of the axial stress–strain curves among the experimental data [7], standard SPH and SPH_SFDI with non-associative flow rule. According to the results of Figure 18a, the stress–strain relation obtained by the SPH_SFDI method is in more agreement with experimental data than the ones of standard SPH, despite some unavoidable discrepancies due to the complication of the physical problem. Figure 18b gives the comparisons of the stress–strain curve in the experimental data with the results obtained by SPH_SFDI with associative and non-associative plastic flow rules. According to the results of Figure 18, there exists a certain difference between the numerical results and the experimental results for real sea ice. The elastic–plastic model of this paper can get a reasonable agreement with the ones of experimental data. However, the nonlinear behavior of stress–strain time histories cannot be captured exactly; there exists many different factors, such the ice viscosity, the anisotropy and the temperature, which should be further investigated to make it more reliable for the numerical simulation of real sea ice.

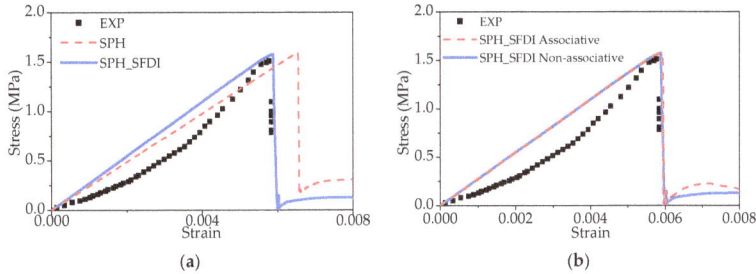

(a)

(b)

Figure 18. Comparisons of the stress–strain curve between experimental data and numerical results: (a) standard SPH and SPH_SFDI with non-associative flow rule; and (b) only SPH_SFDI with different flow rules.

Figure 19 shows the comparisons of the typical fracture pattern among experimental results of Zhang [48] (Figure 19a), the standard SPH and SPH_SFDI with non-associative flow rule. According to the results of Figure 19, the shear failures in the ice sample are predicted by the standard SPH and SPH_SFDI at $t = 0.69$ s and $t = 1.38$ s. The ice specimen exists the brittle failure and there is a main crack in the fracture pattern. The upper part of the body has the trend of sliding along the main crack and falling out of the specimen. Although the standard SPH method can predict the shear failure, the position of fracture crack differs greatly from the experimental result. The results of the SPH_SFDI are in better agreement with the experimental test than the ones of the standard SPH. In addition, some irrational damage occurred where the particle distribution is obviously ill conditioned in the SPH result, which can be seen in Figure 19b1,c1. By comparison, the results of SPH_SFDI are more stable and more regular. In summary, the present simulations also provide a strong indication that the results of SPH_SFDI method could be superior to the standard SPH in predicting the compressive failure process accurately. It should be noted that with the development of shear failure, the lower part of the ice specimen tilts under the downward sliding extrusion of the upper part and deformation occurs at the lower left corner, as shown in Figure 19c.

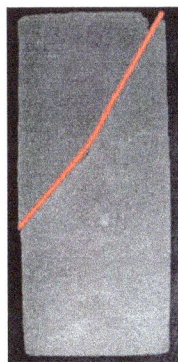

(a)

Figure 19. *Cont.*

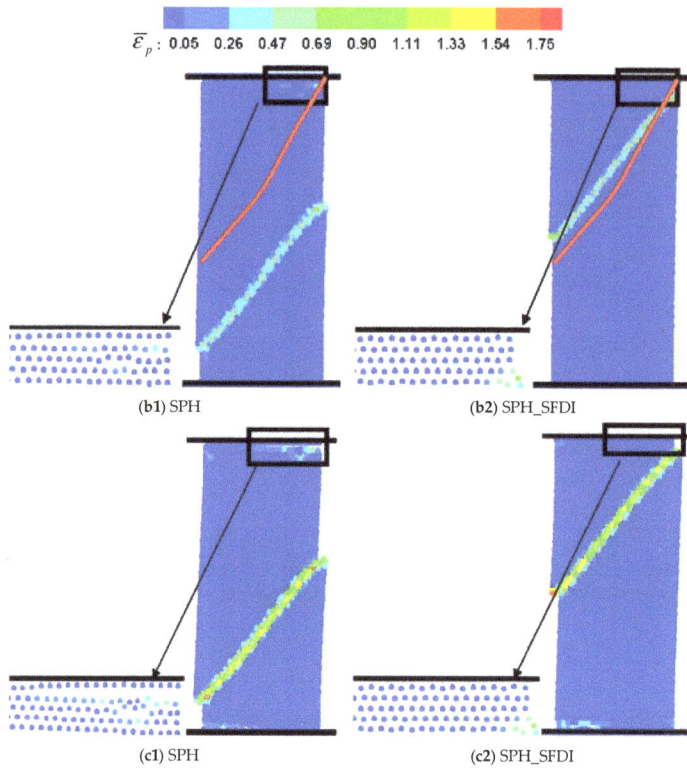

Figure 19. Comparisons of (**a**) the typical fracture pattern among experimental results of Zhang [48] (experimental fracture crack: red lines); and the standard SPH and SPH_SFDI results at different time: (**b**) $t = 0.69$ s; and (**c**) $t = 1.38$ s (contours of accumulated plastic strain).

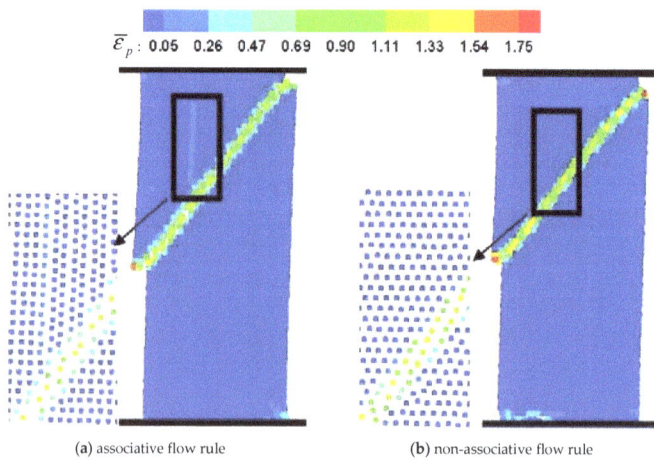

Figure 20. Comparison of the SPH_SFDI results with different flow rules (contours of accumulated plastic strain).

Figure 20 shows a direct comparison for the brittle shear failure simulation by SPH_SFDI with the associative flow rule and non-associative flow rule respectively. Although the stress–strain curve obtained by the associative rule is basically consistent with the experimental data (seen in Figure 18b) and the failure patterns predicted by the associative rule are also consistent with that of the non-associative flow rule, it can be seen in Figure 20a that there is a slight particle strip distribution in the SPH_SFDI results with associative flow rule. In contrast, the distribution of particles for the SPH_SFDI result using non-associative flow rule is more stable and reliable, more details can be found in Figure 20b. Therefore, with the combination of a comparative analysis of the different flow laws in the four points bending, it can be found that the non-associative flow rule can yield the better results for simulating the fractures.

In addition, Figure 21 shows the comparison of the bulge fracture patterns among the experiment results by Zhang [48], the standard SPH and SPH_SFDI. According to the results of Figure 21b, the ice sample exhibits the ductile failure feature and the bulge failure occurs in the bottom part of the ice sample. During the compression process, the failure progress of the ice sample is slow and there is no obvious main crack at the stage of the specimen failure. The bottom part of the ice sample distends to the two outer sides and eventually fractures. It is also shown in Figure 21b that the predicted cracks by SPH_SFDI can make a better agreement with the experimental test than the ones of the standard SPH. Although the standard SPH method can predict the bulge failure, the position of bulge fracture differs greatly from the experimental results. In addition, the particles below the damage position are obviously disordered in the results of standard SPH, which can be shown in Figure 21b1. By comparison, the results of SPH_SFDI are more reliable and the particle distributions of SPH_SFDI are more regular.

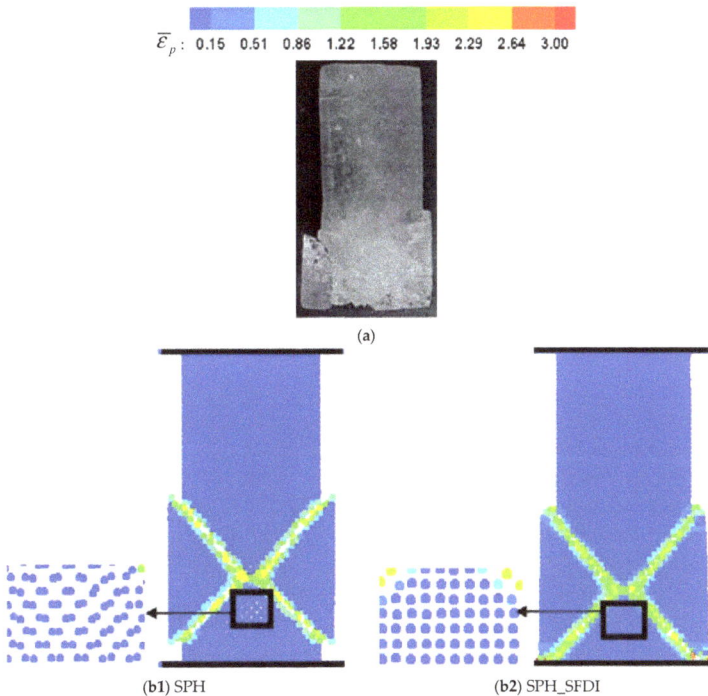

Figure 21. Comparisons of the bulge fracture pattern among (**a**) the bulge fracture patterns among the experiment results by Zhang [48], (**b**) the standard SPH and SPH_SFDI results (The color legend is the accumulated plastic strain).

6. Conclusions

In this paper, the SPH_SFDI model including the elastic–plastic cohesion softening Drucker–Prager failure model is proposed to simulate the bending and compression failure processes of ice. The predicted force in a four-point bending and the axial stress of a uniaxial compressive test are in a good agreement with the experimental data. The simulated fracture patterns are also reasonably close to the reality. The conducted studies disclosed that the elasto-plastic cohesion softening Drucker–Prager failure model, which originated from the soil and rock mechanics, can also be effectively used to simulate the physical destruction phenomena during the failure process of the ice. According to the comparisons between the numerical results conducted by the standard SPH and improved SPH_SFDI, the performance of the latter is found to be much better in view of the numerical accuracy and stability in the study of the bending and compression failure processes of ice.

Acknowledgments: This research work is supported by the National Natural Science Foundation of China (Nos. 51739001, 51279041, 51379051 and 51639004), Foundational Research Funds for the Central Universities (Nos. HEUCF170104 and HEUCDZ1202) and Defense Pre Research Funds Program (No. 9140A14020712CB01158), to which the authors are most grateful. Author Q. Ma also thanks the Chang Jiang Visiting Chair Professorship Scheme of the Chinese Ministry of Education, hosted by HEU.

Author Contributions: N. Zhang made the computations and data analysis; X. Zheng guided the engineering project and provided the data; and Q. Ma made the proof reading and editing. All authors contributed to the work.

Conflicts of Interest: The authors declare no conflict of interest.

References

1. Zhou, L.; Riska, K.; Moan, T.; Su, B. Numerical modeling of ice loads on an icebreaking tanker: Comparing simulations with models tests. *Cold Reg. Sci. Technol.* **2013**, *87*, 33–46. [CrossRef]
2. Lindqvist, G. A straightforward method for calculation of ice resistance of ships. In Proceedings of the 10th International Conference on Port and Ocean Engineering under Arctic Conditions, Luleaa, Sweden, 12–16 June 1989.
3. Valanto, P. The resistance of ships in level ice. *Trans. Soc. Nav. Archit. Mar. Eng.* **2001**, *119*, 53–83.
4. Kujala, P.; Riska, K.; Varsta, P. Results from in situ four point bending tests with Baltic Sea Ice. In Proceedings of the IAHR Symposium on Ice Problems, Helsinki, Finland, 19–23 August 1990; pp. 261–278.
5. Schulson, E.M. Compressive shear faults within arctic sea ice: Fracture on scales large and small. *J. Geophys. Res. Oceans* **2004**, *109*, 101–111. [CrossRef]
6. Schulson, E.M.; Fortt, A.L.; Iliescu, D.; Renshaw, C.E. Failure envelope of first-year arctic sea ice: The role of friction in compressive fracture. *J. Geophys. Res. Oceans* **2006**, *111*, 2209–2223. [CrossRef]
7. Li, Z.J.; Zhang, L.M.; Lu, P.; Lepparanta, M. Experimental study on the effect of porosity on the uniaxial compressive strength of sea ice in bohai sea. *Sci. China Technol. Sci.* **2011**, *54*, 2429–2436. [CrossRef]
8. Polach, R.V.B.U.; Ehlers, S.; Kujala, P. model-scale ice—Part A: Experiments. *Cold Reg. Sci. Technol.* **2013**, *94*, 74–81. [CrossRef]
9. Corasdale, K. Platform shape and ice interaction: A review. In Proceedings of the 21st International Conference on Port and Ocean Engineering under Arctic Conditions, Montreal, QC, Canada, 10–14 July 2011.
10. Lu, W.; Lubbad, R.; Hoyland, K.; Loset, S. Physical model and theoretical model study of level ice and wide sloping structure interactions. *Cold Reg. Sci. Technol.* **2014**, *101*, 40–72. [CrossRef]
11. Zhou, L.; Su, B.; Riska, K.; Moan, T. Numerical simulation of moored structure station keeping in level ice. *Cold Reg. Sci. Technol.* **2012**, *71*, 54–66. [CrossRef]
12. Kolari, K.; Kuutti, J.; Kurkela, J. FE-simulation of continuous ice failure based on model update technique. In Proceedings of the 20th International Conference on Port and Ocean Engineering under Arctic Conditions, Luleå, Sweden, 9–12 June 2009.
13. Hopkins, M.A. On the mesoscale interaction of lead ice and floes. *J. Geophys. Res.* **1996**, *101*, 18315–18326. [CrossRef]
14. Paavilainen, J.; Tuhkuri, J. Parameter effects on simulated ice rubbling forces on a wide sloping structure. *Cold Reg. Sci. Technol.* **2012**, *81*, 1–10. [CrossRef]

15. Paavilainen, J.; Tuhkuri, J.; Polojarvi, A. 2D numerical simulation of ice rubble formation process against an inclined structure. *Cold Reg. Sci. Technol.* **2011**, *68*, 20–34. [CrossRef]

16. Das, J. Modeling and validation of simulation results of an ice beam in four-point bending using smoothed particle hydrodynamics. *Int. J. Offshore Polar Eng.* **2017**, *27*, 82–89. [CrossRef]

17. Deb, D.; Pramanik, R. Failure process of brittle rock using smoothed particle hydrodynamics. *J. Eng. Mech.* **2013**, *139*, 1551–1565. [CrossRef]

18. Lucy, L.B. A numerical approach to the testing of fusion process. *Astron. J.* **1977**, *88*, 1013–1024. [CrossRef]

19. Gingold, R.A.; Monaghan, J.J. Smoothed particle hydrodynamics—Theory and application to non-spherical stars. *Mon. Not. R. Astron. Soc.* **1977**, *181*, 375–389. [CrossRef]

20. Morris, J.; Fox, P.; Zhu, Y. Modeling low Reynolds number incompressible flows using SPH. *J. Comput. Phys.* **1997**, *136*, 214–226. [CrossRef]

21. Cleary, P.; Prakash, M. Discrete-element modelling and smoothed particle hydrodynamics: Potential in the environmental sciences. *Philos. Trans. R. Soc. Lond. Ser. A* **2004**, *362*, 2003–2030.

22. Khayyer, A.; Gotoh, H.; Shao, S.D. Corrected incompressible SPH method for accurate water surface tracking in breaking waves. *Coast. Eng.* **2008**, *55*, 236–250. [CrossRef]

23. Khayyer, A.; Gotoh, H.; Shao, S.D. Enhanced predictions of wave impact pressure by improved incompressible SPH methods. *Appl. Ocean Res.* **2009**, *31*, 111–131. [CrossRef]

24. Zheng, X.; Ma, Q.W.; Duan, W.Y. Incompressible SPH method based on Rankine source solution for violent water wave simulation. *J. Comput. Phys.* **2014**, *276*, 291–314. [CrossRef]

25. Shao, S.D. Incompressible SPH flow model for wave interactions with porous media. *Coast. Eng.* **2010**, *57*, 304–316. [CrossRef]

26. Liu, X.; Xu, H.; Shao, S.; Lin, P. An improved incompressible SPH model for simulation of wave–structure interaction. *Comput. Fluids* **2013**, *71*, 113–123. [CrossRef]

27. Shao, S.D.; Lo, E.Y.M. Incompressible SPH method for simulating newtonian and non-newtonian flows with a free surface. *Adv. Water Resour.* **2003**, *26*, 787–800. [CrossRef]

28. Xu, R.; Stansby, P.; Laurence, D.; Xu, R.; Stansby, P.; Laurence, D. Accuracy and stability in incompressible SPH (ISPH) based on the projection method and a new approach. *J. Comput. Phys.* **2009**, *228*, 6703–6725. [CrossRef]

29. Napoli, E.; Marchis, M.D.; Gianguzzi, C.; Milici, B.; Monteleone, A. A coupled finite volume–smoothed particle hydrodynamics method for incompressible flows. *Comput. Methods Appl. Mech. Eng.* **2016**, *310*, 674–693. [CrossRef]

30. Libersky, L.D.; Petschek, A.G. Smoothed particle hydrodynamics with strength of materials. In *Proceedings of the Next Free Lagrange Conference, Moran, WY, USA, 3–7 June 1990*; Trease, H., Friits, J., Crowley, W., Eds.; Springer: New York, NY, USA, 1991; Volume 395, pp. 248–257.

31. Benz, W.; Asphaug, E. Simulations of brittle solids using smooth particle hydrodynamics. *Comput. Phys. Commun.* **1995**, *87*, 253–265. [CrossRef]

32. Randles, P.; Libersky, L. Smoothed particle hydrodynamics: Some recent improvements and applications. *Comput. Methods Appl. Mech. Eng.* **1996**, *139*, 375–408. [CrossRef]

33. Bui, H.; Fukagawa, R.; Sako, K.; Ohno, S. Lagrangian meshfree particles method (SPH) for large deformation and failure flows of geomaterial using elastic–plastic soil constitutive model. *Int. J. Numer. Anal. Methods Geomech.* **2008**, *32*, 1537–1570. [CrossRef]

34. Douillet-Grellier, T.; Jones, B.D.; Pramanik, R.; Pan, K.; Albaiz, A.; Williams, J.R. Mixed-mode fracture modeling with smoothed particle hydrodynamics. *Comput. Geotech.* **2016**, *79*, 73–85. [CrossRef]

35. Zhang, N.B.; Zheng, X.; Ma, Q.W.; Hao, H.B. Numerical simulation of failure progress of ice using smoothed particle hydrodynamics. In Proceedings of the SOPE 2017, International Ocean and Polar Engineering Conference, San Francisco, CA, USA, 25–30 June 2017.

36. Whyatt, J.K.; Board, M.P. *Numerical Exploration of Shear-Fracture-Related Rock Bursts Using a Strain-Softening Constitutive Law*; US Department of the Interior, Bureau of Mines: Washington, DC, USA, 1991.

37. Ma, Q.W. A new meshless interpolation scheme for MLPG_R method. *CMES Comput. Model. Eng. Sci.* **2008**, *23*, 75–89.

38. Monaghan, J.J.; Lattanzio, J.C. A refined particle method for astrophysical problems. *Astron. Astrophys.* **1985**, *149*, 135–143.

39. Monaghan, J.J. Smoothed particle hydrodynamics. *Annu. Rev. Astron. Astrophys.* **1992**, *30*, 543–574. [CrossRef]

40. Monaghan, J.J. SPH without a tensile instability. *J. Comput. Phys.* **2000**, *159*, 290–311. [CrossRef]

41. Gray, J.; Monaghan, J.; Swift, R. SPH elastic dynamics. *Comput. Methods Appl. Mech. Eng.* **2001**, *190*, 6641–6662. [CrossRef]

42. Swegle, J.; Hicks, D.; Attaway, S. Smoothed particle hydrodynamics stability analysis. *J. Comput. Phys.* **1995**, *116*, 123–134. [CrossRef]

43. Libersky, L.D.; Petschek, A.G.; Carney, T.C.; Hipp, J.R.; Allahdadi, F.A. High strain Lagrangian hydrodynamics: A three-dimensional SPH code for dynamic material response. *J. Comput. Phys.* **1993**, *109*, 67–75. [CrossRef]

44. Zheng, X.; Shao, S.D.; Khayyer, A.; Duan, W.Y.; Ma, Q.W.; Liao, K.P. Corrected first-order derivative ISPH in water wave simulations. *Coast. Eng. J.* **2017**, *59*, 1750010. [CrossRef]

45. Long, S.Y. *Meshless Methods and Their Applications in Solid Mechanics*; Science Press: Beijing, China, 2014; pp. 235–238.

46. Ehlers, S.; Kujala, P. Optimization-based material parameter identification for the numerical simulation of sea ice in four-point bending. *Proc. Inst. Mech. Eng. Part M J. Eng. Marit. Environ.* **2013**, *228*, 70–80. [CrossRef]

47. ITTC. Ice Property Measurements, 7.5-02-04-02. Available online: http://ittc.sname.org (accessed on 19 June 2017).

48. Zhang, L.M. Experimental Study on Uniaxial Compressive Strength and Influencing Factors of Ice. Ph.D. Thesis, Dalian University of Technology, Dalian, China, 2012.

water

MDPI

Article

Seasonal Variation in Sediment Oxygen Demand in a Northern Chained River-Lake System

Eric Akomeah * and Karl-Erich Lindenschmidt

Global Institute for Water Security, University of Saskatchewan, 11 Innovation Boulevard,
Saskatoon, SK S7N 3H5, Canada; karl-erich.lindenschmidt@usask.ca
* Correspondence: era524@mail.usask.ca; Tel.: +1-306-2800-233

Academic Editor: Jiangyong Hu
Received: 21 January 2017; Accepted: 30 March 2017; Published: 5 April 2017

Abstract: Sediment oxygen demand (SOD) contributes immensely to hypolimnetic oxygen depletion. SOD rates thus play a key role in aquatic ecosystems' health predictions. These rates, however, can be very expensive to sample. Moreover, determination of SOD rates by sediment diagenesis modeling may require very large datasets, or may not be easily adapted to complex aquatic systems. Water quality modeling for northern aquatic systems is emerging and little is known about the seasonal trends of SOD rates for complex aquatic systems. In this study, the seasonal trend of SOD rates for a northern chained river-lake system has been assessed through the calibration of a water quality model. Model calibration and validation showed good agreement with field measurements. Results of the study show that, in the riverine section, SOD_{20} rates decreased from 1.9 to 0.79 g/m^2/day as urban effluent traveled along the river while a SOD_{20} rate of 2.2 g/m^2/day was observed in the lakes. Seasonally, the SOD_{20} rates in summer were three times higher than those in winter for both river and lakes. The results of the study provide insights to the seasonal trend of SOD rates especially for northern rivers and lakes and can, thus, be useful for more complex water quality modeling studies in the region.

Keywords: sediment oxygen demand; hydrodynamic; water quality modeling; calibration; validation; seasonal trend

1. Introduction

Oxygen depletion in aquatic ecosystems has received much attention in the world. The importance of sediments in oxygen depletion have been well studied by many authors [1,2]. Oxidation (aerobic decomposition) of settled organic materials and the anaerobic respiration of invertebrates in sediments have been found to consume a large percentage of water column oxygen in surface water bodies [3,4]. Oxygen is used during the decomposition of organic matter by microorganisms and as it reacts with the by-products of respiration. The stabilization of organic material by organisms (e.g., bacteria) can exert high oxygen demand in sediments. High benthic oxygen demand can also result from high primary production by benthic algae (periphyton). Oxygen is consumed when algae respires in the night and it is produced during photosynthesis in the day. Studies by authors [5,6] indicated that oxygen demand by sediments was the major source of water column oxygen depletion. Benthic deposits usually originate from surface runoff, wastewater effluents, and aquatic conditions [7–9]. Once at the receiving surface water, the deposits are transported from these allochthonous materials through the water column to the river bed. The deposits could also be generated from autochthonous materials as aquatic plants and phytoplankton die off [10,11]. Extremely low levels of oxygen (below 2–3 mg/L) in aquatic systems may lead to dead zones [12], and flaura and fauna death. The rate at which water column oxygen is removed during the decomposition of organic matter and respiration of organisms in stream or lake bed sediments is known as sediment oxygen demand (SOD) [13–15].

SOD is influenced by temperature, velocity of water flow, residence time, and sediment composition [16]. An increase in water temperature accelerates, for example, the rate at which benthic bacteria respire, which elevates the SOD rate. Investigations by [17] show that a 10 °C rise in temperature doubled biological activities in sediments. MacPherson et al. [7] also found that SOD rate positively correlates with temperature. At low velocity (<10 cm/s), the SOD rate has been found to increase, but it decreases at higher velocity [18]. Since SOD plays a key role in the dissolved oxygen balance of surface water bodies at the sediment-water interface, determining its value in water quality modeling is paramount to the overall prediction of the health of aquatic ecosystems.

SOD can be estimated by in situ and laboratory measurements, sediment diagenesis modeling or through water quality model calibration. Liu and Chen [19] conducted SOD measurements along the Xindian River of Northern Taiwan for the purpose of water quality modeling. The authors found that SOD rates were directly proportional to the river discharge—higher discharge resulted in lower SOD rates. SOD rates at 20 °C ranged from 0.367 to 1.246 g/m^2/day. In the study by [15], the profile method was used to measure SOD rates in the Millstone River, Georgia. SOD ranged from 0.5 to 2.2 g/m^2/day. Caldwell and Doyle [20] also conducted in situ SOD measurements along a river using SOD chambers. The measured SOD rates corrected to a temperature of 20 °C ranged from 1.3 to 4.1 g/m^2/day. Several sediment diagenesis models have been built since 1996 to estimate the chemical concentrations and reaction rates often lacking from in-situ or laboratory measurements [21]. The study by [22] used a sediment diagenesis model to illustrate the influence of resuspension on the water column, and oxidation and denitrification in sediments. Sohma et al. [23–25] assessed the effect of water column anoxic fluctuations on sediment using a three-dimensional sediment diagenesis model. Calibrating the zero order SOD rate of a water quality model is another approach used to estimate SOD rate. Gualtieri [26] assessed SOD rates of a river by calibrating a water quality model within the minimum and maximum values of reported SOD rates. The author found the influence of SOD on dissolved oxygen to be substantial at effluent outfalls.

Estimating SOD by measurement techniques can become expensive due to the spatial and seasonal variation of SOD. In most surface water systems, SOD rate varies considerably, spatially, and seasonally, due to varying sediment composition. Spatially, the rate of deposition, physical, and chemical composition of sediment beds which affect oxygen consumption for rivers and lakes are substantially different. Steep sections of a river will have more boulder or cobble deposition and fine sediments (silt or clay) settle in low-velocity reaches, and for lakes, cobbles and pebbles settle at the inlet (as the current of a river slows down due to resistance from the rather still body of water in the lake); beyond the inlet and up to obstructing structures, sand, silt, and clay are, respectively, deposited as the current becomes very slow. These different sediment beds result in different SOD rates [27]. Seasonal variation of temperature impacts the composition of benthic and microbial communities, which influences the rate of SOD. Biological and chemical reactions in sediments are also elevated with temperature rise. As a result of these variations, often a large number of measurements are required to fully characterize SOD dynamics. In addition to the different SOD configurations and seasonal variation of temperature, it is often challenging to relate measurements to external loads. Sediment diagenesis models have been developed to handle the biogeochemical reactions that occur at the water column-sediment interface. The models are, however, complex and not easily adaptable to solve site-specific issues, such as complex river systems with varying sediment configurations. In addition, a large amount of data is needed to accurately drive these models, which might not be easily available [28,29]. Except in the case of special studies, SOD data for modeling purposes are not routinely collected [30].

In this paper, the seasonal variation of SOD for a prairie chained river-lake system is examined using the Water Quality Analysis Simulation Program (WASP 7.52) (Athens, GA, USA). Under-ice surface water quality modeling is emerging in the northern climate regions, but not much is known about the seasonal trends of SOD, especially for a complex system like chained river-lakes. One objective of the study is to examine the seasonal trend of SOD through an accurate but inexpensive modeling approach. Another objective is to determine the variation of SOD between the river and lake parts of the system. The estimated SOD values from this study will be used to set up an advanced eutrophication model in the future.

For a complex aquatic ecosystem, constraining processes and calibrating the water quality and transport modules of a low complexity water quality model provide a reasonable SOD estimate for water quality models [31]. At a low water quality model complexity, like the Streeter and Phelps dissolved oxygen balance, other process parameters are constrained, allowing models to be properly calibrated to few parameters. Building a model by gradually adding complexity is a strategy adopted by some modelers [32]. Complex models increase the number of processes being modeled, the output variables, and parameters [32]. Apart from the uncertainties associated with the model structure (as not all physical processes might be adequately represented, or even included) and forcings in a complex model, calibrating to optimum parameters, is often difficult in the face of sparse input data, as is common at most sites. According to [33], building complex models should be accompanied by a corresponding incorporation of relevant model observations. This paper adopts a slightly modified form of the Streeter and Phelps dissolved oxygen balance model, to estimate the SOD rates. All other water quality processes including reaeration are constrained.

2. Materials and Methods

2.1. Overview of Study Area

The study area is the Qu'Appelle River (QR) watershed, which extends eastward from Lake Diefenbaker in Saskatchewan to the Assiniboine River in Manitoba, Canada. Mainstay economic activities within the basin are agriculture (over 75% of total catchment area), fertilizer industries, underground and solution potash mines, oil refineries, and commercial fisheries. This study focused on the central lower half of the QR from the confluence of QR and Wascana Creek to Katepwa Lake (Figure 1). This section of the QR is a chained river-lake system, including four hypereutrophic lakes [34]: Pasqua, Echo, Mission, and Katepwa and an off-channel lake, Last Mountain Lake. A concrete control structure with timber stop-logs and lift gate on Echo Lake are used to regulate the water levels at Pasqua and Echo Lakes for recreational purposes. Water levels at Mission and Katepwa Lakes are maintained for commercial fishing by the regulation of a control structure at Katepwa Lake. The in-stream lakes have an average residence time of approximately nine days. Within the watershed two cities, Moose Jaw and Regina, release treated wastewater effluent into the river upstream of the lakes [34,35]. Tributaries from west to east include Wascana Creek, Last Mountain Creek, Loon Creek, Jumping Deer Creek, and Echo Creek. During floods, backflow from the Qu'Appelle is diverted through Last Mountain Creek into Last Mountain Lake. Echo Lake, Katepwa Lake, and Last Mountain Creek at Craven are equipped with control structures to regulate lake levels and to maintain environmental flow within the QR system.

The temperature in the region ranges from a mean of $-16\ ^\circ$C in the winter to a mean of $19\ ^\circ$C in the summer. Mean annual potential evapotranspiration (600 mm) is twice the precipitation [35].

The Qu'Appelle River discharge is modified by interbasin transfer from the South Saskatchewan River. The flow regime depicts the characteristics of a typical prairie river and lake drainage setting (Figure 2). At the Water Survey of Canada (WSC) hydrometric station 05JF001, Lumsden, the monthly mean discharge below 4.29 m^3/s occurs during the winter (November to March). During the spring, (March to April), monthly mean discharge increases sharply to 24.28 m^3/s as a result of snowmelt. Discharge then decreases sharply until August and then levels off for the rest of the season.

During the winter season (November to March), the QR system is typically covered with ice and snow. Construction of the Qu'Appelle Dam was completed in 1967. Ice freeze-up and break-up conditions after the dam construction for the period 1970 to 2010 at WSC station 05JG006 is shown in Figure 3. The graph shows a steady increase to a late freeze-up and early break-up conditions.

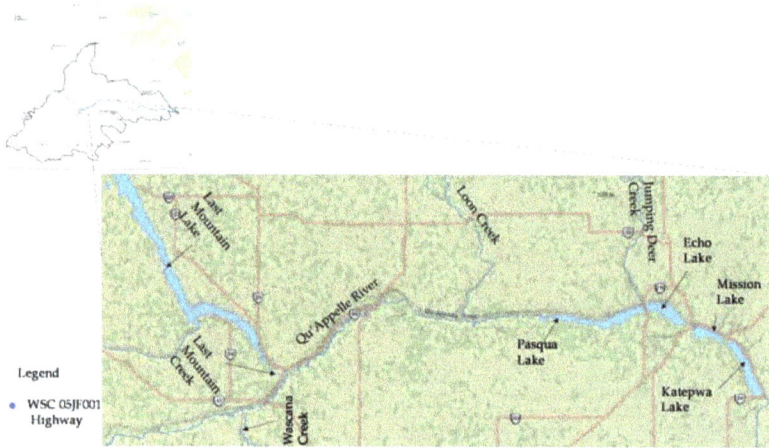

Figure 1. Map of the Qu'Appelle River System.

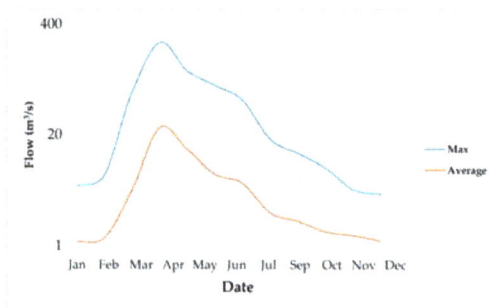

Figure 2. Monthly mean flow at WSC gauge 05JF001, Lumsden, from 1911 to 2016.

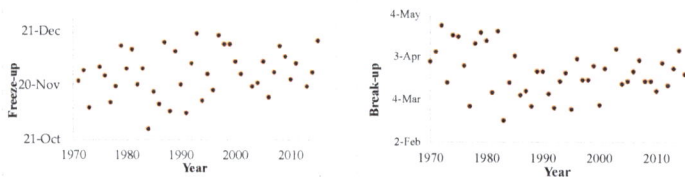

Figure 3. Ice freeze-up (**left**) and break-up (**right**) conditions at the Elbow River gauge between 1970 and 2010.

2.2. Model Setup

The seasonal variation of sediment oxygen demand for the chained river-lake system of Qu'Appelle River valley was assessed using the Water Quality Analysis Simulation Program 7.52 (WASP) (Athens, GA, USA). The WASP program dynamically models the water column and underlying benthos of aquatic systems including rivers, lakes, estuaries and coastal water bodies. The program is used to simulate conventional and toxic water pollution problems based on the principles of conservation of mass, momentum, and energy. Model development includes discretizing the surface water network into one, two or three-dimensional segments and defining model boundary conditions and input parameters. A one-dimensional model was considered for the river and shallow lake system. The system was assumed to be well-mixed laterally and vertically. The transport and kinetics of biochemical oxygen demand and dissolved oxygen (BOD-DO) dynamics of the study area were characterized in WASP's eutrophication module (EUTRO). EUTRO can simulate the dynamics of up to four systems including dissolved oxygen balance, the nitrogen cycle, phosphorus cycle, and phytoplankton kinetics.

The BOD-DO dynamics selected in EUTRO represents the basic complexity (complexity 1) of processes and kinetics involved in the dissolved oxygen balance. The main processes (see Equations (1)–(3) below) involved in the WASP's complexity 1 are atmospheric reaeration as a source of dissolved oxygen (DO) and as sinks, the total oxidation, and settling of oxidizable organic material and SOD. This low level of complexity (a slightly modified form of Streeter-Phelps) was selected to allow reasonable estimation of SOD through model calibration, by driving the model with a verified streamflow transport and constraining all other processes [31]. The total biochemical oxygen demand (TBOD), which represents oxidation of organic matter, is the main kinetic reaction for oxygen demanding materials in the water column. The particulate fraction of biologically oxidizable organic material gets settled out and is deposited in the sediments under low flow conditions. The deposition sometimes influences benthic sediment oxygen demand. In EUTRO, SOD is described for water column segments and seasonal changes are determined by the temperature coefficient [31]. where;

$$\frac{d(TBOD)}{dt} = TBOD_{in} - k_d\theta_d^{T-20}TBOD - \frac{v_{s3}}{D}(1 - f_{D5})TBOD, \qquad (1)$$

$$\text{Total BOD in} \quad \text{total oxidization} \quad \text{settling}$$

$$\frac{d(DO)}{dt} = +k_2\theta_2^{T-20}(DO_s - DO) - k_d\theta_d^{T-20}TBOD - \frac{SOD_T}{D}, \qquad (2)$$

$$\text{Reaeration} \quad \text{total oxidization} \quad \text{sediment oxygen demand}$$

$$SOD_T = SOD_{20}\theta^{T-20}. \qquad (3)$$

$d(TBOD)/dt$	=	Rate of change of the concentration of oxygen required to mineralize organic matter in mg/L/day. This is corrected before comparison with field BOD_5 data
$TBOD_{in}$	=	Oxygen demand due to the oxidization of newly produced organic matter per day, mg/L/day
k_d	=	Deoxygenation rate at 20 °C, 1/day
T	=	Water temperature, °C
θ_d	=	Temperature coefficient
$d(DO)/dt$	=	Rate of change of dissolved oxygen concentration, mg/L/day
θ_2	=	Temperature coefficient
DO	=	Dissolved oxygen concentration, mg/L
$TBOD$	=	Total Biochemical oxygen demand in mg/L
v_{s3}	=	Organic matter settling velocity, m/day
D	=	Average segment depth, m
f_{D5}	=	Fraction dissolved TBOD
k_2	=	Reaeration rate at 20 °C, 1/day
DO_s	=	DO solubility at temperature T, mg/L
SOD_T	=	Sediment oxygen demand rate at T, $g/m^2/day$
θ	=	Temperature coefficient

2.3. Advective Transport

The river network within the study area was modeled using WASP's dynamic stream transport module. Dynamic flow in the model is calculated using kinematic wave flow routing, ponded weir overflow, or backwater flow equations. The module requires river segments and flow information. Segment characteristics including lengths, widths, and depths for average flow conditions, as well as bottom slopes and Manning friction coefficients, are used to define the hydraulics of the system. Flow pathways and inflow time functions for the main channel and tributaries are used to define the river network and boundary conditions. Within the river, flow is established by upstream or tributary inflow boundary conditions. As the flow moves downstream, bottom slope and channel roughness control discharge variation.

The riverine section of the network was modeled as free flowing using the kinematic wave flow routine of the dynamic stream transport module. In WASP (Athens, GA, USA), this routine calculates flow wave propagation, which results in varying depths, volumes, flows, and velocities for the network. Flow is estimated by solving the one-dimensional continuity and momentum equations. River segments between the confluence with Loon Creek and the Qu'Appelle lakes were characterized as backwater flow while the four lakes were represented as ponded segments with weir overflow. WASP 7.52 (Athens, GA, USA) allows up to 25 segments for ponds. River segments upstream of the Loon Creek confluence were characterized as one-dimensional kinematic wave flow.

A calibrated and validated HEC-RAS model (Davis, CA, USA) was used to generate hydraulic parameters used as inputs for the dynamic flow routines in WASP (Athens, GA, USA). HEC-RAS (Davis, CA, USA) performs up to two-dimensional steady and unsteady flow hydraulic calculations for full network of natural and constructed channels [36]. HEC-RAS (Davis, CA, USA) river and lake cross-sections were generated from a 2013 digital elevation model (DEM), river depth survey and lake bathymetry survey of the site. The river network was discretized using an interval of 700 m in the riverine section and up to 2800 m in the lakes. This resulted in 175 one-dimensional horizontal water columns in WASP (Athens, GA, USA). Hydrodynamic parameters generated in HEC-RAS (Davis, CA, USA) included the depth exponent and multipliers, segment widths, and average depths. For the free-flow segments, the depth multiplier was taken as the cross-sectional average segment depth under average flow condition. In WASP (Athens, GA, USA), depth exponent controls channel shape. A depth exponent value of 0.3, representing an irregular cross-section was selected for these segments. Together with segment widths, slopes, and roughness factors, these depths are used to estimate segment depths during simulations. In the same vein, depth multipliers for backwater flow segments and ponded segments were taken as the cross-sectional average segment depth. Continuity is maintained when calculating the total segment depth for the ponded segments, while both continuity and momentum equations are used for the backwater depth calculation [37].

2.4. Initial Condition

Initial water quality concentrations in the model were estimated from closest long-term water quality monitoring stations by interpolating between two conjunctive stations. These included DO and TBOD concentrations for each segment, representing the concentrations at the beginning of the simulation. In situ river and lake DO measurements were undertaken using the sonde instrument.

TBOD was estimated from available BOD_5 data using the classic ultimate BOD equation with a deoxygenation rate of 0.2/day [38]. This rate was selected based on similar studies at nearby catchments and reported values which ranged from 0.1/day to 0.2/day [32,38].

2.5. Boundary Condition

Water quality, river flow, and temperature data for the period 2013 to 2015 were used as model forcings. The closest long-term water quality and river flow station (at Lumsden) downstream of the Wascana Creek and Qu'Appelle River junction was used to define the upstream boundary conditions. Water-quality and flow data on major creeks, including Last Mountain, Loon, and Jumping Deer creeks were used to represent contributions from these tributaries. Flow data for tributaries with either short time series of data or no data were estimated using a continuity equation involving nearest upstream and downstream streamflow stations.

Ice-covers influence reaeration rates during winter months in northern climates. Atmospheric oxygen transfer to the open water surface is blocked as a result of the ice-cover. To account for the effect of ice-cover on reaeration in WASP (Athens, GA, USA), a time function value of ice cover (XICECVR = 1-ice in WASP) fractions is used to multiply a reaeration rate constant. The value represents the ratio of open water surface to ice cover. A value of 1 indicates open surface condition whereas a value of 0 represents ice-covered conditions. In the study area, an ice cover value of 1 was used to represent freeze-up conditions (November to March) and 0 for open-water conditions (April to November) [31].

2.6. Calibration and Validation

The model hydrodynamics, water quality, and SOD rates were all calibrated and validated. SOD rate calibration involved successive parameter space iterations within reasonable ranges of reported values. Parameter values that gave the best curve fitting with field water quality data were selected as optimum parameters. A model run with these values was then compared with a different set of field data to validate the model. Model performance and errors were then estimated using mean error (ME) and mean absolute error (MAE) model evaluation statistics.

3. Results and Discussion

3.1. Model Calibration and Validation

The HEC-RAS (Davis, CA, USA) hydrodynamic model calibration was completed for the periods 2013 to 2014 and validation was conducted from 2014 to 2015. Model calibration involved successive iterations of selected Manning's coefficient values, flow, and seasonal factors within reported ranges [36]. Calibrated Manning's n values ranged from 0.022 to 0.025 for the riverine segments of the study site and 0.025 to 0.05 for river banks, flood plains, and lakes. Figure 4 shows the longitudinal profile of model calibration results for the river and lakes.

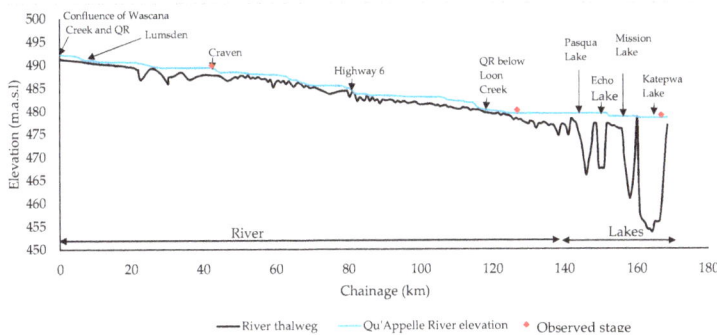

Figure 4. Longitudinal profile showing modeled surface water elevation (blue line), observed stage (red diamond square), and river thalweg (black line).

Optimum temporal and spatial SOD rates and temperature correction coefficient values for each segment were calibrated and verified for the periods 1 May 2013 to 30 April 2014 and 1 May 2014 to 30 April 2015 to cover one full summer (May–October) and one full winter (November–April) season in WASP(Athens, GA, USA). Reported SOD rates and temperature correction coefficients for the river and lake systems were varied successively within their reasonable ranges using the Dynamically Dimensioned Search (DDS) algorithm [39] of OSTRICH software (Boston, MA, USA). Optimum rates and coefficients were selected based on how well simulated DO levels matched observed DO levels. The best-fit curves for long-term stations along the river are as shown in Figure 5. Model bias and average prediction errors are shown in Table 1. Mean error (ME) measures the model bias, while mean absolute error (MAE) measures its predictive power. A value of zero ME is usually desired and less than 1.5 mg/L acceptable for variable (DO) prediction. Generally, predicted DO levels were in close agreement with sampled DO at Lumsden, Craven, upstream of Pasqua, and the lakes, but fair agreement was observed for Highway 6. Specifically, the model overpredicted the sampled low DO, especially during the summer season for Highway 6. As the low DO concentration was sampled during periods of high flow (summer time), it is plausible that high levels of diffused loadings from catchments draining into these sites could be occurring. As in-stream sampling data for loading was sparse, this process could not be well represented in the model. Another contributing factor could be the low frequency of DO and temperature observation readings. The model uses a daily time step while sampling was undertaken, monthly, on average. The model, thus, interpolates between the monthly values during the simulation and in so doing, could miss the low DO values.

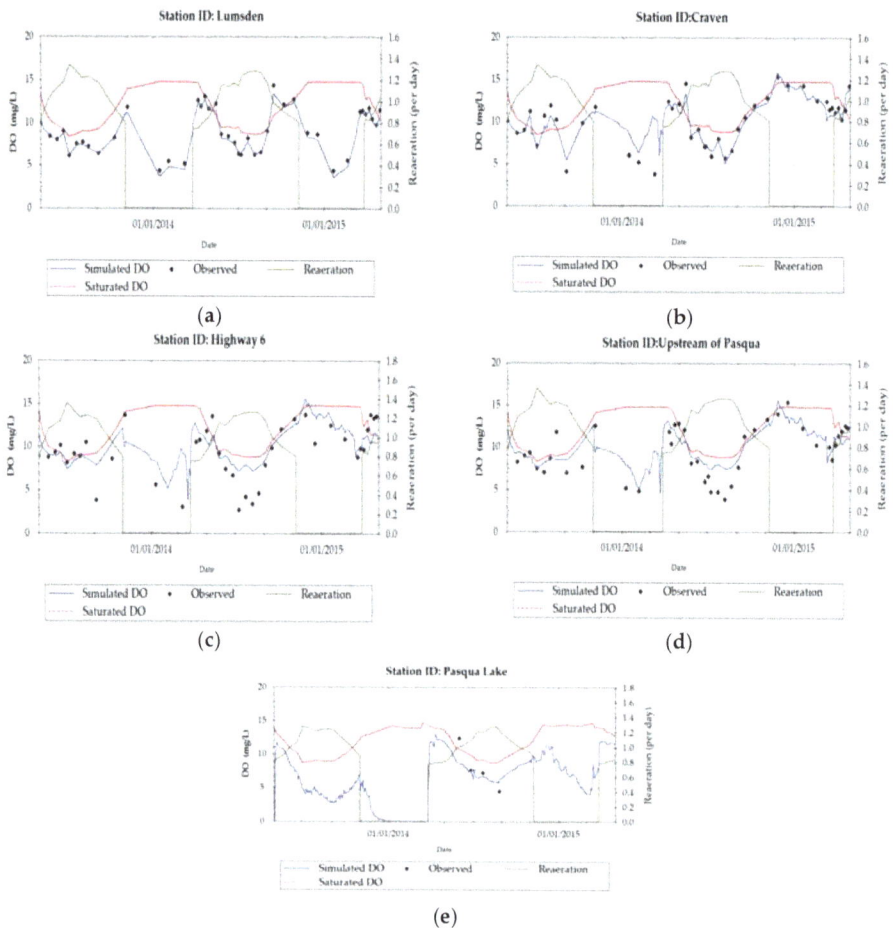

(a)

(b)

(c)

(d)

(e)

Figure 5. Simulated and sampled dissolved oxygen for the calibration and validation periods: the left and right vertical axes represent dissolved oxygen and reaeration coefficient, respectively; from the upstream of the study area, (**a**) Lumsden; (**b**) Craven; (**c**) Highway 6; (**d**) upstream of Pasqua Lake; and (**e**) Pasqua lake.

Table 1. Model bias and prediction error.

Gauge Station	Mean Error	Mean Absolute Error
Lumsden	−0.32	0.35
Craven	−0.03	0.77
Highway 6	0.62	1.78
Upstream of Pasqua	0.36	1.36
Pasqua	−0.62	1.30

3.2. Spatial and Seasonal Variation of Sediment Oxygen Demand

Generally, the estimated areal sediment flux at 20 °C ranged from 0.79 to 2.2 $g/m^2/day$ across the river-lake system (Table 2). These rates are within reported ranges for rivers and lakes [15,19,20,40,41]. A SOD_{20} rate of 1.9 and 1.7 $g/m^2/day$ were estimated for reaches 1 and 2 (over two-fold increase compared to reach 3). Studies by [42] showed a SOD rate range of 0.17 to 0.33 $g/m^2/day$ for the Athabasca River. The authors observed a downward gradient of SOD rate from sites downstream of pulp mill and municipal effluent discharge points. Reaches 1 and 2 form the upstream part of the riverine section of the study site. Algae bloom, siltation, and nutrient export from agricultural land are common phenomena for this part of the study area [43]. Urban effluents from Regina and Moose Jaw are conveyed through this stretch of the river and eventually to the chained lakes. This stretch of the river could be experiencing increased benthic decomposition of settled organic matter from effluent and dead algae. Oxygen is usually consumed faster than it is supplied for locations immediately downstream of point discharges through the oxidation of reduced substances. The SOD_{20} rate, however, decreases as the effluent travels downstream through reaches 3 and 4 (0.79 and 0.98 $g/m^2/day$) with organic matter mineralization, effluent stabilization, and increasing atmospheric reaeration. Background SOD values, however, influence these amounts [44–46]. Steep bank, deep morphology, low velocity, and large sediment surface area may have contributed to the SOD_{20} rate in the lakes [47]. The steep slopes enhance diffuse organic material loadings (from leaves and plant branches), especially during the summer.

Seasonally, across the system (river and lake), the actual model SOD rates during the simulation period increased steadily after ice-breakup and peaked in the summer but leveling off in the winter (Figure 6). The riverine winter modeled SOD rates ranged from 0.30 to 0.94 $g/m^2/day$. In the lakes, winter SOD rates ranged from 0.83 to 0.95 $g/m^2/day$. In the study by [48], SOD rates for 11 eutrophic ice-covered lakes in Alberta ranged from 0.243 to 0.848 $g/m^2/day$. The higher estimated SOD rates for the Qu'Appelle Lakes may be due to their hypereutrophic nature. The lakes' SOD rates increased three-fold in the summer compared to the winter and likewise, the river SOD rates increased by a factor of 3. Breakup and summer rainfall often lead to large deposition of sand, debris and allocthonous organic material especially in pool and riffle areas [49]. Hypolimnetic oxygen consumption thus increases as organic matter settles and are oxidized several days after such events.

Table 2. Calculated Model Segment SOD rates and temperature correction coefficient [1].

River Reach/Lake	Description	Optimum SOD (g/m²/Day) at 20 °C	Reported SOD Range	Reference
Reach 1	Wascana-QR confluence to Lumsden	1.90	(0.1–5.3) (SOD) (1.04–1.13) (Temp. coefficient)	[40] [2] [50]
Reach 2	Lumsden to Craven	1.70	(0.1–5.3)	[40]
Reach 3	Craven to Highway 6	0.79	(0.1–5.3)	[40]
Reach 4	Highway 6 to Qu'Appelle below Loon Creek	0.98	(0.1–5.3)	[40]
Lakes	Pasqua, Echo Mission, Katepwa lakes	2.20	(0.7–8.4)	[41] [3]

Notes: [1] The model was calibrated to an optimum temperature coefficient of 1.05 for all reaches and lakes; [2] The reported range represents SOD measurements between 19 and 25 °C for six Eastern Michigan river stations; [3] Oxygen mass balance for lakes.

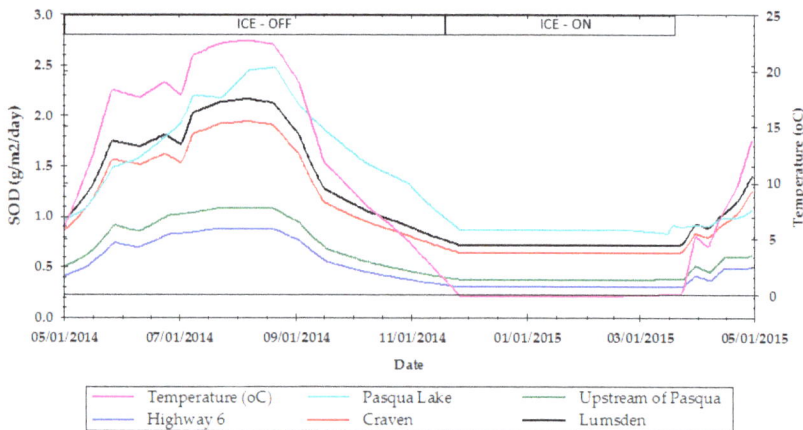

Figure 6. Seasonal variation of SOD rate in the QR system (these rates are model-based results and not normalized to 20 °C).

4. Conclusions

In this study, the SOD rates for a chained river-lake system have been estimated through the calibration of a slightly modified Streeter-Phelps BOD-DO water quality model. Due to the high sampling expenditure, routine sampling of SOD rates are often rare and little is known about the trend in a complex system like a chained river-lake system. River-lake transport and reaeration which influences SOD estimations were calibrated and constrained, respectively. Study results show that SOD_{20} rates gradually decreased from 1.9 to 0.79 $g/m^2/day$ in the riverine section of the system as urban effluent traveled through the river; meanwhile, a SOD_{20} rate of 2.2 $g/m^2/day$ was observed in the lakes. Seasonally, modeled SOD rate increased three-fold in the lakes and the river, respectively, in summer as compared to winter periods. The estimated SOD rates are within reported values of similar studies for rivers and lakes [15,19,20,40–42,48]. A downward gradient of SOD was observed in the riverine section of the study site. This trend is attributed to a plausible increase in organic matter decomposition at reaches 1 and 2 and mineralization, effluent stabilization, and increased reaeration in reaches 3 and 4. Studies by [42] for an ice-covered river reported increases in SOD rates for sites immediately downstream of effluent discharge points. Most ice-covered rivers in the world exhibit a linear relationship between oxygen depletion and river distance [51]. The lakes in the study site are shallow. Large sediment surface area, steep banks, and morphology may have contributed to the highly-estimated SOD rate in the lakes. The study by [48] attributed estimated SOD rates for ice-covered lakes in Alberta to a high lake productivity.

Increased bacterial activities in the sediments during the summer and sediment resuspension as a result of high flows may have contributed to the observed higher summer SOD rates in both rivers and lakes.

The zero-order model adopted to estimate the SOD rates in this study is not expected to predict the absolute values of SOD for the river and lakes due to limitations including the mechanism to capture how sediment organic matter converts to sediment oxygen demand [52]. Increasing the model complexity may improve the model results but this has to be accompanied by relevant observations [33] in order to reduce model uncertainties. In situ SOD measurements may be required in the river reaches and lake in the future to further verify the modeled results. Despite these shortcomings, the results are deemed reasonable for water quality modeling purposes of the complex river system in the absence of measured SOD data. The approach adopted in this study is also better than just selecting literature values to run water quality models.

The results of the study provide insights into the SOD rates and seasonal trends, especially for northern rivers.

Acknowledgments: This project was funded by Environment Canada's Environmental Damages Fund under the project: A water quality modeling system of the Qu'Appelle River catchment for long-term water management policy development.

Author Contributions: Eric Akomeah and Karl-Erich Lindenschmidt conceived and designed the experiments; Eric Akomeah performed the experiments and processed the results; Karl-Erich Lindenschmidt supervised, reviewed, and contributed to the writing of the article.

Conflicts of Interest: Authors declare no conflicts of interest.

References

1. Martin, N.; McEachern, P.; Yu, T.; Zhu, D.Z. Model development for prediction and mitigation of dissolved oxygen sags in the Athabasca River, Canada. *Sci. Total Environ.* **2013**, *443*, 403–412. [CrossRef] [PubMed]
2. Zahraeifard, V.; Deng, Z. Modeling sediment resuspension-induced DO variation in fine-grained streams. *Sci. Total Environ.* **2012**, *441*, 176–181. [CrossRef] [PubMed]
3. Liu, W. Measurement of sediment oxygen demand for modelling dissolved oxygen distribution in tidal Keelung River. *Water Environ. J.* **2009**, *23*, 100–109. [CrossRef]
4. Tian, Y. A Dissolved Oxygen Model and Sediment Oxygen Demand Study in the Athabasca River. Master's Thesis, University of Alberta, Edmonton, AB, Canada, 2005.
5. MacPherson, T.A. Sediment Oxygen Demand and Biochemical Oxygen Demand: Patterns of Oxygen Depletion in Tidal Creek Sites. Master's Thesis, University of North Carolina Wilmington, Wilmington, NC, USA, 2003.
6. Rounds, S.; Doyle, M.C. *Sediment Oxygen Demand in the Tualatin River Basin, Oregon, 1992–1996;* US Department of the Interior, US Geological Survey: Porland, OR, USA, 1997.
7. MacPherson, T.A.; Cahoon, L.B.; Mallin, M.A. Water column oxygen demand and sediment oxygen flux: Patterns of oxygen depletion in tidal creeks. *Hydrobiologia* **2007**, *586*, 235–248. [CrossRef]
8. Hu, W.; Lo, W.; Chua, H.; Sin, S.; Yu, P. Nutrient release and sediment oxygen demand in a eutrophic land-locked embayment in Hong Kong. *Environ. Int.* **2001**, *26*, 369–375. [CrossRef]
9. Matlock, M.D.; Kasprzak, K.R.; Osborn, G.S. Sediment oxygen demand in the Arroyo Colorado River. *JAWRA J. Am. Water Resour. Assoc.* **2003**, *39*, 267–275. [CrossRef]
10. Higashino, M.; Gantzer, C.J.; Stefan, H.G. Unsteady diffusional mass transfer at the sediment/water interface: Theory and significance for SOD measurement. *Water Res.* **2004**, *38*, 1–12. [CrossRef] [PubMed]
11. Haag, I.; Schmid, G.; Westrich, B. Dissolved Oxygen, and Nutrient Fluxes across the Sediment–Water Interface of the Neckar River, Germany: In Situ Measurements and Simulations. *Water Air Soil Pollut. Focus* **2006**, *6*, 413–422. [CrossRef]
12. Diaz, R.J.; Rosenberg, R. Spreading Dead Zones and Consequences for Marine Ecosystems. *Science* **2008**, *321*, 926–929. [CrossRef] [PubMed]
13. Chau, K. Field measurements of SOD and sediment nutrient fluxes in a land-locked embayment in Hong Kong. *Adv. Environ. Res.* **2002**, *6*, 135–142. [CrossRef]
14. Rasheed, M.; Al-Rousan, S.; Manasrah, R.; Al-Horani, F. Nutrient fluxes from deep sediment support nutrient budget in the oligotrophic waters of the Gulf of Aqaba. *J. Oceanogr.* **2006**, *62*, 83–89. [CrossRef]
15. Miskewitz, R.J.; Francisco, K.L.; Uchrin, C.G. Comparison of a novel profile method to standard chamber methods for measurement of sediment oxygen demand. *J. Environ. Sci. Health Part A* **2010**, *45*, 795–802. [CrossRef] [PubMed]
16. Zeledon-Kelly, R.V. *Effects of Landuse on Sediment Oxygen Demand for Streams in Northwest Arkansas and the Validation of a Sediment Oxygen Demand Measure-Plus-Calculate Model;* ProQuest: Fayetteville, AR, USA, 2009.
17. McDonnell, A.J.; Hall, S.D. Effect of environmental factors on benthal oxygen uptake. *J. Water Pollut. Control Fed.* **1969**, *41*, R353–R363.
18. Mackenthun, A.A.; Stefan, H.G. Effect of flow velocity on sediment oxygen demand: Experiments. *J. Environ. Eng.* **1998**, *124*, 222–230. [CrossRef]
19. Liu, W.; Chen, W. Monitoring sediment oxygen demand for assessment of dissolved oxygen distribution in river. *Environ. Monit. Assess.* **2012**, *184*, 5589–5599. [CrossRef] [PubMed]

20. Caldwell, J.M.; Doyle, M.C. *Sediment Oxygen Demand in the Lower Willamette River, Oregon*; Water-Resources Investigations Report; U.S. Geological Survey: Portland, OR, USA, 1995.

21. Paraska, D.W.; Hipsey, M.R.; Salmon, S.U. Sediment diagenesis models: Review of approaches, challenges and opportunities. *Environ. Model. Softw.* **2014**, *61*, 297–325. [CrossRef]

22. Massoudieh, A.; Bombardelli, F.; Ginn, T.; Green, P.; Ferreira, R. Mathematical modeling of the biogeochemistry of dissolved and sediment-associated trace metal contaminants in riverine systems. In Proceedings of the International Conference on Fluvial Hydraulics, Lisbon, Portugal, 6–8 September 2006.

23. Sohma, A.; Sekiguchi, Y.; Yamada, H.; Sato, T.; Nakata, K. A new coastal marine ecosystem model study coupled with hydrodynamics and tidal flat ecosystem effect. *Mar. Pollut. Bull.* **2001**, *43*, 187–208. [CrossRef] [PubMed]

24. Sohma, A.; Sekiguchi, Y.; Nakata, K. Modeling and evaluating the ecosystem of sea-grass beds, shallow waters without sea-grass, and an oxygen-depleted offshore area. *J. Mar. Syst.* **2004**, *45*, 105–142. [CrossRef]

25. Sohma, A.; Sekiguchi, Y.; Kuwae, T.; Nakamura, Y. A benthic–pelagic coupled ecosystem model to estimate the hypoxic estuary including tidal flat—Model description and validation of seasonal/daily dynamics. *Ecol. Model.* **2008**, *215*, 10–39. [CrossRef]

26. Gualtieri, C. Sediment oxygen demand modeling in dissolved oxygen balance. In Proceedings of the International Conference on Environmental engineering and renewable energy, Ulaanbaatar, Mongolia, 7–10 September 1998.

27. Slama, C.A. Sediment oxygen Demand and Sediment Nutrient Content of Reclaimed Wetlands in the Oil Sands Region of Northeastern Alberta. Master's Thesis, University of Windsor, Windsor, ON, Canada, 2010.

28. Boudreau, B.P. A method-of-lines code for carbon and nutrient diagenesis in aquatic sediments. *Comput. Geosci.* **1996**, *22*, 479–496. [CrossRef]

29. Meysman, F.J.; Middelburg, J.J.; Herman, P.M.; Heip, C.H. Reactive transport in surface sediments. I. Model complexity and software quality. *Comput. Geosci.* **2003**, *29*, 291–300. [CrossRef]

30. Bierman, V., Jr.; DePinto, J.; Dilks, D.; Moskus, P.; Slawecki, T.; Bell, C.; Chapra, S.; Flynn, K. *Modeling Guidance for Developing Site-Specific Nutrient Goals*; Water Environment Research Foundation: Alexandria, VA, USA, 2013.

31. Wool, T.A.; Ambrose, R.B.; Martin, J.L.; Comer, E.A.; Tech, T. *Water Quality Analysis Simulation Program (WASP). User's Manual*; U.S Environmental Protection Agency, Center for Exposure Assessment Modeling: Athens, GA, USA, 2006.

32. Terry, J.A.; Sadeghian, A.; Lindenschmidt, K. Modelling Dissolved Oxygen/Sediment Oxygen Demand under Ice in a Shallow Eutrophic Prairie Reservoir. *Water* **2017**, *9*, 131. [CrossRef]

33. Cox, L.A. Internal dose, uncertainty analysis, and complexity of risk models. *Environ. Int.* **1999**, *25*, 841–852. [CrossRef]

34. Quinlan, R.; Leavitt, P.R.; Dixit, A.S.; Hall, R.I.; Smol, J.P. Landscape effects of climate, agriculture, and urbanization on benthic invertebrate communities of Canadian prairie lakes. *Limnol. Oceanogr.* **2002**, *47*, 378–391. [CrossRef]

35. Hall, R.I.; Leavitt, P.R.; Quinlan, R.; Dixit, A.S.; Smol, J.P. Effects of agriculture, urbanization, and climate on water quality in the northern Great Plains. *Limnol. Oceanogr.* **1999**, *44*, 739–756. [CrossRef]

36. United States Army Corps of Engineers, Hydrologic Engineering Center. *HEC-RAS River Analysis System 2D Modelling User's Manual Version 5.0*; Hydrologic Engineering Center, United States Corps of Engineer: Davis, CA, USA, 2016.

37. Ambrose, R.; Wool, T. *WASP7 Stream Transport-Model Theory, and User's Guide, Supplement to Water Quality Analysis Simulation Program (WASP) User Documentation*; National Exposure Research Laboratory, Office of Research and Development, US Environmental Protection Agency: Athens, GA, USA, 2009.

38. Akomeah, E.; Chun, K.P.; Lindenschmidt, K. Dynamic water quality modelling and uncertainty analysis of phytoplankton and nutrient cycles for the upper South Saskatchewan River. *Environ. Sci. Pollut. Res.* **2015**, *22*, 18239–18251. [CrossRef] [PubMed]

39. Tolson, B.A.; Shoemaker, C.A. Dynamically dimensioned search algorithm for computationally efficient watershed model calibration. *Water Resour. Res.* **2007**, *43*, 208–214. [CrossRef]

40. Chiaro, P.S.; Burke, D.A. Sediment oxygen demand and nutrient release. In *Conference on Environmental Engineering: Research Development and Design, Proceedings of the Environmental Engineering Division Specialty Conference, Kansas City, MO, USA, 10–12 July 1978*; American Society of Civil Engineers: Kansas City, MO, USA, 1978; pp. 313–322.

41. James, A. The measurement of benthal respiration. *Water Res.* **1974**, *8*, 955–959. [CrossRef]

42. Zhu, D.; Yu, T. Spatial variation of sediment oxygen demand in Athabasca River: Influence of water column pollutants. In Proceedings of the World Environmental and Water Resources Congress, Kansas City, MO, USA, 17–21 May 2009; pp. 1–12.

43. Clifton Associates Ltd. *Upper Qu'Appelle Water Supply Project: Economic Impact and Sensitivity Analysis*; Water Security Agency: Regina, SK, Canada, 2012.

44. McCulloch, J.; Gudimov, A.; Arhonditsis, G.; Chesnyuk, A.; Dittrich, M. Dynamics of P-binding forms in sediments of a mesotrophic hard-water lake: Insights from non-steady state reactive-transport modeling, sensitivity and identifiability analysis. *Chem. Geol.* **2013**, *354*, 216–232. [CrossRef]

45. Maerki, M.; Müller, B.; Wehrli, B. Microscale mineralization pathways in surface sediments: A chemical sensor study in Lake Baikal. *Limnol. Oceanogr.* **2006**, *51*, 1342–1354. [CrossRef]

46. United States Environmental Protection Agency. *Technical Guidance Manual for Developing Total Maximum Daily Loads, Book 2: Streams and Rivers, Part 1: Biochemical Oxygen Demand and Nutrients/Eutrophication*; U.S. EPA: Washington, DC, USA, 1995.

47. Blais, J.M.; Kalff, J. The influence of Lake Morphometry on sediment focusing. *Limnol. Oceanogr.* **1995**, *40*, 582–588. [CrossRef]

48. Babin, J.; Prepas, E. Modelling winter oxygen depletion rates in ice-covered temperate zone lakes in Canada. *Can. J. Fish. Aquat. Sci.* **1985**, *42*, 239–249. [CrossRef]

49. Elwood, J.W.; Waters, T.F. Effects of floods on food consumption and production rates of a stream brook trout population. *Trans. Am. Fish. Soc.* **1969**, *98*, 253–262. [CrossRef]

50. Thomann, R.V.; Mueller, J.A. *Principles of Surface Water Quality Modeling and Control*; Harper & Row: New York, NY, USA, 1987.

51. Chambers, P.; Scrimgeour, G.; Pietroniro, A. Winter oxygen conditions in ice-covered rivers: The impact of pulp mill and municipal effluents. *Can. J. Fish. Aquat. Sci.* **1997**, *54*, 2796–2806. [CrossRef]

52. Chapra, S.C. *Surface Water-Quality Modeling*; Waveland Press: Long Grove, IL, USA, 2008.

water

MDPI

Article

Modelling Dissolved Oxygen/Sediment Oxygen Demand under Ice in a Shallow Eutrophic Prairie Reservoir

Julie A. Terry *, Amir Sadeghian and Karl-Erich Lindenschmidt

Global Institute for Water Security, University of Saskatchewan, Saskatoon, SK S7N 3H5, Canada;
amir.sadeghian@usask.ca (A.S.); karl-erich.lindenschmidt@usask.ca (K.-E.L.)
* Correspondence: julie.terry@usask.ca; Tel.: +1-306-966-2825

Academic Editor: Jiangyong Hu
Received: 17 December 2016; Accepted: 10 February 2017; Published: 17 February 2017

Abstract: Dissolved oxygen is an influential factor of aquatic ecosystem health. Future predictions of oxygen deficits are paramount for maintaining water quality. Oxygen demands depend greatly on a waterbody's attributes. A large sediment–water interface relative to volume means sediment oxygen demand has greater influence in shallow systems. In shallow, ice-covered waterbodies the potential for winter anoxia is high. Water quality models offer two options for modelling sediment oxygen demand: a zero-order constant rate, or a sediment diagenesis model. The constant rate is unrepresentative of a real system, yet a diagenesis model is difficult to parameterise and calibrate without data. We use the water quality model CE-QUAL-W2 to increase the complexity of a zero-order sediment compartment with limited data. We model summer and winter conditions individually to capture decay rates under-ice. Using a semi-automated calibration method, we find an annual pattern in sediment oxygen demand that follows the trend of chlorophyll-a concentrations in a shallow, eutrophic Prairie reservoir. We use chlorophyll-a as a proxy for estimation of summer oxygen demand and winter decay. We show that winter sediment oxygen demand is dependent on the previous summer's maximum chlorophyll-a concentrations.

Keywords: ice-cover; chlorophyll-a; shallow lakes; modelling; dissolved oxygen; sediments

---

1. Introduction

Oxygen is essential for a healthy aquatic system. The Canadian water quality guidelines for the protection of aquatic life state that dissolved oxygen (DO) is the most important parameter in water [1]. Severe oxygen depletion can lead to fish kills [2,3], deformities in fish larvae [1], and changes in community composition and lake trophic state [3–5]. The prediction of DO concentrations is vital for fisheries, and for aquatic managers responsible for maintaining ecosystem health [3].

The shallow lakes and reservoirs of the Canadian Prairies are naturally mesotrophic to eutrophic [6], and display severe fluctuations in DO [2]. Large phytoplankton blooms can occur, and the waterbodies are subject to an extreme climate with hot summers and ice-covered winters. DO is additionally important in drinking water reservoirs as dissolved gas supersaturation can be an issue in water treatment [7]. Low oxygen can also induce release of nutrients, and sulphide production.

Phytoplankton contribute greatly to DO in reservoirs by photosynthesis, as will macrophytes if present in large volumes [3,8]. Periphyton may also contribute [9]. Additional DO will enter as inflows and reaeration from the atmosphere. As well as replenishing DO, inflowing waters also transport organic matter into a reservoir. This matter will settle in the sediments along with dead plants and algae. When this material decomposes both chemical oxidation and biological respiration exert a significant oxygen demand to the water column [10], known as biochemical oxygen demand (BOD), and to the sediments, known as sediment oxygen demand (SOD). Both BOD and SOD have a positive relationship with reservoir productivity in this case. Nitrification also contributes to oxygen demand.

In open water oxygen deficits are replenished through reaeration [11] to the surface and mixed to the bottom by wind and turbulence. Reaeration is the exchange of gases at the air–water interface. In contrast, ice-covered conditions bring significant changes to the DO dynamics. Under ice-cover atmospheric gas exchange is removed from the oxygen balance [12]. If sufficient light penetrates through the ice, plants and algae continue to photosynthesise and produce oxygen [13]. The cooler winter water temperatures slow the decomposition of organic matter and reduce the consumption of oxygen through bacterial activity [1,5]. Breaks in the ice can increase the oxygen balance by allowing gas exchange.

Conversely, heavy snow loads reduce light penetration to a point where photosynthesis is greatly reduced [5,14,15]. The resultant decomposition of dying biota consumes further oxygen supplies [12]. Inflow volumes are often low in winter with less new oxygen inflow to offset consumptive processes [16]. There may be no breaks in the ice and extended ice-cover. The absence of wind on the water surface reduces the chance of oxygen mixing through the water column to deeper waters. Oxygen levels can reach the point of anoxic conditions at the bottom of reservoirs with high oxygen demands [3].

Low winter DO concentrations have been linked to shallower lakes with sizeable littoral zones and prolonged ice-cover [17]. Shallow waterbodies have a large sediment–water interface relative to water volume. This interface is where the organic matter and bacterial activity tends to be concentrated [17]. The relative influence of bottom decomposition on the water column is therefore greater in shallow systems [11]. While open waters are often well-mixed from wind action, under ice-cover the shallow water depth means that the anoxic zone could potentially thicken along the bottom sediments. SOD is highly sensitive to small temperature fluctuations at lower water temperatures, with small increases intensifying oxygen depletion [18]. When modelling DO in a shallow, eutrophic system the ability to simulate SOD and the rate of SOD decay is important.

Water quality modellers usually work in a series of steps: first is a water balance model followed by a water temperature and mixing model to set-up the hydrodynamics for the system. Some modellers then choose to move to a full nutrient and phytoplankton model, and their DO predictions are part of the overall sources and sinks of the model. The danger with greatly increasing the complexity at once is that each additional state variable will require additional parameters and functions to control the escalating number of interacting processes. The result is a large number of parameters in relation to output variables and objective functions. An over-parameterised model is difficult to calibrate due to the greater number of parameter combinations that may provide *non-unique* optima as described by the *equifinality* thesis [19].

Another strategy is to approach the nutrient and phytoplankton modelling with a stepwise approach: building the model complexity in stages rather than adding all the water-quality data at once. This method allows parameters to be constrained at a lower complexity (fewer output variables) before enabling further state variables, parameters and functions.

One of the simplest methods to begin a DO model would be the Streeter–Phelps model, a long-standing model with the state variables BOD and DO [11]. In practice, the relative importance of BOD depends on the system being investigated. In Europe, for example, rivers have high loading of waste water BOD in areas of dense population and industry [20–22]. The Prairie reservoirs in Canada are often in rural areas and BOD inflows can be small. For these shallow, eutrophic systems it is far more important to include SOD when modelling DO.

Water quality models generally fall into two categories for modelling SOD: a full sediment diagenesis model, or a much simplified year-round SOD rate that varies in response to water temperature. A diagenesis model has the advantage that it can be calibrated for specific applications such as wastewater studies. The disadvantage is that, in reality, SOD is fairly difficult to measure in the field. The diagenesis model is useful when sediment core analyses are available, yet few aquatic managers and fisheries would have access to this kind of information.

A full water-quality model is currently being built for Buffalo Pound Lake (BPL), a shallow eutrophic Prairie reservoir in the Canadian province of Saskatchewan. BPL has insufficient sediment data to properly parameterise a diagenesis model. A constant SOD rate, however, is unrepresentative of the processes in a shallow, eutrophic system that spends approximately half of the year under-ice.

Our objective in this study is to test an alternative approach that allows us to increase the complexity in the constant rate SOD formulation with limited data. Our method extends the year-round constant rate by building an empirical model for SOD that considers both ice-on and ice-off periods. Modelling both winter and summer allows us to constrain certain parameters during certain seasons in order to better calibrate other parameters. For instance, setting reaeration to zero under ice-covered conditions allows us to better describe the SOD parameterisation.

For the DO model, we use CE-QUAL-W2 (W2) (Portland, OR, USA)—a two-dimensional (vertical and longitudinal) coupled hydrodynamic and water quality model. W2 is a complex model suitable for reservoirs. W2 is chosen due to its suitability for BPL as a long, narrow waterbody, and the inclusion of an ice model. A full description of the hydrodynamics and transport processes of W2 is given in the user manual [23].

The results obtained by our simulations will allow us to constrain our baseline SOD within a sensible range for BPL. We will be able to maintain appropriate SOD rates as the model becomes more complex on incorporating algal-nutrient dynamics.

2. Materials and Methods

2.1. Site Description

Buffalo Pound Lake (BPL) is an impounded natural lake located on the Upper Qu'Appelle River in the Saskatchewan Province of Canada (Figure 1). The reservoir supplies the water demands of the cities of Moose Jaw, Regina, surrounding communities, and an expanding industrial corridor and potash mines. The reservoir forms part of the glacially formed upper Qu'Appelle River system described in detail in Hammer [24]. Annual mean precipitation is 365.3 mm and approximately 30% falls as snowfall [25]. Ice cover is typically November to late April. Air temperatures range between a daily minimum of $-17.7\,^{\circ}\text{C}$ in January to a daily maximum of $26.2\,^{\circ}\text{C}$ in July [25]. Over 95% of the drainage basin is agricultural land [26] suggesting that non-point nutrient sources (diffuse pollution, overland run-off) may factor significantly in nutrient loading to BPL. Water quality issues such as eutrophication remain a challenge, and the reservoir has persistent problems with taste, odour, and algal blooms [27,28].

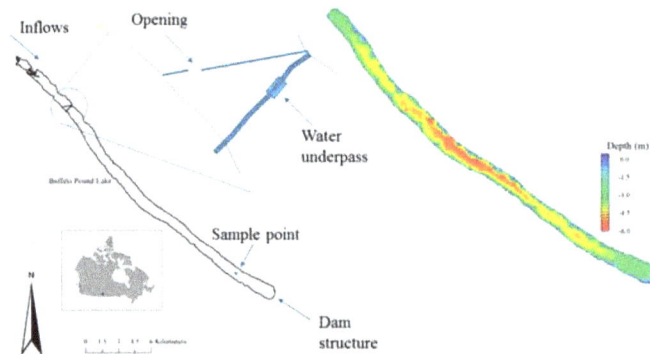

Figure 1. Buffalo Pound Lake, Saskatchewan, Canada. Mean depth is 3.8 m with a maximum depth of 5.98 m. Mean residence time is highly variable (6 to 30 months). Flow is in a southeast direction. The black reservoir outline is to the provided scale. The digital elevation model (DEM) shows bathymetry for the main body of the lake downstream of the underpass.

2.2. Model Setup

W2 needs full geometric data to operate. A digital elevation model (DEM) was prepared in ArcGIS 10.2.2 (ESRI Inc., Redlands, CA, USA). The DEM includes sonar data collected by boat in 2014, and a reservoir extent polygon and shoreline digital elevation data provided by the Saskatchewan Water Security Agency (WSA). The combined GIS data are interpolated using a spline barrier method at 30 m resolution. The Upper Qu'Appelle flows into the northwest end of the reservoir with the dam located at the southeast end. In essence, the upstream area of BPL is split into separate waterbodies by Highway 2, which dams the reservoir down to the reservoir bed (Figure 1). The first obstacle that the inflows meet is the breaker built to protect the highway. Once the flows are through the breaker they are then squeezed through a gap of 45 m (three connected 15 m sections) under the bridge, and into the main body of the reservoir. The top section of the reservoir is extremely shallow and weed choked, and sonar data could not be collected by boat. The top and main body of the reservoir will likely experience some differences in reservoir conditions making it less realistic to model the reservoir as just one waterbody. Water quality data were available for under Highway 2 and so these are set as boundary data. The DEM and water quality model covers the whole main body of BPL downstream of Highway 2.

W2 discretises the waterbody into a finite grid of longitudinal segments, vertical layers and cross-sectional widths. The user specifies the space steps in the longitudinal and vertical directions. The cross-sectional widths are determined by the shoreline bathymetry as each cell spans the width of the waterbody. The prepared DEM has been segmented into a numerical grid in the Watershed Modelling System (WMS) (Aquaveo, Provo, UT, USA) for final output as a bathymetry text file for W2. Longitudinal segments average 100.9 m with a total length for all 256 segments of 25,834 m. Vertical layers are 0.25 m with the maximum number of layers being 26 at the deepest part of the reservoir. W2 requires boundary layers and segments that are all zero meters and these are included in these totals. Average width at the surface is 890 m.

2.3. Data Collection and Analysis

Hourly meteorological forcing data have been downloaded from Environment Canada (EC) for the Moose Jaw station located approximately 30 km south from BPL. In order to estimate the wind conditions at the reservoir surface, comparisons have been made of recent EC data against data from an in situ high-frequency data collecting buoy. This buoy has been deployed on BPL by the Global Institute for Water Security since 2014 for open water field seasons. Snowfall figures are also taken from the EC Moose Jaw station, and are monthly totals. The "snow on the ground" measurement is the physical quantity of snow-cover on the last day of each month.

Gauged averaged daily inflows have been downloaded directly from the EC website. Accurate inflow data is not available for the BPL boundary of Highway 2, and flows are from the nearest gauge 19 km upstream on the Upper Qu'Appelle River. This is land distance—the flows will travel further as the channel meanders. Monthly mean estimates of ungauged inflows are provided by the WSA and include minor tributaries located after the EC gauge, as well as overland run-off estimates.

The main outflows from BPL are dam releases and piped withdrawals. The dam releases have been derived using EC data for two downstream flow gauges. The withdrawal volumes are provided by the on-site Buffalo Pound Water Treatment Plant (WTP) and by SaskWater. Daily averaged water-level measurements are provided by the WSA for an in-reservoir gauge.

Monthly inflow DO and BOD measurements are provided by the WSA for a sample site at the Highway 2 boundary. The in-reservoir observed data are taken from a substantial weekly dataset provided by the WTP laboratory. The WTP weekly samples are normally taken around 07:20 a.m. at a sample site midway between the north and south shorelines near the downstream end of the reservoir, and approximately one meter off the reservoir bed. The reservoir is expected to be well-mixed at the sampling point. Water is withdrawn through an intake pipe at this location to the WTP's pumping station, on the south shore, where sampling takes place before the water is pumped to the WTP itself.

These samples are transported to the WTP laboratory for analyses. This procedure is performed weekly in both open water and under-ice conditions. Some spot sample water quality data are available for other locations across the lake, although not all constituents are measured regularly at these additional sites, and they have not been included in this study.

Weekly inflow temperatures are estimated through a linear regression ($R^2 = 0.861$; equation $y = 1.0598x - 2.7747$; 59 samples; no outliers removed) between WTP spot sample temperature measurements over 34 years at the site of the inflow gauge upstream, and the WTP weekly temperature data for the reservoir. Precipitation temperatures are set at dew-point temperature, or zero if the dew-point is negative.

Initial conditions for water temperature and DO are also taken from the WTP weekly dataset. Sediment temperature is set at the mean annual air temperature over the simulation period as per the W2 manual recommendation [23]. Parameter coefficients are set according to knowledge of the reservoir, or are left at W2 default values where data are not available to support a change. The kinetic coefficients for BOD and SOD are W2 defaults (Table 1).

Table 1. W2 Default kinetic coefficients used in this study for the sediment oxygen demand (SOD) and biochemical oxygen demand (BOD) calculations.

Coefficient	Description	Value	Units
TSED	Sediment temperature	10.3 [1]	°C
CBHE	Coefficient of bottom heat exchange	0.3	$W{\cdot}m^{-2}{\cdot}°C^{-1}$
KBOD	5-day BOD decay rate at 20 °C	0.1 [2]	day^{-1}
TBOD	Temperature coefficient (decay rate)	1.02 [2]	
RBOD	Ratio of 5-day BOD to ultimate BOD	1.85 [2]	
CBODS	BOD settling rate	0.0 [2]	$m{\cdot}day^{-1}$
SODT1	Lower temperature for zero-order SOD or first-order sediment decay	4.0	°C
SODT2	Upper temperature for zero-order SOD or first-order sediment decay	25.0	°C
SODK1	Fraction of SOD or sediment decay at lower temperature	0.1	
SODK2	Fraction of SOD or sediment decay at upper temperature	0.99	
REAERAT	Reaeration formulation	LAKE, 6	

Notes: [1] Where the value is different to the W2 default; [2] W2 uses CBOD as the model group; we are assuming that CBOD makes up the majority of our BOD.

For quality assurance, the WTP data span the complete simulation period and undergo strict quality control sample procedures. The flow data, water-level data and meteorological data downloaded from the WSA and EC websites are expected to have undergone quality control prior to commencement of the study. Metadata is available for the WSA water-quality database that details the source and perceived accuracy of the measurements.

2.4. Model Customisation

We have customised two components of the W2 model: SOD and the ice algorithm. This study uses W2 version 3.72, which includes a zero-order, or a limited first-order, sediment compartment for estimating SOD. The latest version of W2 (v4.0) also includes a new sediment diagenesis model; however, with no sediment data to drive a full diagenesis compartment, there would be considerable uncertainty at the large scale of a reservoir. We opted for v3.72 as the complete source code for v4.0 was not available for download on commencement of our study, and we were unable to customise the later version for our specific objective.

W2 uses three different types of data for model calibration: the first group are set prior to the model run and remain constant throughout the simulation—examples being latitude for the calculation of solar radiation, bathymetry, and parameter coefficients. The second group are the time-varying state variables such as inflows, outflows, and meteorological data. The third group are the variables changing internally in the model at each time step; temperature, shear stress, and horizontal and vertical velocities are examples of this group.

DO is calculated in W2 as per Equation (1). The complete set of DO equations in W2 are more complex as the model recognises up to thirteen sources and sinks of DO [23]. We present here the W2 equations we use in our own reservoir DO/SOD model.

$$S_{DO} = \underbrace{A_{sur}K_L\left(\Phi'_{DO} - \Phi_{DO}\right)}_{aeration} - \underbrace{\boldsymbol{SOD}\gamma_{OM}\frac{A_{sed}}{V}}_{zero-order\ SOD} - \underbrace{\sum K_{BOD}R_{BOD}\Theta^{T-20}\Phi_{BOD}}_{BOD\ decay} \qquad (1)$$

where:

A_{sur}	water surface area, m^2
K_L	interfacial exchange rate for oxygen, m·s^{-1}
Φ'_{DO}	saturation DO concentration, g·m^{-3}
Φ_{DO}	dissolved oxygen concentration, g·m^{-3}
$\boldsymbol{SOD}$	sediment oxygen demand, g·m^{-2}·s^{-1}
γ_{OM}	temperature rate multiplier for organic matter decay
A_{sed}	sediment surface area, m^2
V	volume of computational cell, m^{-3}
K_{BOD}	BOD decay rate, s^{-1}
R_{BOD}	conversion from BOD in the model to BOD ultimate
Θ	BOD temperature rate multiplier
Φ_{BOD}	BOD concentration, g·m^{-3}

The zero-order SOD is a user-defined constant rate that is temperature dependant. In the original source code the model reads the SOD at the start of the simulation, and uses the same rate in the equation for the whole simulation period. The zero-order SOD is displayed in bold text in Equation (1). In W2, BOD is imported as a time-varying variable in the inflow constituent file. We modified the W2 code to treat SOD in a similar manner and read SOD as a time-varying temperature-dependent input file. The model checks for new values of SOD during each iteration and updates the zero-order SOD in Equation (1). The original constant SOD rate in W2 is now a variable rate in the DO equations, although the DO module itself is unchanged.

For the ice model W2 calculates the formation and melting of ice during simulations, and the relevant processes (e.g., light, wind, heat fluxes) are adjusted accordingly by the model. Snow is not considered in the algorithm. Snow depth at BPL is often between 0.1 and 0.3 m as per the supplied WSA long-term data. To account for this lack of snow the ice model has been extended to include two empirical coefficients to the existing W2 algorithms, as have been previously applied [29]. The first coefficient α extends the ice growth and thickness equations and reduces the heat lost through back radiation from black surfaces. The second coefficient β extends the ice melt equations and reduces the heat conduction between air and ice. Both coefficients are assigned a value between zero and one to be multiplied by the appropriate equation parameter. For BPL, no ice thickness data are available for calibration of α. A 39-year data set of ice-on and ice-off dates has been provided by the WTP, and it is found that W2 predicts the ice-on dates to be closely matched with the observed dates. For this study, the coefficient α is set to have no contribution to the ice growth equation (given the value 1). Ice-off dates were difficult to match as the ice melts too quickly in the W2 simulations—up to a period of several weeks. The optimum value of coefficient β is found to be 0.24 to predict the best spring ice-off dates over the simulation period.

2.5. Model Setup and Application

The model simulates a continuous seven-year period (1 April 1986–31 March 1993). This period is chosen due to the availability of daily flow data recorded by two WSA gauges just above and below BPL that were subsequently discontinued.

The water balance, ice-on and ice-off dates, and the water temperature model were calibrated. The final temperature model shows good results (Figure 2). Some discrepancy occurs in the winter of

1989/90 and 1991/92 with the model under-predicting the winter bottom temperature and possibly the stratification. The temperature profile can depend on the meteorological conditions at freeze-over. In addition, many of the temperature sensitive parameters and coefficients in W2 (e.g., sediment temperature, bottom heat exchange, surface albedo) are fixed in the model. It is likely that there is some temporal variance in these in-reservoir.

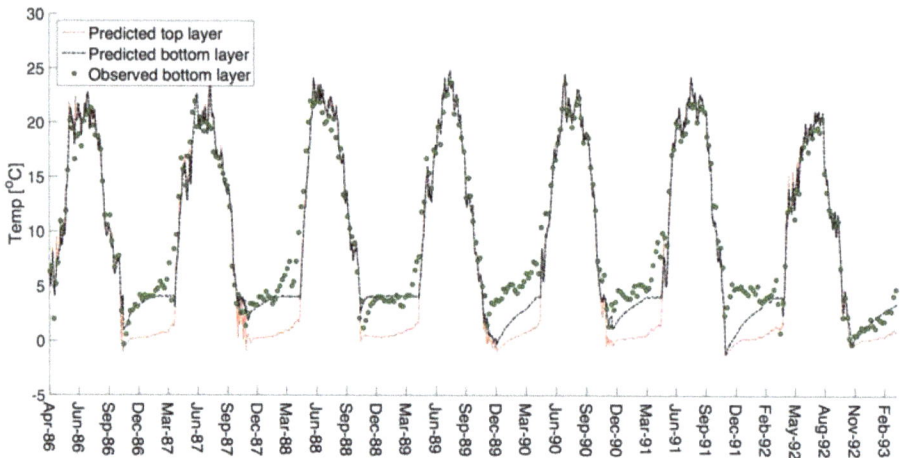

Figure 2. Results of the water temperature model. Compares predicted temperatures in the same grid cell as the Buffalo Pound Water Treatment Plant weekly observations. Note that CE-QUAL-W2 converts the negative water temperature modelled at the start of each winter to equivalent ice thickness. Root mean square error = 1.46 (to 2 dp); mean absolute error = 1.12 (to 2 dp).

The monthly ungauged inflow estimates provided by the WSA are created to close their own water-balance for BPL, and our respective water-balances differ as a result of methodology and data. We chose not to use the provided estimates due to the uncertainty. Another limitation is that the downstream flow data, which we have included, have room for error due to the presence of wetlands, and the potential for backwater flows during the freshet from a tributary confluence downstream of the reservoir. To close our water-balance we have incorporated a distributary tributary (DT) using the W2 in-built water-balance tool. The total contribution of the DT flows is approximately 1.4% of total inflows and precipitation over the eight-year simulation period, although there are seasonal fluctuations. An exception is the winter of 1992/93 where the maximum contribution of DT flows to total inflows under ice reach 22%. This is likely due to uncertainty attributed to error in the withdrawals to the industrial corridor as they are reported on yearly totals. These final year DT flows equate to an approximate 6.5% of BPL volume based on our initial reservoir volume in the DEM (BPL water levels are controlled within a few cm). We aim to assign the DT flows to ungauged inflows and/or outflows once we calibrate the full water-quality model—based on our chemical and nutrient data. For this study, we are assuming that constituent concentrations are primarily introduced in the main river inflows, and that DO and BOD inputs are zero in the DT flows.

We first simulated a simple DO model of BPL with a constant SOD, for comparative purposes. We extended the calibrated temperature model by enabling the water quality variables DO and BOD (BOD as one group) in W2. We proceeded to calibrate the SOD rate as part of a Monte Carlo analyses for several coefficients. We used MATLAB (MathWorks, Natick, MA, USA) to run W2 for these calibration iterations and attempted to fit the predicted DO to observed DO concentrations. Using a constant SOD we were only able to produce a moderately good fit (Figure 3): with both underestimations and overestimations of DO throughout the simulation period.

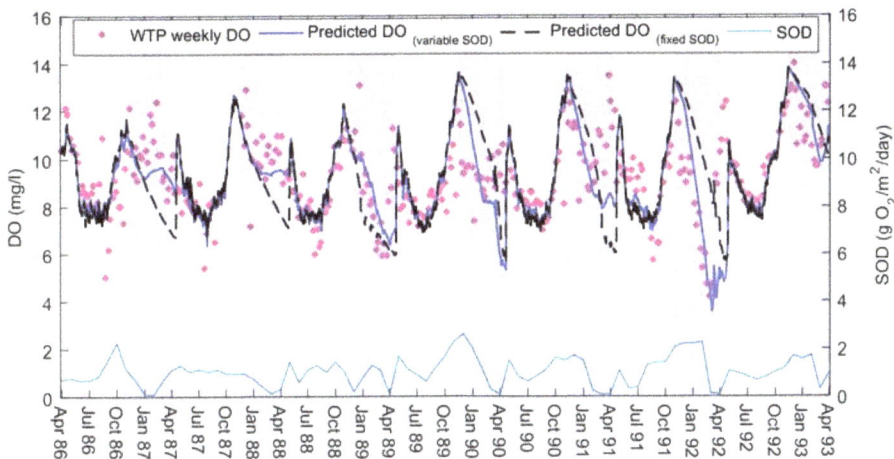

Figure 3. The dissolved oxygen (DO) model using variable sediment oxygen demand (SOD) rates found through a semi-automated calibration procedure to match weekly predicted and observed DO concentrations (WTP weekly DO). These SOD rates are maximum values, as used by CE-QUAL-W2. The black line represents the best fit we could achieve by Monte Carlo analyses using a constant SOD rate (root mean square error = 1.94 (to 2 dp); mean absolute error = 1.43 (to 2 dp). Ice-cover days shown here in blue stripes are observed data from the Buffalo Pound Water Treatment Plant. Predicted DO concentrations using the variable SOD have root mean square error = 1.58 (to 2 dp); mean absolute error = 1.1 (to 2 dp).

To introduce a variable SOD we took the DO model of Figure 3 and implemented a semi-automated calibration through MATLAB. The code attempted to match W2's predicted DO to the observed data by changing the SOD at weekly intervals. We used simple rules: for each weekly period, if the predicted DO concentrations were overestimated then the MATLAB code increased the SOD to increase consumption. If the predicted DO concentrations were underestimated then the SOD decreased that week. All weeks were changing simultaneously during the iterations and we ran the model until the SOD rates reached a stable condition.

We found that the DO model performed better with the variable SOD rates (Figure 3). On examining the results of this new model, we noted that SOD followed a relatively consistent seasonal trend. SOD was high over summer, peaking towards the end of the season, and then gradually depleting over winter. The rates of SOD were different in magnitude each year, yet similar in behaviour.

We compared the new SOD results against observed in-reservoir water-quality data to look for trends. We noticed that the predicted SOD appeared to follow a similar pattern to the observed weekly summer chlorophyll-a (Chla) concentrations (Figure 4) over the first few years: with SOD peaking not long after Chla. In light of this, we investigated if any relationships could be found between Chla abundance, and SOD. Our aim was to determine if Chla might be useful as an alternative measurement for estimating SOD. We approached the open water and under-ice periods differently due to the restriction of ice-cover on reaeration. We wanted to maintain the assumption of having limited data with which to build a model, and we aimed for simple strategies.

For open water seasons reaeration can replenish oxygen as it consumed, and we elected to keep our SOD constant over these periods. We allowed the model to have interannual variability by using individual SOD rates for each year. We began by averaging each summer variable SOD presented in Figure 3. Taking these averages, we found that the lowest and highest seasonal SOD occurred in the respective years of the lowest and highest maximum summer Chla concentrations in the reservoir. This made it simpler to assume the two SOD values as being our SOD range. We then used an equation based on the apparent relationship of these two variables (summer SOD = 0.0042 × max summer Chla + 0.9345) to set the remaining summer SOD rates based on the maximum summer Chla concentrations each year. By this method, we used the previous summer's maximum Chla concentrations as a proxy of the magnitude of biomass production that settled to the bottom sediments by the end of the open water season.

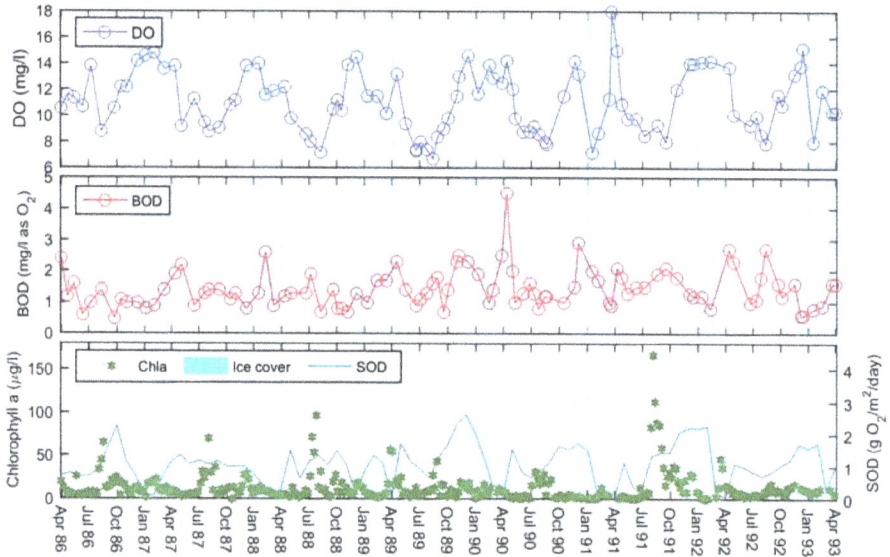

Figure 4. Observed dissolved oxygen (DO) and biochemical oxygen demand (BOD) inflow data, and in-reservoir Chlorophyll-a (Chla) concentrations in BPL. The DO and BOD data are monthly measurements at the upstream boundary (BOD as the standard five-day BOD at 20 °C), and the Chla data are from the long-term weekly dataset, provided by the Buffalo Pound Water Treatment Plant, at the downstream sample point.

To simulate end of season algal bloom mortality and winter decay we again used MATAB to adjust the SOD rate, so that predicted DO fit to observed DO, from one-month before ice-on occurred until ice-off the following spring. We implemented the same weekly semi-automated calibration process as before, and the SOD rates generally peaked before the ice-on event. Under ice-cover W2 automatically stops any gas exchange, and reaeration equals zero. This allows us to imitate a first-order decay rate during this time.

Once we had both the end of season peak SODs and winter SODs we were then able to back-calculate the winter SOD decay rates (k) for each year based on Equation (2):

$$SOD = peak\ SOD \times weeks^{-decay(k)}, \tag{2}$$

where the predicted winter SOD in W2 is assumed to be a function of the predicted peak SOD, and the number of weeks since the start of ice-cover to the decay rate k; With k being the unknown in this equation. These back-calculated decay rates were then plotted against summer Chla.

3. Results

3.1. Dissolved Oxygen Simulation

The final DO model shows good overall results (Figure 5). The predicted DO observations follow the pattern of the observed DO measurements in most years. There is some underestimation in the winters of 1986/87, 1987/88 and 1992/93. SOD follows a similar trend for each year with an end-of-season peak, and winter decay. The SOD remains high in the winter of 1991/92 due to a greater than average oxygen depletion that year. There is a clear connection in the model between the predicted DO, and the observed ice-on and ice-off dates provided by the WTP.

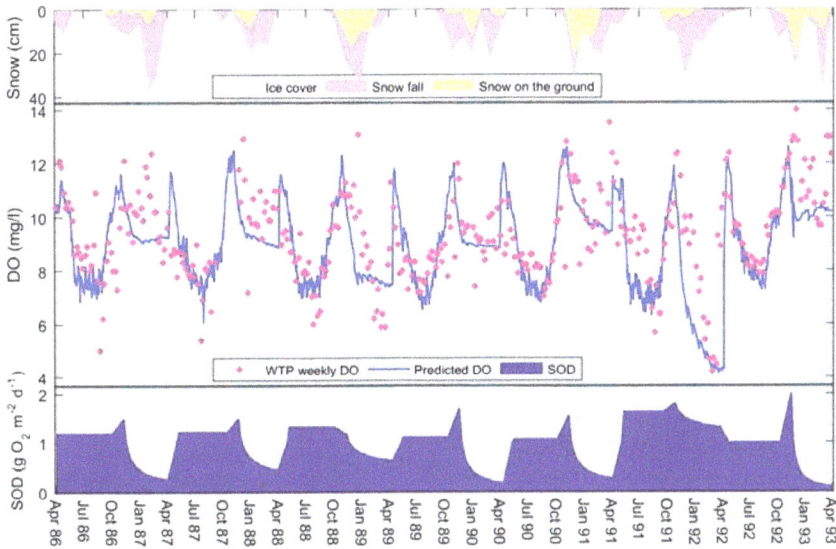

Figure 5. Dissolved oxygen (DO) model using summer sediment oxygen demand (SOD) rates based on the maximum summer Chlorophyll-a. The end of season peak and winter decay are found through a semi-automated calibration procedure to match weekly observed DO concentrations (WTP weekly DO). Ice-cover days shown here are the observed data from the Buffalo Pound Water Treatment Plant, and snow data are from Environment Canada. Snow on the ground is measured on the last day of each month. Predicted DO have root mean square error = 1.47 (to 2 dp); mean absolute error = 1.09 (to 2 dp).

3.2. Sediment Oxygen Demand Relationships

The peak SOD does not fit particularly well with the maximum or average summer Chla. Interestingly, in comparison with observed data for BPL, the peak SODs appear to have a high correlation ($R^2 = 0.85$) with the average BOD inflows included in our model for the open water period. (Figure 6a). The winter SOD decay rates have a negative, exponential relationship with both the average and maximum summer Chla concentrations of the previous summer. The relationship between SOD decay and the maximum Chla (Figure 6b) is slightly stronger at $R^2 = 0.88$ (average Chla: $R^2 = 0.84$).

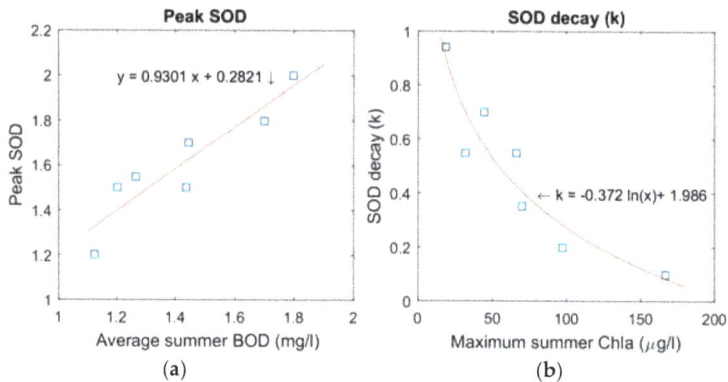

Figure 6. Relationships between SOD (day^{-1}), and observed BPL measurements, after the final DO model simulation: (**a**) Predicted peak SOD and average open water BOD inflows ($R^2 = 0.85$); (**b**) Back-calculated winter SOD decay, and observed maximum summer Chla concentrations of the previous summer ($R^2 = 0.88$).

4. Discussion

A zero-order approach for SOD that treats the demand as an input variable rather than a calculated one does not reflect the conversion of organic matter settling during the simulation [10]. Our original intention was to allow the model to find the changing SOD values through matching to observed DO concentrations. For this, the model was calibrated in MATLAB using a semi-automated iterative process that allowed the model to change the SOD weekly. The resulting SOD values were then to be read by W2 as an input file; this would imitate a first-order compartment, in essence, by varying through the simulation. The model indeed predicted DO concentrations more closely with this variable SOD file than with the W2's original fixed rate option.

Calibrating the SOD with the purpose that the model's DO predictions match with observed DO measurements can be an unsound technique as it assumes that other parameters such as reaeration and settling rates are already well-known [11]. This method of calibration also combines several reservoir processes that contribute to DO into one net value that is assumed to be SOD. While these points suggest that there are limitations to this approach, there remains the problem that few aquatic managers have sufficient data available to run the full diagenesis model. In view of this, we assessed the initial results to see if there were other trends that matched what we know about BPL, and that might suffice as a proxy measurement or explanatory variable.

The reservoir is a highly eutrophic system with high incidences of algal blooms. Deoxygenation can occur after the collapse of a summer algal bloom due to additional bacterial activity [2]. In general, the more enriched the system then the higher the rates of productivity and ultimately the greater the oxygen depletion from decomposition [3]. Chla has previously been used as a proxy for estimating in-lake BOD [30]. Based on this principal, and our knowledge of the reservoir, we are assuming that most of the autochthonous contributions to oxygen demand within BPL are related to algal activity (apart from nitrification and chemical oxygen demand). Any allochthonous inputs to oxygen demand are already included in our BOD time-series data in the inflow constituent file—with the caveat that we are using the W2 default BOD settling rate between upstream and our sample site on the reservoir.

We have found this approach to be successful, as shown in Figure 5. The summer SOD rates based on a correlation with the summer Chla concentrations act effectively as a substitute to our weekly variable SOD. This suggested link between oxygen depletion and productivity also agrees with our findings relating winter SOD decay to the Chla concentrations of the previous summer. This is noticeable in the results for the winters of 1988/89 and 1991/92 where the SOD remains high under ice-cover following large summer algal blooms.

In contrast, a result of no correlation between Chla and DO consumption is found in other shallow prairie lake sites [3]. The study in question suggests that the most important predictor of DO consumption is macrophyte biomass due to the large contribution to particulate organic matter (POM). This is found to be also true in sites with abundant phytoplankton, although the authors point out that the algal-derived carbon averages 150 times less than the macrophyte-derived carbon in their study. In BPL, apart from the top section outside of the model boundary, the reservoir is not thought to have many macrophytes. This may explain why we are able to find a relationship between summer Chla and winter SOD decay as the macrophyte contribution to POM is not important in this reservoir.

Our winter SOD decay pattern declines in an exponential manner with a rapid reduction at the onset of winter. This theory fits with the suggestion that the first three months of ice-cover have the greatest oxygen consumption due to the rapid oxidation of certain organic materials over others, for example [31].

The apparent connection between peak SOD and the average BOD inflows for the open water season is more surprising as we had suspected that the low values of BOD would have little impact in the reservoir. SOD and BOD are, in fact, often combined into one demand known as hypolimnetic oxygen demand [18]. In an ideal model BOD and SOD would be kept separate. Our internal BOD is included with our SOD, as is reaeration. In addition, BOD inflows are based on monthly samples. The result is that we cannot ascertain for certain the relative importance of BOD flowing into the reservoir.

In the winters of 1986/87 and 1987/88, the model under-predicted the DO concentrations. However, in these years, it can be seen that although snowfall was still high in both years, there was little snow left on the ground at the end of each month. It is possible that the reservoir winter albedo is relatively low in these years and light can penetrate the ice to allow photosynthesis to take place. This is evident in the observed winter Chla concentrations (Figure 4). This is a winter phenomenon that we are unable to capture in the model as we do not simulate primary productivity, and our SOD equations are founded on summer Chla. While snow on the ground is also minimal in the winter of 1991/92, this year is different as the intense summer algal bloom preceding this year results in the DO concentrations falling and the SOD remaining high.

What is interesting in the winter of 1992/93 is that the snow cover is high throughout the winter, suggesting low light availability, and yet DO concentrations are high. On examination of the temperature model (Figure 2), it can be seen that the observed bottom water temperatures are much lower in this year than in previous years. Cold water holds more oxygen than warm water [5], and estimating DO inputs based on monthly samples may be missing occasional elevated DO concentrations in cold river inflows during periodic snowmelts. The winter of 1992/93 also has the greatest contribution of the distributary tributary inflows, which indicates that there may be a larger amount of ungauged inflows contributing to the water balance in W2 with no corresponding DO input file.

Another point to consider is that oxygen consumption rates are shown to be temperature dependent with lower consumption rates at lower temperatures [12]. Both the summer temperatures of 1992 and winter temperatures of 1992/93 were colder than previous years. This suggests that less heat may be stored in the sediments. The sediment temperature in W2 is a fixed user parameter, and in our model has been set at the average annual air temperature over the simulation period (where temperature >0 °C). The observed sediment temperature may be colder than this average value in the final winter, and, in reality, the sediment oxygen requirements are lower than we are modelling.

An equation for the effect of water temperature on SOD shows how temperature and SOD have a positive relationship with each other: SOD starts to decline more rapidly after temperatures go lower than 10 °C, and reduces towards zero as water temperatures drop below 5 °C [11]. Part of the pattern of SOD changes in BPL will also be a function of bottom water temperature. Our SOD decays rapidly to near zero levels quite early in the colder water temperatures of winter 1992/93. The temperatures in BPL are fairly consistent year-on-year until the summer of 1992, and the winter decreases and summer increases of SOD in response to temperature are expected to be comparable up to this point. This leaves Chla explaining the SOD variance between the individual years.

W2 uses four SOD temperature-rate multipliers to adjust the rate of SOD decay as a function of temperature in the model. They are model calibration parameters, and can be helpful in reproducing the changing rate of consumption of DO. We have used the default settings in W2 (Table 1). The variable SOD values that we include in W2 as an input file are maximum SOD rates—the same as the model format for a constant SOD rate. Figure 7 shows the temperature adjusted rates that W2 is actually using during the simulations based on these default calibration parameters. A contributing reason for the end of summer SOD peaks, for example, might be that the rapid decreases in the temperature adjusted SOD are too extreme. The model may potentially increase the maximum SOD to compensate for this temperature effect when calibrating SOD to match the predicted DO with observed DO. We had no data with which to justify changing the default values. Likewise, we did not wish to increase our model uncertainty by expanding the number of parameters with no additional data. We instead chose to adjust just one parameter (SOD) for our purposes.

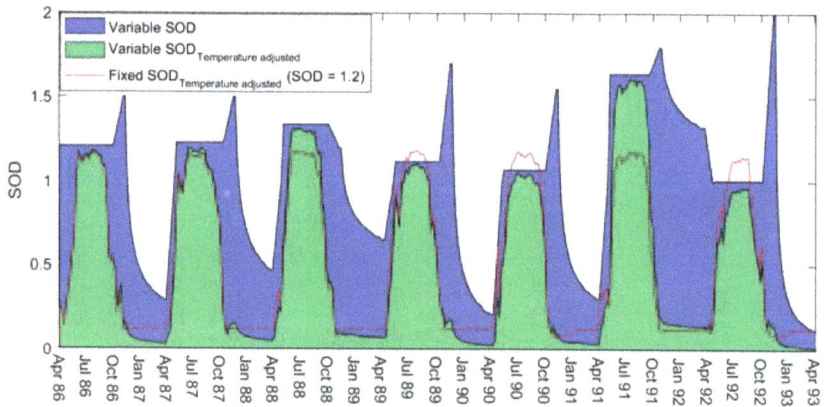

Figure 7. A comparison of the maximum SOD rates that we input into the model (blue) against the temperature adjusted rates that the model is actually using based on W2 default values for the four temperature-rate multipliers (green). Also shown is the temperature adjusted rates for the fixed SOD simulations.

There are a few differences between SOD found by using MATLAB to vary SOD each week to fit the predicted and observed DO to each other (weekly model—Figure 3), and SOD found through basing the demand on summer Chla (final model—Figure 5). The winter decay in the final model agrees well with the drops in SOD in the weekly model except for the winter of 1991/92. In this year, the weekly model drops the SOD to zero when the DO levels are extremely low and then increases SOD as the oxygen levels rise. This is the opposite behaviour to what we would expect in a limnological sense. The behaviour of the final model in this time period is more realistic. The relatively large peak in SOD at the end of this summer may possibly be due to the BOD inputs being higher this year (Figure 4).

In reality, the rapid peaks in SOD shown in Figure 5 are also likely a result of our holding the SOD at a constant rate over the summer period. The final model uses average summer rates, and will miss some of the variability that would naturally occur. In the semi-automated calibration of Figure 3, we show that when the model uses variable weekly SOD rates there is a general (with some fluctuations) increase over the course of the summer. This agrees with our assumption that SOD in

BPL cumulates to a peak due to biota dying towards the end of the season. By holding the model at an average seasonal SOD, instead, we are likely overestimating the SOD in spring, and underestimating the SOD in autumn—if we are to assume that SOD increases as suggested by Figure 3. We found that our winter decay relationship with Chla was stronger if we allowed MATLAB to adjust SOD so that the predicted DO fits to the observed DO at the onset of ice-on. In order to achieve this, we had to release the model from the fixed SOD rate at the end of the season—thereby simulating the end of season peak. We chose a one-month period prior to ice-on as a sensitivity analysis showed that this gradient of SOD adjustment gave us the best overall results for summer and winter.

This summer averaging method is most likely responsible for the sudden increase in SOD at ice-off in April. There will be natural processes leading to an increase in SOD (e.g., warmer water, spring blooms, and spring turnover) that will be exaggerated by the need for the SOD to instantly increase to the average summer rate at the start of spring. Other methods may be to allow the model to increase gradually over the whole summer duration, either linearly or exponentially, although we would need to consider some way of verifying the manner in which it increased. Our strategy was to approach the problem as if having little to no data to verify the predicted SOD rates (except using DO data). We decided to constrain the model to using an average summer rate based on a relationship found with Chla, and then allow the SOD to decay over winter dependent on a fixed equation. Thus, the aquatic manager would only need to find (and ideally verify) limited points, such as the end of season peak rate, rather than weekly SOD rates.

One limitation to our study is the disconnection between the top and main section of the reservoir. While our inflow constituent file is based on observed data from under Highway 2, and the boundary of our model, the inflows themselves relate to a gauge further upstream on The Upper Qu'Appelle River. It is uncertain at present what effect the top section of the reservoir, the wave-breaker, and the 45 m gap under the highway have on our inflow boundary data.

Finally, while our extended W2 ice model is a suitable model for seasonally ice and snow-covered waterbodies, the W2 model has a fixed albedo coefficient through the simulation. The extension to the ice model stops the ice melting too quickly by keeping the snow on the ground, yet does not help with modelling the correct amount of light penetrating the ice. This will make it difficult to calibrate DO, when primary production is added to the model, with just one value for high and low snow years, and different ice structures. This is due to the influence of light on photosynthesis and oxygen production, as indicated in the observed data in the winters of 1986/87 and 1987/88. Our future plans include modifying the ice model further to include a function for a variable albedo. We think that this will be an interesting step to take forward and is a missing-link in modelling DO/SOD relationships in ice-covered reservoirs.

5. Conclusions

From the modelling, we show that winter SOD decay is inversely dependent on the previous summer's maximum Chla concentrations. The decay rate is faster when less algae are produced. A constant SOD value suffices during the summer half-year; however, a better DO simulation is obtained in winter when the SOD rate decays during the course of the winter. This implies that the biomass supply during winter is limited and much of the draw on DO is diminished by the end of the winter. This result is backed by several field studies. We have shown that for a Prairie shallow reservoir with few macrophytes and BOD inputs variable SOD can be used in a water quality model to represent additional oxygen demand after an algal bloom. The summer SOD and winter SOD decay can be estimated by treating the open water and under-ice period individually in the model. This variable SOD over-time can be estimated for both summer and winter conditions based on summer concentrations of Chla. This concept can be widely applied to similar systems that do not have data to support a full diagenesis model, yet would benefit from a more representative estimation of SOD than is provided by a zero-order constant SOD rate.

Acknowledgments: This work is funded through the Natural Sciences and Engineering Research Council of Canada Strategic Project Grant, the Buffalo Pound Water Treatment Plant (WTP), the Water Security Agency (WSA), the Global Institute for Water Security (GIWS), and the School of Environment and Sustainability (SENS). We thank Heather Wilson (DEM preparation), and Helen Baulch (buoy data + funding support) from GIWS, and Paul Jones (boat sonar data) from SENS. We also thank John-Mark Davies (water-quality data), Dave MacDonald (GIS data), and Andrew Thornton (water demands) from the WSA. We are grateful to Curtis Hallborg from the WSA for hydrological data and his time in explaining the hydrology of the reservoir system. Water-quality data provided by Dan Conrad at the WTP is gratefully acknowledged. We thank the anonymous reviewers for valuable feedback.

Author Contributions: Julie A. Terry, Amir Sadeghian and Karl-Erich Lindenschmidt conceived and designed the experiments under the latter's supervision. Julie A. Terry set-up and calibrated the Buffalo Pound reservoir model including the water balance, water temperature dynamics and initial DO-BOD kinetics. Amir Sadeghian adapted CE-QUAL-W2 for snow-cover, and, after establishing trends between SOD, Chla and BOD, extended the model to include variable SOD. Julie A. Terry wrote the bulk of the paper with text inputs and conceptual edits from Karl-Erich Lindenschmidt. All authors proofread and approved the manuscript.

Conflicts of Interest: The authors declare no conflict of interest.

References

1. Canadian Council of Ministers of the Environment. Canadian water quality guidelines for the protection of aquatic life: Dissolved oxygen (freshwater). In *Canadian Environmental Quality Guidelines*; Canadian Council of Ministers of the Environment: Winnipeg, MB, Canada, 1999.
2. Robarts, R.D.; Waiser, M.J.; Arts, M.T.; Evans, M.S. Seasonal and diel changes of dissolved oxygen in a hypertrophic prairie lake. *Lakes Reserv. Res. Manag.* **2005**, *10*, 167–177. [CrossRef]
3. Meding, M.E.; Jackson, L.J. Biotic, chemical, and morphometric factors contributing to winter anoxia in prairie lakes. *Limnol. Oceanogr.* **2003**, *48*, 1633–1642. [CrossRef]
4. Ruuhijärvi, J.; Rask, M.; Vesala, S.; Westermark, A.; Olin, M.; Keskitalo, J.; Lehtovaara, A. Recovery of the fish community and changes in the lower trophic levels in a eutrophic lake after a winter kill of fish. *Hydrobiologia* **2010**, *646*, 145–158. [CrossRef]
5. Wetzel, R.G. *Limnology: Lake and River Ecosystems*, 3rd ed.; Academic Press: San Diego, CA, USA, 2001.
6. Finlay, K.; Leavitt, P.R.; Patoine, A.; Wissel, B. Magnitudes and controls of organic and inorganic carbon flux through a chain of hard-water lakes on the northern great plains. *Limnol. Oceanogr.* **2010**, *55*, 1551–1564. [CrossRef]
7. Scardina, P.; Edwards, M. Prediction and measurement of bubble formation in water treatment. *J. Environ. Eng.* **2001**, *127*, 968. [CrossRef]
8. Hosseini, N.; Johnston, J.; Lindenschmidt, K.-E. Impacts of climate change on the water quality of a regulated prairie river. *Water* **2017**, submitted.
9. Thornton, K.W.; Kimmel, B.L.; Payne, F.E. *Reservoir Limnology: Ecological Perspectives*; Wiley: New York, NY, USA, 1990.
10. Cross, T.; Summerfelt, R. Oxygen demand of lakes: Sediment and water column bod. *Lake Reserv. Manag.* **1987**, *3*, 109–116. [CrossRef]
11. Chapra, S.C. *Surface Water-Quality Modeling*; McGraw-Hill: Boston, MA, USA; New York, NY, USA, 1997.
12. Golosov, S.; Maher, O.; Schipunova, E.; Terzhevik, A.; Zdorovennova, G.; Kirillin, G. Physical background of the development of oxygen depletion in ice-covered lakes. *Oecologia* **2007**, *151*, 331–340. [CrossRef] [PubMed]
13. Vehmaa, A.; Salonen, K. Development of phytoplankton in Lake Pääjärvi (Finland) during under-ice convective mixing period. *Aquat. Ecol.* **2009**, *43*, 693–705. [CrossRef]
14. Salonen, K.; Leppäranta, M.; Viljanen, M.; Gulati, R. Perspectives in winter limnology: Closing the annual cycle of freezing lakes. *Aquat. Ecol.* **2009**, *43*, 609–616. [CrossRef]
15. Fang, X.; Stefan, H.G. Projected climate change effects on winterkill in shallow lakes in the northern United States. *Environ. Manag.* **2000**, *25*, 291–304. [CrossRef]
16. Martin, N.; McEachern, P.; Yu, T.; Zhu, D.Z. Model development for prediction and mitigation of dissolved oxygen sags in the Athabasca River, Canada. *Sci. Total Environ.* **2013**, *443*, 403–412. [CrossRef] [PubMed]
17. Leppi, J.; Arp, C.; Whitman, M. Predicting late winter dissolved oxygen levels in arctic lakes using morphology and landscape metrics. *Environ. Manag.* **2016**, *57*, 463–473. [CrossRef] [PubMed]

18. Kirillin, G.; Leppäranta, M.; Terzhevik, A.; Granin, N.; Bernhardt, J.; Engelhardt, C.; Efremova, T.; Golosov, S.; Palshin, N.; Sherstyankin, P.; et al. Physics of seasonally ice-covered lakes: A review. *Aquat. Sci.* **2012**, *74*, 659–682. [CrossRef]

19. Beven, K. A manifesto for the equifinality thesis. *J. Hydrol.* **2006**, *320*, 18–36. [CrossRef]

20. Williams, R.; Keller, V.; Vo, A.; Barlund, I.; Malve, O.; Riihimaki, J.; Tattari, S.; Alcamo, J. Assessment of current water pollution loads in europe: Estimation of gridded loads for use in global water quality models. *Hydrol. Process.* **2012**, *26*, 2395–2410. [CrossRef]

21. Lindenschmidt, K.-E. The effect of complexity on parameter sensitivity and model uncertainty in river water quality modelling. *Ecol. Model.* **2006**, *190*, 72–86. [CrossRef]

22. Lindenschmidt, K.-E.; Pech, I.; Baborowski, M. Environmental risk of dissolved oxygen depletion of diverted flood waters in river polder systems—A quasi-2d flood modelling approach. *Sci. Total Environ.* **2009**, *407*, 1598–1612. [CrossRef] [PubMed]

23. Cole, T.M.; Wells, S.A. Ce-qual-w2: A two-dimensional, laterally averaged, hydrodynamic and water quality model, version 3.72. In *Department of Civil and Environmental Engineering*; Portland State University: Portland, OR, USA, 2015.

24. Hammer, U. Limnological studies of the lakes and streams of the upper qu'appelle river system, saskatchewan, Canada. *Hydrobiologia* **1971**, *37*, 473–507. [CrossRef]

25. Environment Canada. Climate Normals 1981–2010 Station Data for Moose Jaw a Station. Available online: http://climate.weather.gc.ca (accessed on 21 November 2014).

26. Hall, R.I.; Leavitt, P.R.; Dixit, A.S.; Quinlan, R.; Smol, J.P. Limnological succession in reservoirs: A paleolimnological comparison of two methods of reservoir formation. *Can. J. Fish. Aquat. Sci.* **1999**, *56*, 1109–1121. [CrossRef]

27. Kehoe, M.; Chun, K.; Baulch, H. Who smells? Forecasting taste and odor in a drinking water reservoir. *Environ. Sci. Technol.* **2015**, *49*, 10984–10992. [CrossRef] [PubMed]

28. Slater, G.P.; Blok, V.C. Isolation and identification of odourous compounds from a lake subject to cyanobacterial blooms. *Water Sci. Technol.* **1983**, *15*, 229–240.

29. Sadeghian, A.; de Boer, D.; Hudson, J.J.; Wheater, H.; Lindenschmidt, K.-E. Lake diefenbaker temperature model. *J. Gt. Lakes Res.* **2015**, *41*, 8–21. [CrossRef]

30. Fang, X.; Stefan, H.G. Simulations of climate effects on water temperature, dissolved oxygen, and ice and snow covers in lakes of the contiguous U.S. Under past and future climate scenarios. *Limnol. Oceanogr.* **2009**, *54*, 2359–2370. [CrossRef]

31. Babin, J.; Prepas, E.E. Modelling winter oxygen depletion rates in ice-covered temperate zone lakes in canada. *Can. J. Fish. Aquat. Sci.* **1985**, *42*, 239–249. [CrossRef]

water

MDPI

Article

How Does Changing Ice-Out Affect Arctic versus Boreal Lakes? A Comparison Using Two Years with Ice-Out that Differed by More Than Three Weeks

Kate A. Warner [1,*] , Rachel A. Fowler [1], Robert M. Northington [1] , Heera I. Malik [2], Joan McCue [1] and Jasmine E. Saros [1]

[1] Climate Change Institute and School of Biology and Ecology, University of Maine, Orono, ME 04469, USA; fowlerrachelanne@gmail.com (R.A.F.); robert.northington@maine.edu (R.M.N.); joan.mccue@maine.edu (J.M.); jasmine.saros@maine.edu (J.E.S.)
[2] Rhithron Associates, Inc., 33 Fort Missoula Rd., Missoula, MT 59804, USA; heera.malik@maine.edu
*   Correspondence: kathryn.warner@maine.edu

Received: 15 December 2017; Accepted: 14 January 2018; Published: 17 January 2018

Abstract: The timing of lake ice-out has advanced substantially in many regions of the Northern Hemisphere, however the effects of ice-out timing on lake properties and how they vary regionally remain unclear. Using data from two inter-annual monitoring datasets for a set of three Arctic lakes and one boreal lake, we compared physical, chemical and phytoplankton metrics from two years in which ice-out timing differed by at least three weeks. Our results revealed regional differences in lake responses during early compared to late ice-out years. With earlier ice-out, Arctic lakes had deeper mixing depths and the boreal lake had a shallower mixing depth, suggesting differing patterns in the influence of the timing of ice-out on the length of spring turnover. Differences in nutrient concentrations and dissolved organic carbon between regions and ice-out years were likely driven by changes in precipitation and permafrost thaw. Algal biomass was similar across ice-out years, while cell densities of key *Cyclotella sensu lato* taxa were strongly linked to thermal structure changes in the Arctic lakes. Our research provides evidence that Arctic and boreal regions differ in lake response in early and late ice-out years, however ultimately a combination of important climate factors such as solar insolation, air temperature, precipitation, and, in the Arctic, permafrost thaw, are key drivers of the observed responses.

Keywords: climate change; lakes; early ice-out; late ice-out; Arctic; boreal; mixing depth; phytoplankton

1. Introduction

Lakes throughout the Northern Hemisphere are experiencing changes in the timing of ice-on, ice-out and the duration of ice cover [1–5]. Changes in the timing of ice-out are of particular interest for understanding plankton dynamics, as ice-out marks the onset of spring conditions and the period leading to the peak of the growing season. Ice-out timing also has stronger direct connection to climate change than ice-on because individual lake properties influence the freezing process more strongly than the thawing process [6,7]. The timing of ice-out has advanced substantially, occurring up to 21 days earlier over the past 40 to 100 years at mid-latitudes [8–11] and up to 13 days earlier since 2000 in the Arctic [12].

Correlations suggest that the timing of ice-out is an important driver of phytoplankton community structure and biomass. Paleolimnological studies have inferred that earlier ice-out has triggered changes in lake properties that caused shifts in diatom communities at both high and mid-latitudes and that the taxon-specific shifts occurred earlier in Arctic lakes (ca. 1870) than in boreal lakes (ca. 1970) due to expansion of planktonic diatom habitat and lengthening of the growing season [13–15]. Specifically, it has been hypothesized that shorter periods of ice cover induced by warming air temperatures favor small *Cyclotella* taxa due to increased water column stability throughout the growing season [15,16]. However, based on neo- and paleolimnological approaches, small *Cyclotella sensu lato* taxa can be more abundant during early [14,17] or late [18,19] ice-out years. Similarly, monitoring of a boreal lake over a 14-year period and a temperate lake over a 15-year period both revealed that the timing of ice-out does not clearly influence total phytoplankton biomass during the growing season [20,21]. Collectively, these studies reveal that the links between ice-out and phytoplankton dynamics vary in pattern and strength across systems.

This regional variation is well illustrated by comparing Arctic and boreal lakes. The rate of warming is at least twice the global average at high Arctic latitudes above 60° North compared to other latitudes [22–24], which will influence seasonal light patterns, the length of the growing seasons, timing of ice-out relative to phytoplankton blooms [25] and the onset of stratification [26] differently than at lower latitudes that contain boreal regions. The relationship between air temperature and the actual timing of ice-out is also not linear among different latitudes and differs greatly between Arctic and boreal regions within the Northern Hemisphere [27]. With both Arctic and boreal regions experiencing rapid climate change, questions remain regarding the magnitude of effect between the regions. In Arctic lakes, ice-out occurs between May to July depending on latitude, while in boreal lakes it occurs between March to May. Therefore, Arctic lakes experience a shorter ice-free season during which there is higher light exposure and rapid onset of stratification shortly after ice-out compared to boreal lakes, which have a longer spring turnover period, longer growing season and a gradual increase in light exposure and temperatures. These differences suggest that the strength of effects of changes in the timing of ice-out may differ between Arctic and boreal lakes.

Changes in ice-out are an important physical change in lake ecosystems and there are several potential pathways by which the timing of ice-out can affect phytoplankton ecology (Figure 1). These pathways, however, are not only affected by the timing of ice-out but also by other climatic factors including precipitation, wind and cloud cover (i.e., incoming solar radiation). For example, while there are assertions in much of the limnological literature that earlier ice-out will lead to earlier onset and strengthening of thermal stratification [28–30], there is not extensive evidence to support an exclusive relationship. Dependent on elevation, precipitation can be more influential than temperature in driving ice-out [31]. However, the timing of ice-out is also strongly related to air temperatures in the month or two prior to ice breakup [3,32], and these months vary by region, with ice-out dates in mid-latitudes reflecting February to March air temperatures and at higher latitudes April to May air temperatures. While air temperatures during those months will be important for lake stratification via effects on ice-out timing, many additional factors (e.g., air temperatures during open water months, wind, cloud cover, water clarity) will affect thermal stratification patterns, potentially weakening any links with ice-out timing. Changes in the length of spring turnover and the length of the open water season are additional physical changes in lake ecosystems that are altered by climatic factors and affect phytoplankton ecology through similar pathways (Figure 1). Earlier ice-out will likely lengthen spring turnover and increase the length of the open water season, potentially altering phytoplankton growth and succession [19]. A subsequent physical implication from earlier ice-out and changes in thermal stratification and the length of the open water season, is a change in light exposure (Figure 1). The light environment plays an important role in phytoplankton abundance and composition [21] and will change variably in boreal and Arctic regions based on changes in ice-out timing, thus clear links between ice-out timing and light climate are still being investigated. It is also important to note that under-ice algal growth is greater than previously understood [33], raising questions about the extent to which earlier ice-out will strongly affect seasonal phytoplankton dynamics.

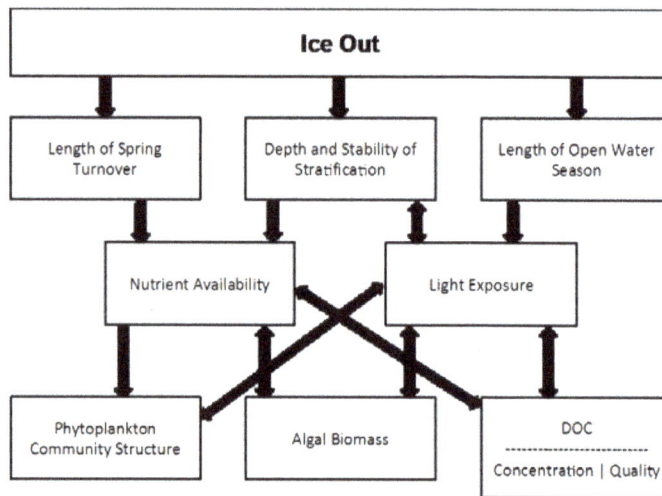

Figure 1. Conceptual diagram of a subset of the potential effects of ice-out on lake ecosystems.

These pathways that drive physical changes in lake ecosystems may also contribute to chemical changes that influence phytoplankton (Figure 1). Earlier ice-out may lead to increased nutrient loading [34], or conversely, reductions in the duration of winter ice cover may contribute to reduced under-ice nitrate production [35], thus links between ice-out and changes in nutrients remain unclear. In addition, increased light exposure from earlier ice-out can alter dissolved organic carbon concentrations and quality [36]. It is important to note that changes in ice-out have effects on chemical pathways in lakes, but that climate factors also influence chemical pathways independent of ice-out. For example, in the Arctic, warming promotes thawing of permafrost, which may increase nutrient loading to lakes [37] and further affect aquatic ecosystems, making it important to distinguish links between ice-out and phytoplankton to better resolve how future climate will alter aquatic ecosystems.

To address the extent to which ice-out affects phytoplankton dynamics requires a better understanding of how spring and summer lake conditions vary between early and late ice-out years and how they compare in different regions. How different are lake conditions in an early versus late ice-out year? To improve mechanistic understanding of the influence of ice-out on Arctic and boreal lake ecosystems, we evaluated the effects of ice-out timing on thermal stratification and differences in biological and biogeochemical characteristics in an early and late ice-out regime. We analyzed data from two inter-annual monitoring datasets, one from the Arctic (a set of 3 lakes in West Greenland) and one from the boreal zone (a lake in Maine, USA). These datasets were collected over multiple years to assess changing lake conditions over time and were originally collected for two different studies. We chose two years from each of these datasets for which monitoring data were available and that had the largest differences in ice-out dates (Table A1). Ice-out timing differed by at least three weeks and we compared a suite of physical, chemical and phytoplankton metrics between the years in each area.

2. Materials and Methods

2.1. Study Design

To compare the responses of Arctic and boreal lakes to the timing of ice-out, we used data from two inter-annual monitoring datasets that were originally collected for two different studies. In the Arctic, a set of three lakes was monitored, while in the boreal region, one lake was monitored. For the boreal lake, we chose two years from the dataset in which ice-out timing differed by 41 days (2012 early ice-out and 2015 late ice-out; Table 1). Data were available for both years to compare lake parameters during late spring (hereafter referred to simply as spring), as well as during the peak of summer stratification (hereafter referred to as summer). For the Arctic lakes, data were available to compare spring lake parameters during two years in which ice-out timing differed by 30 days (2016 early ice-out and 2015 late ice-out; Table 1). Summer data were not available for 2015 but were available for 2013, a year in which ice-out was 22 days later than in 2016 (Table 1). As a result, for the Arctic lakes, the comparisons of spring lake parameters are from one set of years (2016 versus 2015) and for a different set of years (2016 versus 2013) for summer responses. This limits our ability in the Arctic lakes to address questions about whether ice-out effects on spring conditions are sustained into summer.

Table 1. Dates of comparison for early ice-out versus late ice-out years in Arctic and boreal ecosystems. Comparisons were also made in the late spring (denoted Spring) and in mid-summer during peak thermal stratification (denoted Summer). Range of dates for Arctic includes sampling at all three lakes.

Region		Spring		Summer	
		Early Ice-Out	Late Ice-Out	Early Ice-Out	Late Ice-Out
Arctic	Year	2016	2015	2016	2013
	Ice-out date	18 May	17 June	18 May	9 June
	Sampling dates	28–30 June	27 June–1 July	15–17 July	19–21 July
Boreal	Year	2012	2015	2012	2015
	Ice-out date	19 March	29 April	19 March	29 April
	Sampling dates	11 June	11 June	12 July	10 July

2.2. Site Description

The Arctic lakes in this study are located adjacent to Kangerlussuaq, southwest Greenland, which is situated within the Arctic Circle and spans from the Greenland Ice Sheet to midway to the coast (Figure 2). Soils are derived from weathered granidoritic gneisses [38] and vegetation is variable but consists largely of woody shrubs around the lakes in this study. Continuous permafrost underlies the region [38] and surface inflow and outflow are not typically apparent [39]. Mean summer temperature is 10.2 °C from June to August and precipitation averages 173 mm per year [40]. Ice-out typically occurs between late May and late June with thermal stratification occurring very quickly thereafter [41]. This region contains approximately 20,000 lakes that are mostly chemically dilute and oligotrophic [42]. The three lakes selected for this study are all located in the Kellyville region to the east of Kangerlussuaq (Table 2). The lakes are generally small and similar in depth and surface area (Table 2). These lakes are not fed by the Greenland Ice Sheet, therefore turbidity is low.

Table 2. Select characteristics of the four study lakes.

Region	Lake	Lat	Long	Elevation (m)	Surface Area (km^2)	Volume ($\times 10^6$ m^3)	Max Depth (m)
	SS2	66.99	−50.96	190	0.368	2.49	12
Arctic	SS85	66.98	−51.06	195	0.246	0.94	11
	SS1590	67.01	−50.98	200	0.243	1.16	18
Boreal	Jordan	44.33	−68.26	83	0.800	17.4	45

The boreal lake in this study, Jordan Pond, is located in Acadia National Park in Maine, USA (Figure 2; Table 2). Lakes in Acadia National Park cover 2600 acres of the approximately 35,000-acre park. Soils in Acadia are derived from granite and schist tills, and granite dominates the landscape throughout the park [43]. Representative of northern boreal forest, spruce-fir forests persist in Acadia with stands of oak, maple and beech dominant in some areas that were burned in a fire in 1947. Data from Acadia National Park's weather station suggests average summer temperature from June through

August is 19 °C and average annual precipitation is 1455 mm. Ice-out timing is variable but typically occurs between late March and late April. Jordan Pond is an oligotrophic lake with a maximum depth of 45 m and is somewhat larger than the Arctic lakes in this study.

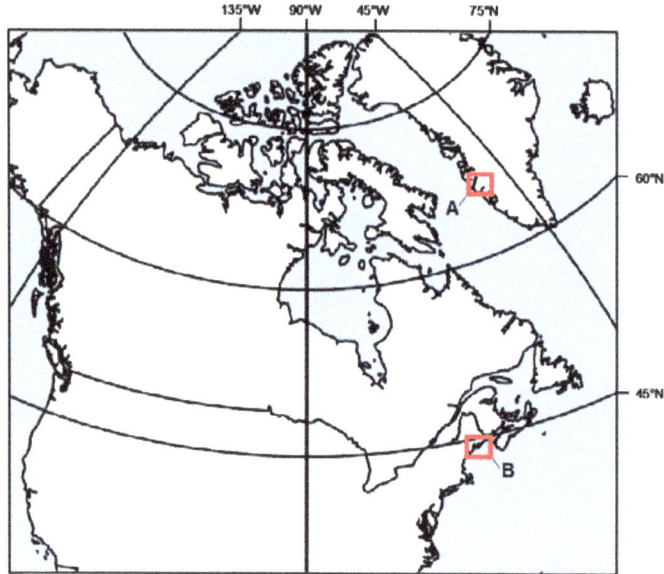

Figure 2. Map depicting the location of the (**A**) Arctic and (**B**) boreal study sites.

2.3. Climate Variables

Air temperature and precipitation data for Jordan Pond were collected from the Acadia National Park McFarland Hill (ACAD-MH) weather station. Air temperature and precipitation data for Kangerlussuaq and the Arctic lakes were collected from the Kangerlussuaq airport (DMI 04231) weather station.

2.4. Comparative Lake Sampling

2.4.1. Physical

Sampling across all four of the study lakes was conducted using the same methods during each of the dates listed in Table 1. Secchi depth was measured on the shady side of the boat using a black and white disc. Temperature profiles consisted of measurements at each meter down to 25 m using a YSI EXO2 Sonde (Xylem Inc., Yellow Springs, OH, USA). Epilimnion thickness was calculated based on temperature profiles and defined as the first depth at which there was ≥ 1 °C change per meter. Water column stability (Schmidt stability) was calculated from temperature profiles and lake bathymetry using the rLakeAnalyzer package in R [44]. The onset of stratification for the boreal lake was determined as the first day there was a ≥ 1 °C difference per meter in the water column.

2.4.2. Chemical

Water was collected from the epilimnion, metalimnion and hypolimnion using a van Dorn bottle at each lake for analysis of total phosphorus (TP) and dissolved inorganic nitrogen (DIN), which is the sum of nitrate (NO_3^-) and ammonium (NH_4^+). For analysis of DIN, NO_3^- and NH_4^+, samples were filtered through Whatman GF/F filters pre-rinsed with deionized water. Flow injection analysis using the phenate (NH$_4$-N) and cadmium reduction (NO$_3$-N) methods [45] on a Lachat Quikchem 8500 (Hach Company, Loveland, CO, USA) flow injection analyzer (FIA) were used to quantify NO$_3^-$ and NH$_4^+$. TP was determined from whole-water samples using persulfate digestion followed by the ascorbic acid method on a Lachat Quickchem 8500 (Hach Company, Loveland, CO, USA) flow injection analyzer [45]. After analysis, TP and DIN samples from the epilimnion, metalimnion and hypolimnion were averaged for comparison. Nutrient limitation status was identified by the ratio of DIN:TP, with DIN:TP < 1.5 indicating N limitation, DIN:TP > 3.4 indicating P limitation and values from 1.5 to 3.4 suggesting co-limitation [46].

Water from the epilimnion was used for analysis of dissolved organic carbon (DOC) concentrations and specific ultraviolet absorbance at 254 nm (SUVA$_{254}$). All DOC concentration and SUVA$_{254}$ samples were filtered through Whatman GF/F filters pre-rinsed with deionized water. A Shimadzu Total Organic Carbon Analyzer (Shimadzu Corporation, Kyoto, Japan) was used to analyze DOC concentrations and a Varian Carey UV-VIS spectrophotometer (Agilent Technologies, Santa Clara, CA, USA) was used to analyze SUVA$_{254}$ by measuring dissolved absorbance property at 254 nm. To provide corrected dissolved absorbance values, a Milli-Q deionized water blank was subtracted from the raw absorbance values and Naperian dissolved absorption coefficients were calculated using the following equation [47]:

$$a_d = \frac{2.303 \times D}{r}$$

where D is the decadal optical density value from the spectrophotometer and r (measured in meters) is the path length of the quartz cuvette. The DOC-specific absorption coefficient, SUVA$_{254}$, was calculated by dividing a_d (254 nm) by the DOC concentration (mg C L^{-1}).

2.4.3. Biological

Water was also collected from the epilimnion, metalimnion and hypolimnion at each lake using a van Dorn bottle to determine phytoplankton biomass (as chlorophyll *a*). Chlorophyll samples from each depth were filtered through 25 mm Whatman GF/F filters, wrapped in aluminum foil and frozen until analysis. All chlorophyll *a* samples were analyzed within three weeks of filtration and processed using standard methods [45]. Filters were ground and 90% acetone was used to extract chlorophyll overnight, then samples were centrifuged and a Varian Cary UV-VIS spectrophotometer (Agilent Technologies, Santa Clara, CA, USA) was used to analyze chlorophyll *a* concentrations. After analysis, chlorophyll *a* values from all three depths were averaged on each date to capture a water column average.

We also assessed the response of key diatom taxa that are demonstrated indicators of climate-driven lake ecosystem changes. The relative abundances of *Cyclotella sensu lato* taxa are often correlated with changes in the timing of ice-out [15] and mechanistically have been linked to thermal structure [48]. Two 50-mL centrifuge tubes were collected from the epilimnion, metalimnion and hypolimnion from each of the four study lakes on all samples dates. In the boreal lake, phytoplankton samples were available for many dates over the two years of interest; we present results across the entire study period for this lake to demonstrate how the two focal spring and summer dates fit into the full seasonal pattern for this lake. All samples were preserved with Lugol's solution, settled in Utermohl chambers and counted using a Nikon Eclipse TS-100 (Nikon Instruments Inc., Tokyo, Japan) inverted microscope at 400× magnification.

2.5. *Data Analysis*

To evaluate patterns in lake metrics in each region, responses of the three Arctic lakes were averaged (mean ± standard error) on each date. Qualitative comparisons were made across all data, as the limited sample size and the unequal number of sites between the two regions did not provide enough power to conduct more advanced statistical analyses.

3. Results

3.1. Arctic Region

In the Arctic lakes, ice-out occurred 30 days earlier in 2016 (18 May) compared to 2015 (17 June) and 22 days earlier compared to 2013 (9 June). Air temperatures differed between early and late ice-out years. During the early ice-out year, monthly average air temperatures were 6.9 to 13.7 °C higher from January to April with the biggest temperature differences in March (10.5 °C higher) and April (13.7 °C higher) compared to the 2015 late ice-out year (Figure 3). Average May temperature was 3.7 °C higher in the early ice-out year compared to the 2015 late ice-out year (Figure 3). Air temperatures in the early ice-out year were 0 to 1.5 °C higher from January to March compared to the 2013 late ice-out year (Figure 3). The largest temperature differences between the early ice-out year and the 2013 late ice-out year were in April (4.7 °C higher) and May (5.5 °C higher). Air temperatures were similar in June and July between the early ice-out year and the 2013 and 2015 late ice-out years (Figure 3).

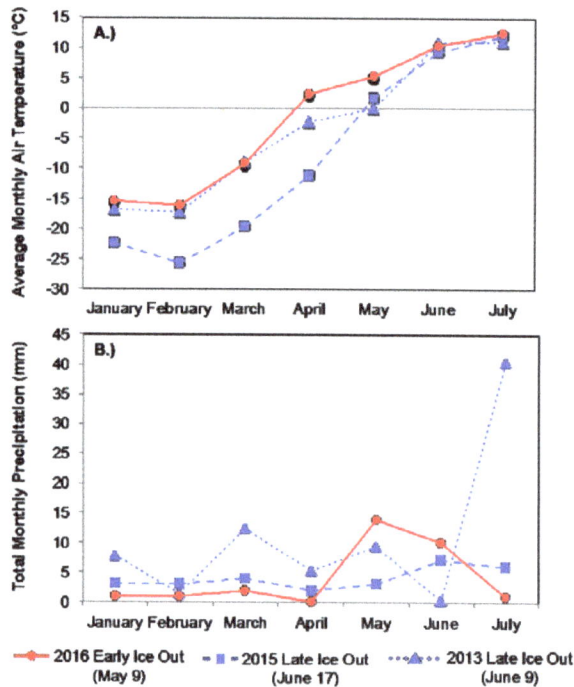

Figure 3. Arctic (**A**) average monthly air temperature in °C and (**B**) total monthly precipitation in mm for early and late ice-out years.

Precipitation varied among early and late ice-out years. Precipitation during the early ice-out year was 2 mm lower in each month from January to April, however in May precipitation was 11 mm higher in the early compared to the 2015 late ice-out year (Figure 3). During the early ice-out year, precipitation in June was 3 mm higher and precipitation in July was 5 mm lower than the 2015 late ice year (Figure 3). Precipitation from January to April was lower in the early ice-out year compared to the 2013 late ice-out year with precipitation differences ranging from 0 to 10 mm less (Figure 3). During the early ice-out year, May precipitation was 5 mm higher and in June precipitation was 10 mm higher compared to the 2013 late ice-out year. In July, precipitation was 39 mm lower in the early ice-out year compared to the 2013 late ice-out year (Figure 3).

3.1.1. Comparison of Spring Response across Early and Late Ice-out Years

Physical variables of lakes differed in spring between the two years. Water temperature at 2 m was 1.4 °C lower during the early ice-out year compared to the late ice-out year (2015; Figure 4). In the early ice-out year, mixing depths were deeper and water clarity was greater compared to the late ice-out year, with epilimnion thickness 2.3 m greater and Secchi depth 2.3 m deeper in the early ice-out year compared to the late ice-out year (Figure 4). Water column stability was 20 J m^{-2} lower during the early ice-out compared to late ice-out year.

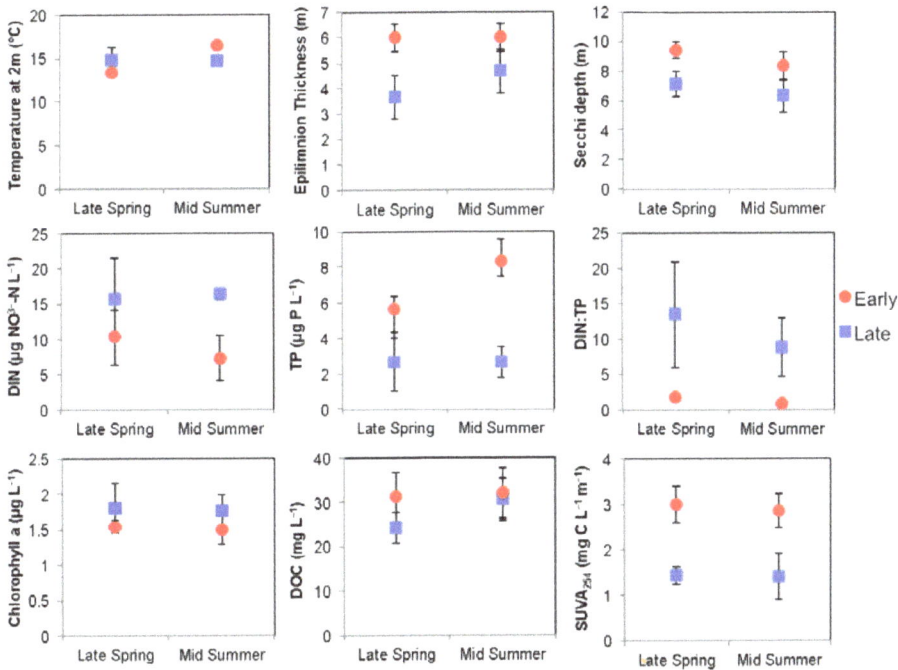

Figure 4. Comparison of lake metrics for early and late ice-out conditions during the spring and summer in Arctic lakes. Responses of the three Arctic lakes are averaged (mean ± standard error) on each date. For the 2016 early ice-out year, spring sampling occurred from 28–30 June and summer sampling was conducted from 15–17 July. For late ice-out years, spring sampling occurred from 27 June–1 July 2015 and summer sampling was conducted from 19–21 July 2013.

Differences across biogeochemical metrics in Arctic lakes in the spring season were variable across early and late ice-out years. DIN and TP had opposite responses in the spring for the two ice-out years. DIN was 5 µg N L^{-1} lower and TP was 3 µg P L^{-1} greater in the early ice-out year compared to the late ice-out year. DIN:TP was 1.7 (indicative of co-limitation by N and P) in the early ice-out year compared to 13 (indicative of P limitation) in the late ice-out year (Figure 4). DOC concentration was higher in the early ice-out year by 6.8 mg L^{-1} and SUVA$_{254}$ was higher by 1.6 mg C L^{-1} m^{-1} during early ice-out compared to late ice-out (Figure 4).

In terms of algal response, algal biomass was similar in early and late ice-out years. Average integrated chlorophyll *a* concentration was 0.3 µg L^{-1} lower in the early ice-out compared to the late ice-out year (Figure 4). Diatom cell densities of the three centric species were different in spring for early and late ice-out years. *D. stelligera* was three times lower in the early ice-out year (by 55 cells mL^{-1}) compared to late ice-out. *L. bodanica* was 1.6 cells mL^{-1} higher in the early ice-out year compared to the late ice-out year. *L. radiosa* was five times lower (29 cells mL^{-1}) in the early ice-out year compared to late ice-out (Figure 5).

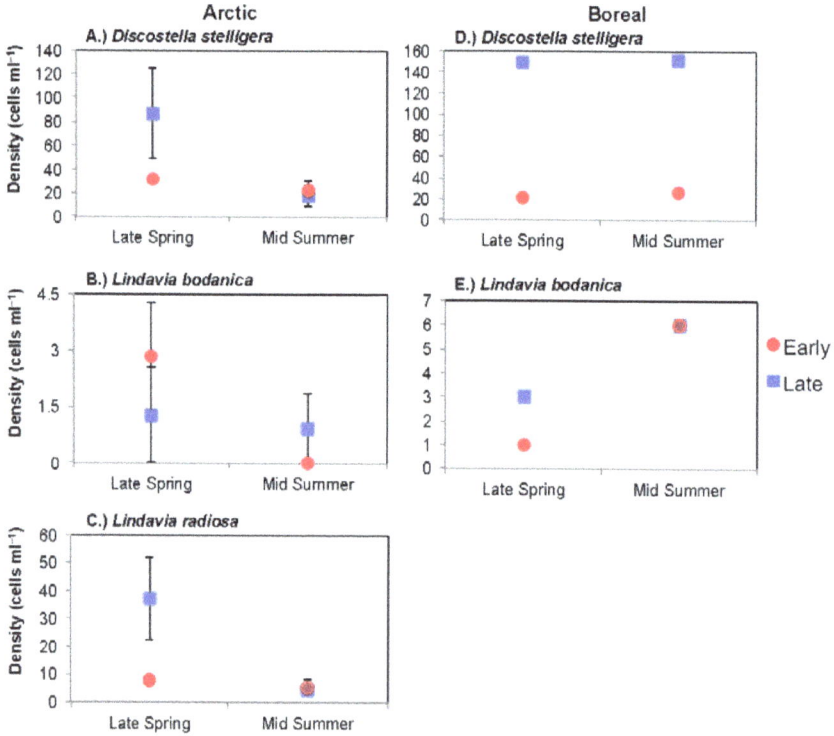

Figure 5. Comparison of (**A**) *Discostella stelligera*; (**B**) *Lindavia bodanica*; and (**C**) *Lindavia radiosa* in Arctic lakes and (**D**) *Discostella stelligera* and (**E**) *Lindavia bodanica* in a boreal lake for early and late ice-out years. Phytoplankton collection occurred at the time of sampling for all lake metrics for spring and summer and early and late ice-out years. Responses of the three Arctic lakes are averaged (mean ± standard error) on each date. Purple points indicate overlapping results for early and late ice-out years.

3.1.2. Comparison of Summer Response across Early and Late Ice-Out Years

Water temperature at 2 m was 1.8 °C higher in the early ice-out year compared to the late ice-out year (2013), the opposite of spring conditions (Figure 4). The deeper mixing depths and greater water clarity in the early ice-out year were sustained from spring, with epilimnion thickness 1.3 m greater and Secchi depth 2 m deeper in the early ice-out year compared to the late ice-out year (Figure 4). Stability was 14 J m^{-2} higher in the early ice-out year compared to the late ice-out year; the opposite of spring conditions (Figure 4).

Biogeochemical metrics were variable in the summer season between early and late ice-out years. DIN and TP responded the same as during spring conditions. DIN was 9 µg N L^{-1} lower and TP was 6 µg P L^{-1} greater in the early ice-out year compared to the late ice-out year and DIN:TP was 0.9 (indicating N limitation) in the early ice-out year compared to 8.8 (indicating P limitation) in the late ice-out year (Figure 4). DOC concentration was higher in the early ice-out year by 1.3 mg L^{-1} and SUVA$_{254}$ was higher by 1.5 mg C L^{-1}m^{-1} during early ice-out compared to late ice-out (Figure 4).

For algal biomass, average integrated chlorophyll *a* was 0.3 µg L^{-1} lower in the early ice-out year compared to late ice-out, the same as during spring conditions (Figure 4). Diatom cell densities of the three centric species were similar between early and late ice-out years in summer, demonstrating a different response from spring conditions (Figure 5).

3.2. Boreal Region

In Jordan Pond, ice-out occurred 41 days earlier in the early ice-out year, on 19 March 2012 compared to the late ice-out year in which ice-out occurred on 29 April 2015. Air temperature differences between the two years were largest in February and March and the largest precipitation differences occurred in May. In the early ice-out year, air temperatures were 9.3 °C higher in February and 6.2 °C higher in March compared to the late ice-out year (Figure 6). In January and from April to May, air temperature was 1.9 °C higher and ranged from 1.5 °C to 2.3 °C higher in the early ice-out year in comparison to the late ice-out year (Figure 6). Precipitation in April and May was 25 mm and 116 mm higher in the early ice-out year compared to the late ice-out year (Figure 6). In January and March, precipitation was similar during both early and late ice-out years and in February, June and July, precipitation was slightly lower in the early ice-out year compared to the late ice-out year (Figure 6).

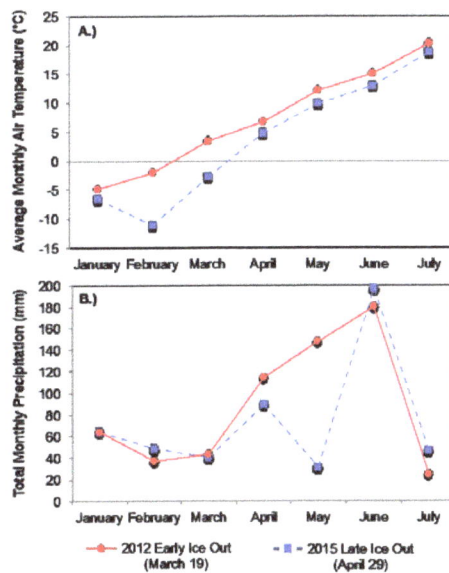

Figure 6. Boreal (**A**) average monthly air temperature in °C and (**B**) total monthly precipitation in mm for early and late ice-out years.

3.2.1. Comparison of Spring Response across Early and Late Ice-Out Years

Physical parameters of Jordan Pond varied between early and late ice-out years in spring. Water temperature at 2 m was 0.5 °C higher in the early ice-out year compared to the late ice-out year (Figure 7). In the early ice-out year, mixing depths were shallower and water clarity was greater. Epilimnion thickness was 2 m shallower and Secchi depth was 5.9 m deeper in the early ice-out year compared to the late ice-out year (Figure 7). Water column stability was 51 J m^{-2} higher in the early ice-out year compared to the late ice-out year (Figure 7). The onset of stratification in the 2012 early ice-out year was on 18 May and on 20 May during the 2015 late ice-out year.

Biogeochemical metrics were variable in the spring between the two years. DIN concentration was higher by 15 µg N L^{-1} and TP concentration was the same (2 µg P L^{-1}) in the early ice-out year compared to the late ice-out year and DIN:TP was 11 (indicating P limitation) in the early ice-out year compared to 3.5 (also P limitation) in the late ice-out year (Figure 7). DOC concentrations were equal for early and late ice-out (1.7 mg L^{-1}) and SUVA$_{254}$ was higher by 0.2 mg C L^{-1} m^{-1} for early ice-out compared to late ice-out (Figure 7).

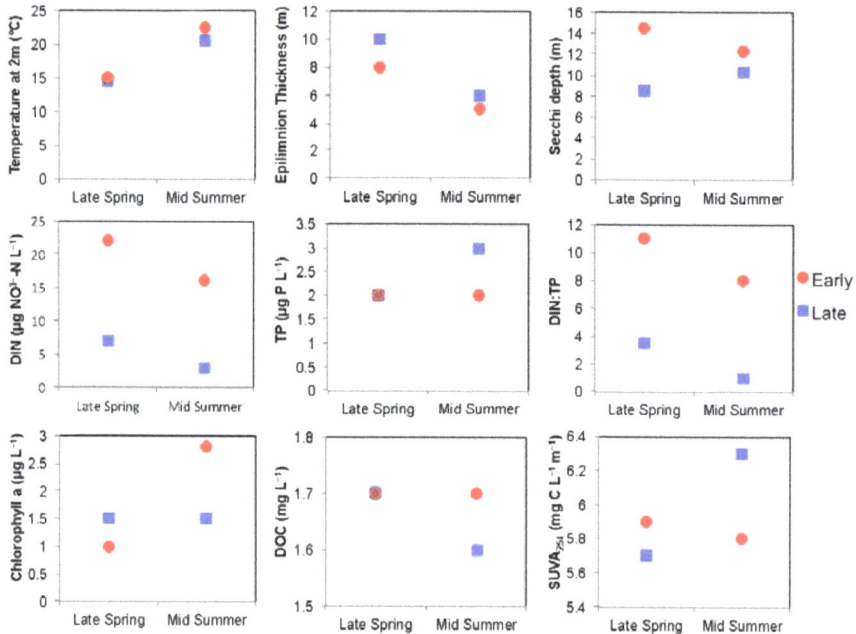

Figure 7. Comparison of lake metrics for early and late ice-out conditions during the spring and summer in the boreal lake. Responses represent one sampling for each of the time periods. For the 2012 early ice-out year, spring sampling occurred on 11 June and summer sampling was conducted on 12 July. For the 2015 late ice-out year, spring sampling occurred on 11 June and summer sampling was conducted on 10 July. Purple points indicate overlapping results for early and late ice-out years.

Algal biomass was similar in early and late ice-out years during the spring. The average integrated chlorophyll *a* concentration was 1.0 μg L^{-1} in the early ice-out year compared to 1.5 μg L^{-1} in the late ice-out year (Figure 7). Diatom cell densities in spring of the two centric species present, *D. stelligera* and *L. bodanica*, were both lower in the early ice-out compared to late ice-out year, however the magnitude of response of the two species varied. *D. stelligera* was seven times lower in the early ice-out year (129 cells mL^{-1}) compared to late ice-out. *L. bodanica* was 2 cells mL^{-1} or three times lower in the early ice-out year compared to the late ice-out year (Figure 5).

3.2.2. Comparison of Summer across Early and Late Ice-Out Years

In summer, conditions of physical lake metrics were sustained from spring. In the early ice-out year, mixing depths remained shallower and water clarity was greater. Epilimnion thickness was 1 m shallower and Secchi depth was 2 m deeper in the early ice-out year compared to the late ice-out year (Figure 7). Stability in summer was 227 J m^{-2} higher in the early ice-out year compared to the late ice-out year (Figure 7).

Biogeochemical metrics varied in response between early and late ice-out years and also with season. DIN, TP and DIN:TP were similar across seasons. DIN was 13 μg N L^{-1} higher and TP was 1 μg P L^{-1} lower in the early ice-out year compared to the late ice-out year and DIN:TP was 8 (indicating P limitation) in the early ice-out year compared to 1 (indicating N limitation) in the late ice-out year (Figure 7). DOC quantity and quality differed across seasons during early and late ice-out years. DOC concentration was 0.1 mg L^{-1} higher and SUVA was lower by 0.5 mg C L^{-1}m^{-1} for early ice-out compared to late ice-out (Figure 7).

Patterns in algal biomass switched from spring to summer during the early and late ice-out years. In contrast to spring, integrated summer chlorophyll *a* concentration was 1.3 µg L^{-1} higher in the early ice-out year compared to late ice-out (Figure 7). Cell density patterns of *D. stelligera* were sustained across seasons and were six times lower in the early ice-out year (by 125 cells mL^{-1}) compared to late ice-out. Summer cell densities of *L. bodanica* were equal when comparing early and late ice-out years with concentrations of 6.6 cells mL^{-1} (Figure 5). Overall seasonal patterns of *D. stelligera* and *L. bodanica* suggest that the spring and summer measurements were representative of seasonal patterns. Figure 8 demonstrates similar changes in the two phytoplankton species throughout the spring and summer seasons. *D. stelligera* and *L. bodanica* had lower cell densities during the early ice-out compared to the late ice-out year from May to mid June. *D. stelligera* remained lower from mid June to mid July while *L. bodanica* became more similar between early and late ice-out years. *D. stelligera* were consistently lower throughout the spring and summer seasons in the early ice-out year compared to the late ice-out year, differences throughout the season ranged from 88 to 148 cells mL^{-1} (Figure 8). *L. bodanica* were consistently lower from early May to mid June with differences ranging from 0.5 to 4.4 cells mL^{-1}, slightly higher in early July by 0.96 cells mL^{-1} and lower by 0.37 cells mL^{-1} in mid July during the late ice-out year compared to the early ice-out year (Figure 8).

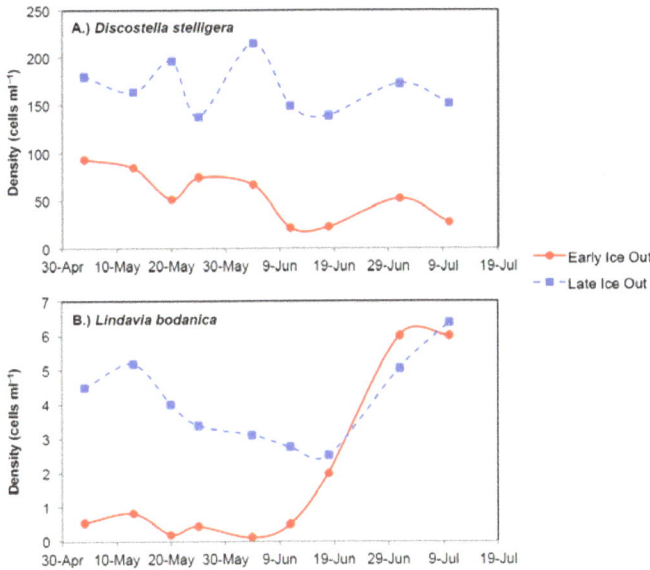

Figure 8. Seasonal comparison of (**A**) *Discostella stelligera* and (**B**) *Lindavia bodanica* in the boreal lake from May to mid-July during early and late ice-out years.

4. Discussion

Our results reveal differences in the response of certain lake metrics in Arctic and boreal regions between early and late ice-out years. During early compared to late ice-out years, Arctic lakes had deeper mixing depths while the boreal lake had a shallower mixing depth. This supports an influence of the timing of ice-out on the length of spring turnover as well as the strength and stability of stratification but with differing effects between the two regions. Nutrient concentrations and inferred limitation patterns also differed across years and regions, though the effects of other factors that determine nutrient loading to lakes (precipitation, permafrost thaw) likely played a stronger role in driving these patterns than the timing of ice-out. Biological responses in the two years across the two regions also differed, with no differences in algal biomass in the Arctic lakes in relation to ice-out and variable effects over seasons in the boreal lake. The cell densities of key *Cyclotella sensu lato* taxa that respond to thermal structure also varied across the years and regions. Collectively, our results indicate that the timing of ice-out is one important driver among many that influence the physical, chemical and biological responses of lake ecosystems to climate, and that the effects of ice-out timing differ between the two regions.

Stratification patterns differed between ice-out years and regions, likely owing to how the timing of ice-out relates to solar insolation patterns. Ice-out occurs between May and June in Arctic lakes, when solar insolation is near its peak ([49], Figure 9) and air temperatures are higher, relative to the year, thus Arctic lakes stratify quickly after ice-out. The length of spring turnover is generally short but important for the timing, depth and stability of stratification [50]. The rapid warming of surface layers in the late ice-out year, when ice off occurred only four days before the annual peak insolation, likely led to the observed shallower stratification depths across Arctic lakes. In contrast, ice-out occurs between March and May in boreal lakes, when solar insolation is lower relative to peak insolation ([49], Figure 9), leading to longer spring turnover periods with extended homothermal mixing of the water column compared to that in Arctic lakes. In the boreal lake, earlier ice-out led to a longer period of spring turnover compared to late ice-out, as the date of the onset of stratification in Jordan Pond for both years was similar. Shallower mixing depths during the early ice-out year correspond with stronger stability, stronger stratification and warmer water temperature at 2 m, similar to observations from King et al. [51]. Compared to the Arctic lakes, the effects of ice-out on the depth and stability of stratification were not as large in the boreal lake, even though the length of the spring turnover period in the boreal lake was 39 days longer. This finding is supported by other work that suggests the timing of the onset of stratification is not directly linked to ice-out timing [52,53].

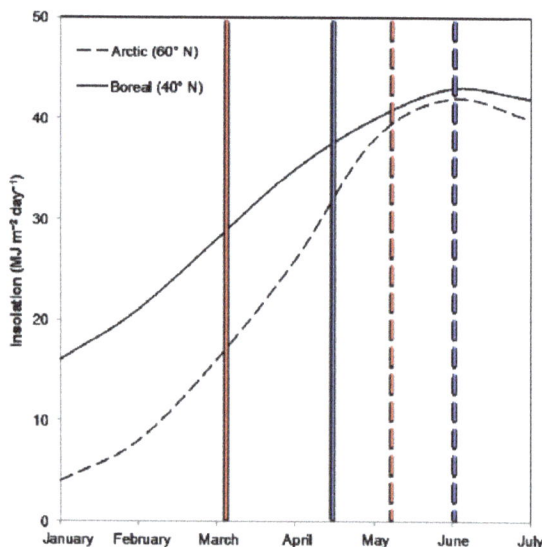

Figure 9. Change in daily solar insolation from January through July for 60° N (representative of the Arctic region) and 40° N (representative of the boreal region). Vertical dashed lines indicate early (red) and late (blue) ice-out dates for the Arctic region and vertical solid lines indicate early (red) and late (blue) ice-out dates for the boreal region. Late ice-out is averaged between the 2013 and 2015 ice-out years. Data are plotted from Buffo et al. [54].

Precipitation amounts were greater in the Arctic and boreal regions during the spring months in the early ice-out years, likely contributing to increased lake water nutrient concentrations. In the early ice-out years, the Arctic region had higher precipitation in May and June and the boreal region had higher precipitation from March through May. Spring precipitation in the Arctic region falls predominantly as snow, including the high precipitation in May during the early ice-out year, which was 82% snow. In the boreal region, precipitation mostly falls as snow from January through March and falls as mostly rain for the remaining spring and summer months. The increased precipitation in May during the early ice-out year fell as rain. Precipitation is a strong driver of increased nutrient inputs to lakes [55,56] and has important effects on terrestrial-aquatic linkages. In both Arctic and boreal lakes, nutrient concentrations and ratios in lakes are affected by alterations in terrestrial export related to climate influences on weathering, precipitation and runoff [57,58]. A key variable further influencing terrestrial-aquatic linkages and consequently nutrient limitation patterns in the Arctic, is permafrost thawing. Permafrost thawing is accelerating the delivery of P to many Arctic lakes [59,60], in part owing to mobilization of P stored in thawing permafrost as well as to changes in groundwater flow paths. Patterns in nutrient concentrations across years in our study differed regionally. In Arctic lakes, DIN concentrations were lower and TP concentrations were higher during the early ice-out year compared to the late ice-out year. In contrast, DIN concentrations in the boreal lake were higher in the early ice-out year and TP concentrations were the same during the two years. These differences in nutrient concentrations led to varying spring nutrient limitation patterns across the regions. Arctic lakes were N and P co-limited in the early ice-out year and P limited in the late ice-out year, while the boreal lake was P limited during both early and late ice-out years. Overall, climate differences between the ice-out years likely drove changes in terrestrial-aquatic linkages that dominated the different lake nutrient conditions, independent of direct effects of ice-out.

While precipitation and permafrost thaw are primary drivers of nutrients in lakes, internal processes related to changes in thermal structure can also influence nutrient availability [61,62]. Ice-out occurs closer to peak solar insolation in the Arctic lakes, likely contributing to short, perhaps incomplete, turnover periods and rapid stratification with late ice-out, with reduced entrainment of P into the photic zone. In contrast, regardless of ice-out timing, the boreal lake has a longer period of spring turnover than Arctic lakes, leading to complete turnover. These differences, in addition to changes in precipitation and permafrost, may influence nutrient cycling and nutrient availability within the lakes. Changes in the depth of the mixed surface layer, or epilimnion, can also alter nutrient cycling [28,62]; however, our results do not provide direct links between nutrient availability and thermal structure or the timing of ice-out. In our study, more precipitation occurred during the early ice-out period, after ice-out and before stratification, which may have influenced DIN and TP concentrations due to runoff. It is possible that precipitation, temperature and epilimnion thickness all contributed to varying DIN concentrations and N:P ratios across all lakes, but direct links between nutrients and the timing of ice-out remain unclear. With continued changes in climate, the relationships between nutrient availability and length of spring turnover and lake thermal structure warrant further study.

Secchi depth was deeper during the early ice-out year in all lakes, while DOC concentrations and SUVA$_{254}$ were variable in the Arctic and boreal regions. In the Arctic lake, DOC concentrations and SUVA$_{254}$ were higher in the early ice-out year compared to the late ice-out year and in the boreal lake, DOC concentrations and SUVA$_{254}$ showed little change between ice-out years. DOC strongly influences transparency in lakes and, similar to nutrients, is altered by many factors in addition to ice-out. In the Arctic region, these factors may include precipitation and permafrost thaw and the deepening of soil active layers [63,64], as well as photodegradation [34], which may increase with earlier ice-out. Cory et al. [36] found changes in DOC may be driven by photochemical oxidation of organic carbon and that sunlight may control the fate of DOC in Arctic surface waters. Our results are inconsistent with photodegradation as a primary mechanism controlling DOC, as DOC concentrations and SUVA$_{254}$ were higher during the early ice-out year. Higher DOC and SUVA$_{254}$ in the early ice-out year suggest that precipitation and permafrost thaw are likely important drivers in explaining our results. Precipitation was higher in May and June during the early ice-out year, which could increase inputs from terrestrial-aquatic linkages. It is important to note that the Arctic lakes in this study have low color DOC [65], therefore deep Secchi depths may be accompanied by high DOC concentrations. In the boreal region, DOC is usually dominated by allochthonous material and lake water DOC concentrations often increase with precipitation [66]. Similar DOC and SUVA$_{254}$ values in early and late ice-out years do not provide evidence to support links between DOC and ice-out, nor do we have enough evidence to elucidate mechanisms in links between similar DOC and deeper Secchi depth in the early ice-out year based on our results. Based on our evidence, differences in climate have strong controls on changes in DOC, which are likely key contributors to the differences observed in this study, rather than direct effects from ice-out.

Algal biomass varied little between early and late ice-out years in both the Arctic lakes and the boreal lake, with algal biomass generally being slightly lower in the early ice-out years compared to the late ice-out years. An exception to this finding occurred during the summer season in the boreal lake, in which algal biomass was higher in the early ice-out year. This result contrasts with other work that suggests increases in algal biomass due to warming [67–69] and earlier ice-out regimes [34]; however, Kraemer et al. [70] found that there is not a direct relationship between warming and algal biomass. Instead, lake surface temperature and trophic state are important in determining algal biomass, thus nutrients and light may be key contributors in algal biomass response and not only lake warming or direct ice-out effects.

The responses of key diatom taxa that are often indicators of thermal structure conditions varied across the two regions. In the Arctic lakes, differences in thermal stratification depths across ice-out years affected cell densities of key diatom taxa in the spring. Cell densities of *D. stelligera* and *L. radiosa* were lower during the early ice-out year, with deeper mixing depths, compared to the late ice-out year. *Discostella stelligera* is more abundant in lakes with shallower mixing depths [40], and *L. radiosa* is more abundant under high light conditions typical of shallower mixing depths [71]. In contrast, cell densities of *L. bodanica* were higher during the early ice-out year with deeper mixing depths; this taxon has a deeper mixing depth optimum than other *Cyclotella* taxa [46]. Patterns for these species in Arctic lakes indicated a strong relationship with mixing depth, resulting in differences in cell densities across differing ice-out years. In contrast, links between these taxa and thermal structure were less clear in the boreal lake. *Discostella stelligera* was more abundant in the late ice-out year, which had deeper mixing depths; this pattern was sustained over the entire open-water season. The same pattern was observed for *L. bodanica*, even though mixing depths showed only small differences across the two ice-out years. Boeff et al. [18] also found that *D. stelligera* was more abundant in some Maine lakes in late ice-out years, in contrast to the early ice-out patterns found in some other areas [14,17]. The effects of the complex interactions between light and nutrients on *Cyclotella* taxa are well known and reviewed by Saros & Anderson [72], and are likely behind the weaker links between thermal structure and taxon responses in this boreal lake compared to those observed in Arctic lakes.

Identifying seasonal effects throughout the open water season provides important insights into the differences between lake responses in Arctic and boreal regions. In Arctic lakes, with the exception of temperature at 2 m, spring and summer response of lake metrics (Figure 4) are sustained between the seasons from the early ice-out year to both the 2013 and 2015 late ice-out years. The biggest difference between spring and summer was a decrease in overall cell densities of phytoplankton (Figure 5). The change in water temperature across the Arctic lakes was likely due to differences between the different late ice-out years used in this study. The boreal lake had larger differences in the lake metric values between spring and summer and a switch in algal biomass and SUVA$_{254}$ concentrations between the early and late ice-out years (Figure 7). The use of two different late ice-out years for the Arctic region make comparisons between spring and summer difficult, however the variation in lake metric values, cell densities of phytoplankton and changes in lake characteristics between the Arctic and boreal regions are likely due to climate conditions at ice-out, which include differences in solar insolation, precipitation and temperature, as well as differences in the timing of stratification relative to ice-out between the two regions. Further investigation of how changes to lake variables are sustained throughout the season relative to ice-out and climate factors could provide important insights about drivers of change in phytoplankton community structure.

Our research provides evidence that lake responses in Arctic and boreal regions differ between early and late ice-out years. However, it is ultimately a combination of climate factors, importantly solar insolation, air temperature, precipitation, and, in the Arctic, permafrost thaw, that are key drivers of the observed responses. Key findings of this study include regional differences in mixing depths and the relationships between length of spring turnover and the strength and stability of stratification. These differences, in concert with climate factors, have further implications for nutrient and light availability and subsequent effects on phytoplankton community structure and biomass. Future work that explicitly examines the pathways and links between the physical and biological effects would strengthen the understanding of how the timing of ice-out influences the biological properties within lakes. Regional differences within the Northern Hemisphere can elicit contrasting lake responses, which will be altered with future climate changes, thus underscoring the importance of this research.

Acknowledgments: Collection of the Arctic dataset was funded by two grants from the US National Science Foundation (grants #1203434 and #1144423) and the Dan & Betty Churchill Fund. Collection of the boreal dataset was funded by the Maine Agricultural and Forest Experiment Station Project ME021409. We are grateful to Benjamin Burpee, Carl Tugend, Emily Rice, Steve Juggins and Andrea Nurse for field assistance and to CH2M Hill Polar Services for providing logistical support in the Arctic. We are also grateful to William Gawley who provided observed ice out dates for Jordan Pond, and also provided air temperature and precipitation data as well as logistical support in Acadia National Park.

Author Contributions: J.E.S. conceived the concept of the paper and organized sampling at all locations. K.A.W., J.E.S., R.A.F., R.M.N. and H.I.M. conducted fieldwork. J.M. conducted all Arctic phytoplankton identifications and counts and H.I.M. conducted all boreal phytoplankton analysis. All authors contributed initial data compilation and analyses. K.A.W. conducted further analyses and K.A.W. and R.A.F. generated figures. K.A.W. and J.E.S. wrote the paper. All authors read and approved the final manuscript.

Conflicts of Interest: The authors declare no conflict of interest. The founding sponsors had no role in the design of the study; in the collection, analyses, or interpretation of data; in the writing of the manuscript and in the decision to publish the results.

Appendix A

Table A1. Ice-out dates from 2010 to 2016 for the Arctic lakes and the boreal lake. Ice-out is defined as the first date that the lake is completely ice-free.

Year	Arctic (Greenland)	Boreal (Jordan Pond)
2010	24 May	22 March
2011	14 June	16 April
2012	3 June	19 March
2013	9 June	4 April
2014	13 June	14 April
2015	17 June	29 April
2016	17 May	17 March

References

1. Kuusisto, E. An analysis of the longest ice observation series made on Finish lakes. *Aqua Fenn* **1987**, *17*, 123–132.
2. Schindler, D.W.; Beaty, K.G.; Fee, E.J.; Cruikshank, D.R.; DeBruyn, E.R.; Findlay, D.L.; Linsey, G.A.; Shearer, J.A.; Stainton, M.P.; Turner, M.A. Effects of climatic warming on lakes of the central boreal forest. *Science* **1990**, *250*, 967–970. [CrossRef] [PubMed]
3. Livingstone, D.M. Large-scale climatic forcing detected in historical observations of lake ice break-up. *Verh. Int. Ver. Theor. Angew. Limnol.* **2000**, *27*, 2775–2783. [CrossRef]

4. Magnuson, J.J.; Robertson, D.M.; Benson, B.J.; Wynne, R.H.; Livingstone, D.M.; Arai, T.; Assel, R.A.; Barry, R.G.; Card, V.; Kuusisto, E.; et al. Historical trends in lake and river ice cover in the Northern Hemisphere. *Science* **2000**, *289*, 1743–1746. [CrossRef] [PubMed]

5. Futter, M.N. Pattern and trends in Southern Ontario lake ice phenology. *Environ. Monit. Assess.* **2003**, *88*, 431–444. [CrossRef] [PubMed]

6. Sporka, F.; Livingstone, D.M.; Stuchlik, E.; Turek, J.; Galas, J. Water temperatures and ice cover in lakes of the Tatra Mountains. *Biologia* **2006**, *61*, S77–S90. [CrossRef]

7. Adrian, R.; O'Reilly, C.M.; Zagarese, H.; Baines, S.B.; Hessen, D.O.; Keller, W.; Livingstone, D.M.; Sommaruga, R.; Straile, D.; Van Donk, E.; et al. Lakes as sentinels of change. *Limnol. Oceanogr.* **2009**, *54*, 2283–2297. [CrossRef] [PubMed]

8. Weyhenmeyer, G.A.; Meili, M.; Livingstone, D.M. Systematic differences in the trend towards earlier ice-out on Swedish lakes along a latitudinal temperature gradient. *Verh. Int. Ver. Theor. Angew. Limnol.* **2005**, *29*, 257–260. [CrossRef]

9. Jensen, O.P.; Benson, B.J.; Magnuson, J.J.; Card, V.M.; Futter, M.N.; Soranno, P.A.; Stewart, K.M. Spatial analysis of ice phenology trends across the Laurentian Great Lakes region during a recent warming period. *Limnol. Oceanogr.* **2007**, *52*, 2013–2026. [CrossRef]

10. Beier, C.M.; Stella, J.C.; Dovciak, M.; McNulty, S.A. Local climatic drivers of changes in phenology at a boreal-temperate ecotone in eastern North America. *Clim. Chang.* **2012**, *115*, 399–417. [CrossRef]

11. Benson, B.J.; Magnuson, J.J.; Jensen, O.P.; Card, V.M.; Hodgkins, G.; Korhonen, J.; Livingstone, D.M.; Stewart, K.M.; Weyhenmeyer, G.A.; Granin, N.G. Extreme events, trends and variability in Northern Hemisphere lake-ice phenology (1855–2005). *Clim. Chang.* **2012**, *112*, 299–323. [CrossRef]

12. Smejkalova, T.; Edwards, M.; Jadunandan, D. Arctic lakes show strong decadal trend in earlier spring ice-out. *Sci. Rep.* **2016**, *6*, 38449. [CrossRef] [PubMed]

13. Rühland, K.; Smol, J.P. Diatom shifts as evidence for recent Subarctic warming in a remote tundra lake, NWT, Canada. *Paleogeogr. Paleoclimatol. Paleoecol.* **2005**, *226*, 1–16. [CrossRef]

14. Rühland, K.; Paterson, A.M.; Smol, J.P. Hemispheric-scale patterns of climate-related shifts in planktonic diatoms from North American and European lakes. *Glob. Chang. Biol.* **2008**, *14*, 2740–2754. [CrossRef]

15. Rühland, K.; Paterson, A.M.; Smol, J.P. Lake diatom response to warming: Reviewing the evidence. *J. Paleolimnol.* **2015**, *54*, 1–35. [CrossRef]

16. Smol, J.P.; Douglas, M.S. Crossing the final ecological threshold in high Arctic ponds. *Proc. Natl. Acad. Sci. USA* **2007**, *104*, 12395–12397. [CrossRef] [PubMed]

17. Wiltse, B.; Paterson, A.M.; Findlay, D.L.; Cumming, B.F. Seasonal and decadal patterns in discostella (Bacillariophyceae) species from bi-weekly records of two boreal lakes (Experimental Lakes Area, Ontario Canada). *J. Phycol.* **2016**, *52*, 817–826. [CrossRef] [PubMed]

18. Boeff, K.A.; Strock, K.E.; Saros, J.E. Evaluating planktonic diatom response to climate change across three lakes with differing morphometry. *J. Paleolimnol.* **2016**, *56*, 33–47. [CrossRef]

19. Kienel, U.; Kirillin, G.; Brademann, B.; Plessen, B.; Lampe, R.; Brauer, A. Effects of spring warming and mixing duration on diatom deposition in deep Tiefer See, NE Germany. *J. Paleolimnol.* **2017**, *57*, 37–49. [CrossRef]

20. Meis, S.; Thackeray, S.J.; Jones, I.D. Effects of recent climate change on phytoplanton phenology in a temperate lake. *Freshwater Biol.* **2009**, *54*, 1888–1898. [CrossRef]

21. Peltomaa, E.; Ojala, A.; Holopainen, A.L.; Salonen, K. Changes in phytoplankton in a boreal lake during a 14-year period. *Boreal Environ. Res.* **2013**, *18*, 387–400.

22. McBean, G. *Arctic Climate Impact Assessment—Scientific Report*; Cambridge University Press: Cambridge, UK, 2005.

23. Screen, J.A.; Simmonds, I. The central role of diminishing sea ice in recent Arctic temperature amplification. *Nature* **2010**, *464*, 1334–1337. [CrossRef] [PubMed]

24. Jeong, J.H.; Kug, J.S.; Linderholm, H.W.; Chen, D.; Kim, B.M.; Jun, S.Y. Intensified Arctic warming under greenhouse warming by vegetation-atmosphere-sea ice interaction. *Environ. Res. Lett.* **2014**, *9*, 94007. [CrossRef]

25. Peeters, F.; Straile, D.; Lorke, A.; Livingstone, D.M. Earlier onset of the spring phytoplankton bloom in lakes of the temperate zone in a warmer climate. *Glob. Chang. Biol.* **2007**, *13*, 1898–1909. [CrossRef]

26. Livingstone, D.M. A change of climate provokes a change of paradigm: Taking leave of two tacit assumptions about physical lake forcing. *Int. Rev. Hydrobiol.* **2008**, *93*, 404–414. [CrossRef]
27. Weyhenmeyer, G.A.; Meili, M.; Livingstone, D.M. Nonlinear temperature response of lake ice breakup. *Geophys. Res. Lett.* **2004**, *31*, L07203. [CrossRef]
28. DeStasio, B.T.; Hill, D.K.; Kelinhans, J.M.; Nibbelink, N.P.; Magnuson, J.J. Potential effects of global climate change on small north-temperate lakes: Physics, fish and plankton. *Limnol. Oceanogr.* **1996**, *41*, 1136–1149. [CrossRef]
29. Peeters, F.; Livingstone, D.M.; Goudsmit, G.H.; Kipfer, R.; Forster, R. Modeling 50 years of historical temperature profiles in a large central European lake. *Limnol. Oceanogr.* **2002**, *47*, 186–197. [CrossRef]
30. Douglas, M.S.V.; Smol, J.P.; Pienitz, R.; Hamilton, P. Algal indicators of environmental change in Arctic and Antarctic lakes and ponds. In *Long-Term Environmental Change in Arctic and Antarctic Lakes*; Pienitz, R., Douglas, M.S.V., Smol, J.P., Eds.; Springer: Dordrecht, The Netherlands, 2004; pp. 117–157.
31. Preston, D.L.; Caine, N.; McKnight, D.M.; Williams, M.W.; Hell, K.; Miller, M.P.; Hart, S.J.; Johnson, P.T.J. Climate regulates alpine lake ice cover phenology and aquatic ecosystem structure. *Geophys. Res. Lett.* **2016**, *43*, 5353–5360. [CrossRef]
32. Beyene, M.T.; Jain, S. Wintertime weather-climate variability and its links to early spring ice-out in Maine lakes. *Limnol. Oceanogr.* **2015**, *60*, 1890–1905. [CrossRef]
33. Hampton, S.E.; Galloway, A.W.E.; Powers, S.M.; Ozersky, T.; Woo, K.H.; Batt, R.D.; Labou, S.G.; O'Reilly, C.M.; Sharma, S.; Lotting, N.R.; et al. Ecology under lake ice. *Ecol. Lett.* **2017**, *20*, 98–111. [CrossRef] [PubMed]
34. De Senerpont Domis, L.N.; Elser, J.J.; Gsell, A.S.; Huszar, V.L.M.; Ibelings, B.W.; Jeppesen, E.; Kosten, S.; Mooij, W.M.; Roland, F.; Sommer, U.; et al. Plankton dynamics under different climatic conditions in space and time. *Freshwater Biol.* **2013**, *58*, 463–482. [CrossRef]
35. Powers, S.M.; Labou, S.G.; Baulch, H.M.; Hunt, R.J.; Lottig, N.R.; Hampton, S.E.; Stanley, E.H. Ice duration drives winter nitrate accumulation in north temperate lakes. *Limnol. Oceanogr. Lett.* **2017**, *2*, 177–186. [CrossRef]
36. Cory, R.M.; Ward, C.P.; Crump, B.C.; Kling, G.W. Sunlight controls water column processing of carbon in Arctic freshwaters. *Science* **2014**, *345*, 925–928. [CrossRef] [PubMed]
37. Levine, M.A.; Whalen, S.C. Nutrient limitation of phytoplankton production in Alaskan Arctic foothill lakes. *Hydrobiologia* **2001**, *455*, 189–201. [CrossRef]
38. Nielsen, A.B. Present Conditions in Greenland and the Kangerlussuaq Area, Working Report. Geological Survey of Denmark and Greenland. 2010. Available online: http://www.posiva.fi/files/1244/WR_2010-07web.pdf (assessed on 30 October 2017).
39. Hasholt, B.; Anderson, N.J. On the formation and stability of oligosaline lakes in the arid, low arctic area of Kangerlussuaq, south-west Greenland. In Proceedings of the Northern Research Basins-14th International Symposium and Workshop, Greenland, Denmark, 25–29 August 2003; C.A. Reitzel: Kobenhavn, Denmark, 2003; pp. 41–48.
40. Saros, J.E.; Northington, R.M.; Anderson, D.S.; Anderson, N.J. A whole-lake experiment confirms a small centric diatom species as an indicator of changing lake thermal structure. *Limnol. Oceanogr. Lett.* **2016**, *1*, 27–35. [CrossRef]
41. Brodersen, K.P.; Anderson, N.J. Subfossil insect remains (Chironomidae) and lake-water temperature inference in the Sisimiut-Kangerlussuaq region, southern West Greenland. *Geol. Greenland Surv. Bull.* **2000**, *186*, 78–82.
42. Anderson, N.J.; Harriman, R.; Ryves, D.B.; Patrick, S.T. Dominant factors controlling variability in the ionic composition of West Greenland lakes. *Arct. Antarct. Alp. Res.* **2001**, *33*, 418–425. [CrossRef]
43. Gilman, R.A.; Chapman, C.A.; Lowell, T.V.; Borns, H.W., Jr. *The Geology of Mount Desert Island: A Visitors Guide to the Geology of Acadia National Park*; Maine Geol. Surv., Department of Conservation: Augusta, GA, USA, 1988; pp. 1–50.
44. Read, J.S.; Hamilton, D.P.; Jones, I.D.; Muraoka, K.; Winslow, L.A.; Kroiss, R.; Wu, C.H.; Gaiser, E. Derivation of lake mixing and stratification indices from high-resolution lake buoy data. *Environ. Model Softw.* **2011**, *25*, 1325–1336. [CrossRef]
45. APHA (American Public Health Association). *Standard Methods for the Examination of Water and Wastewater*, 20th ed.; APHA: Washington, DC, USA, 2000.

46. Bergström, A.K. The use of TN:TP and DIN:TP ratios as indicators for phytoplankton nutrient limitation in oligotrophic lakes affected by N deposition. *Aquat. Sci.* **2010**, *72*, 277–281. [CrossRef]

47. Helms, J.R.; Stubbins, A.; Ritchie, J.D.; Minor, E.C.; Kieber, D.J.; Mopper, K. Absorption spectral slopes and slope ratios as indicators of molecular weight, source and photobleaching of chromophric dissolved organic matter. *Limnol. Oceanogr.* **2008**, *53*, 955–969. [CrossRef]

48. Saros, J.E.; Stone, J.R.; Pederson, G.T.; Slemmons, K.E.H.; Spanbauer, T.; Schliep, A.; Cahl, D.; Williamson, C.E.; Engstrom, D.R. Climate-induced changes in lake ecosystem structure inferred from coupled neo- and paleoecological approaches. *Ecology* **2012**, *93*, 2155–2164. [CrossRef] [PubMed]

49. Kirk, J.T.O. *Light and Photosynthesis in Aquatic Ecosystems*, 2nd ed.; Cambridge University Press: New York, NY, USA, 1994; pp. 40–45, ISBN 0521459664.

50. Prowse, T.D.; Wrona, F.J.; Reist, J.D.; Gibson, J.J.; Hobbie, J.E.; Levesque, L.M.J.; Vincent, W.F. Climate change effects on hydroecology of Arctic freshwater ecosystems. *AMBIO* **2006**, *35*, 347–358. [CrossRef]

51. King, J.R.; Shuter, B.J.; Zimmerman, A.P. Signals of climate trends and extreme events in thermal stratification pattern of multibasin Lake Opeongo, Ontario. *Can. J. Fish. Aquat. Sci.* **1999**, *56*, 847–852. [CrossRef]

52. Weyhenmeyer, G.A.; Blenckner, T.; Pettersson, K. Changes of the plankton spring outburt related to the North Atlantic Oscillation. *Limnol. Oceanogr.* **1999**, *44*, 1788–1792. [CrossRef]

53. Arvola, L.; George, G.; Livingstone, D.M.; Järvinen, M.; Blenckner, T.; Dokulil, M.T.; Jennings, E.; Aonghusa, C.N.; Nõges, P.; Nõges, T.; et al. The impact of the changing climate on the thermal characteristics of lakes. In *The Impact of Climate Change on European Lakes*; Aquatic Ecology Series; George, G., Ed.; Springer: Dordrecht, The Netherlands, 2009; Volume 4, pp. 85–102, ISBN 978-90-481-2944-7.

54. Buffo, J.; Fritschen, L.J.; Murphy, J.L. *Direct Solar Radiation on Various Slopes from 0 to 6 Degrees North Latitude*; USDA Forest Service Research Paper, PNW-142; Pacific Northwest Forest and Range Experiment Station: Portland, OR, USA, 1972.

55. Jeppesen, E.; Kronvang, B.; Olesen, J.E.; Larson, S.E.; Audet, J.; Søndergaard, M.; Hoffmann, C.C.; Andersen, H.E.; Lauridsen, T.L.; Liboriussen, L.; et al. Climate Change effects on nitrogen loading from cultivated catchments in Europe: Implications for nitrogen retention, ecological state of lakes and adaptation. *Hydrobiologia* **2011**, *633*, 1–21. [CrossRef]

56. Fulton, R.S., III; Godwin, W.F.; Schaus, M.H. Water quality changes following nutrient loading reduction and biomanipulation in a large shallow subtropical lake, Lake Griffin, Florida, USA. *Hydrobiologia* **2015**, *753*, 243–263. [CrossRef]

57. Bergström, A.-K.; Jannson, M. Atmospheric nitrogen deposition has cause nitrogen enrichment and eutrophication of lakes in the Northern Hemisphere. *Glob. Chang. Biol.* **2006**, *12*, 635–643. [CrossRef]

58. Rip, W.J.; Ouboter, M.R.L.; Los, H.J. Impact of climatic fluctuations on Characeae biomass in a shallow, restored lake in the Netherlands. *Hydrobiologia* **2007**, *584*, 415–424. [CrossRef]

59. Hobbie, J.E.; Peterson, B.J.; Bettez, N.; Deegan, L.; O'Brien, W.J.; Kling, G.W.; Kipphut, G.W.; Bowden, W.B.; Hershey, A.E. Impact of global change on the biogeochemistry and ecology of an Arctic freshwater system. *Polar Res.* **1999**, *18*, 207–214. [CrossRef]

60. Frey, K.E.; McClelland, J.W. Impacts of permafrost degradation on Arctic river biogeochemistry. *Hydrol. Process.* **2009**, *23*, 169–182. [CrossRef]

61. Jeppesen, E.; Søndergaard, M.; Jensen, J.P.; Havens, K.E.; Anneville, O.; Carvalho, L.; Coveney, M.F.; Deneke, R.; Dokulil, M.T.; Foy, B.; et al. Lake responses to reduced nutrient loading-an analysis of contemporary long-term data from 35 case studies. *Freshwater Biol.* **2005**, *50*, 1747–1771. [CrossRef]

62. Wilhelm, S.; Adrian, R. Impact of summer warming on the thermal characteristics of a polymictic lake and consequences for oxygen, nutrients and phytoplankton. *Freshwater Biol.* **2008**, *53*, 226–237. [CrossRef]

63. Frey, K.E.; Siegel, D.I.; Smith, L.C. Geochemistry of west Siberian streams and their potential response to permafrost degradation. *Water Resour. Res.* **2007**, *43*, W03406. [CrossRef]

64. Tank, S.E.; Frey, K.E.; Striegl, R.G.; Raymond, P.A.; Holmes, R.M.; McClelland, J.W.; Peterson, B.J. Landscape-level controls on dissolved organic carbon flux from diverse catchments of the circumboreal. *Glob. Biogeochem. Cycles* **2012**, *26*, GB0E02. [CrossRef]

65. Saros, J.E.; Northington, R.M.; Osburn, C.L.; Burpee, B.T.; Anderson, N.J. Thermal stratification in small Arctic lakes of southwest Greenland affected by water transparency and epilimnetic temperatures. *Limnol. Oceanogr.* **2016**, *61*, 1530–1542. [CrossRef]

66. Parker, B.R.; Vinebrooke, R.D.; Schindler, D.W. Recent climate extremes alter alpine lake ecosystems. *Proc. Natl. Acad. Sci. USA* **2008**, *105*, 12917–12931. [CrossRef] [PubMed]
67. Persson, L. Trophic interactions in temperate lake ecosystems: A test of food chain theory. *Am. Nat.* **1992**, *140*, 59–84. [CrossRef]
68. Jeppesen, E. The impact of nutrient state and lake depth on top down control in the pelagic zone of lakes: A study of 466 lakes from the temperate zone to the Arctic. *Ecosystems* **2003**, *6*, 313–325. [CrossRef]
69. Hansson, L. Food-chain length alters community responses to global change in aquatic systems. *Nat. Clim. Chang.* **2012**, *2*, 1–6. [CrossRef]
70. Kraemer, B.M.; Mehner, T.; Adrian, R. Reconciling the opposing effects of warming on phytoplankton biomass in 188 large lakes. *Sci. Rep.* **2017**, *7*, 10762. [CrossRef] [PubMed]
71. Malik, H.I.; Saros, J.E. Effects of temperature, light and nutrients on five *Cyclotella sensu lato* taxa assessed with in situ experiments in Arctic lakes. *J. Plankton Res.* **2016**, *38*, 431–442. [CrossRef]
72. Saros, J.E.; Anderson, N.J. The ecology of the planktonic diatom *Cyclotella* and its implications for global environmental change studies. *Biol. Rev.* **2015**, *90*, 522–541. [CrossRef] [PubMed]

Article

The Winter Environmental Continuum of Two Watersheds

Benoit Turcotte [1,*] **and Brian Morse** [2]

[1] Research Scientist, Department of Civil and Water Engineering, Université Laval, Quebec City, QC G1V 0A6, Canada

[2] Department of Civil and Water Engineering, Université Laval, Quebec City, QC G1V 0A6, Canada; brian.morse@gci.ulaval.ca

* Correspondence: benoit.turcotte@gci.ulaval.ca; Tel.: +1-418-656-7857

Academic Editor: Kevin B. Strychar

Received: 16 December 2016; Accepted: 26 April 2017; Published: 9 May 2017

Abstract: This paper examines the winter ecosystemic behavior of two distinct watersheds. In cold-temperate regions, the hydrological signal and environmental parameters can fluctuate dramatically over short periods of time, causing major impacts to aquatic habitats. This paper presents the results of the 2011–2012 winter field campaign in streams and rivers near Quebec City, QC, Canada. The objective was to quantify water quantity and quality parameters and their environmental connectivity from headwater creeks above to the larger rivers below over the entire freeze-up, mid-winter and breakup periods with a view toward exploring the watershed continuum. The paper presents how aquatic pulses (water level, discharge, temperature, conductivity, dissolved oxygen and turbidity, measured at seven sites on an hourly basis along channels of different sizes and orders) evolve through the aquatic environment. Ice conditions and the areal ice coverage were also evaluated (on a daily time step along each instrumented channel). Some findings of the investigation revealed that water temperatures remained well above 0 °C during winter in headwater channels, that dissolved oxygen levels during winter were relatively high, but with severe depletions prior to and during breakup in specific settings, that high conductivity spikes occurred during runoff events, that annual turbidity extremes were measured in the presence of ice and that dynamic ice cover breakup events have the potential to generate direct or indirect mortality among aquatic species and to dislodge the largest rocks in the channel. The authors believe that the environmental impact of a number of winter fluvial processes needs to be further investigated, and the relative significance of the winter period in the annual environmental cycle should be given additional attention.

Keywords: river ice; water quality; watershed; aquatic environment; winter fluvial processes

1. Introduction

The term "aquatic ecosystem" is commonly used, but what is really known about how a watershed works as a system, especially during the cold season? Life subsists under the ice cover of cold regions' river systems. The stress, physical restrictions and environmental conditions endured by aquatic species, directly or indirectly caused by cold air temperatures and consequent freshwater ice processes, have been studied, and key publications on the topic have been completed, e.g., [1–3]. However, so far, continuous aquatic environment monitoring in the presence of ice has seldom been done; the potentially dynamic river ice breakup period generates aquatic habitat constraints that have not been accurately investigated despite the relative importance in the annual hydrological cycle; and the ecological impact of common and less common river ice processes is often only superficially described.

In winter, multiple parameters can vary with greater amplitude and more quickly than during any other season, sending pulses downstream that can either attenuate, amplify or transform.

From a spatial point of view and although scientists and engineers have sampled multiple sites simultaneously, the headwater-to-large-channel dynamics of ice-affected river systems has not been sufficiently documented. This is defendable from a biological point of view since project managers are often emphasizing specific aquatic habitats and channel morphologies. However, the authors believe that measuring the water quality along multiple channel orders is relevant since (1) they are equally important from an aquatic habitat cumulative area or volume; (2) confluences often represent heterogeneous water quality habitats and (3) the water quality at any point in the watershed is substantially influenced by the water flowing from all upstream tributaries.

This paper present continuous environmental data monitored along channels of increasing orders [4] and evolving morphologies in two, geographically-contrasting, watersheds of the Quebec City region, QC, Canada. The data, measured or estimated during the entire 2011–2012 winter season, includes air temperature, channel discharge, ice coverage and ice types, water temperature, specific conductivity, dissolved oxygen and turbidity. The authors propose that this spatiotemporal dataset represents the "winter environmental continuum" of the two watersheds. Here, the continuum can be defined as the multiple, physical links between various parameters at evolving time and space scales. This concept has been presented previously [5,6], but it only included a limited number of monitored parameters in a single watershed.

The objectives of this paper are (1) to demonstrate the relevance of the concept of a winter environmental continuum applied to two independent watersheds and (2) to highlight the local-to-system-scale environmental impact of specific hydrothermal events and ice processes that affect cold regions' watersheds. From an aquatic habitat point of view, the consequences of these events and processes at different channel orders are described or hypothesized, and evident research needs are proposed. The different sections of the paper present a similar structure built on each monitored environmental parameter.

2. Background

2.1. Channel Discharge

The channel discharge (Q), although only indirectly meaningful to ecosystems, represents an important stream parameter because it directly and indirectly affects multiple aquatic characteristics. In sub-arctic and arctic regions, Q is expected to decline uninterruptedly throughout winter because of the absence of runoff from rain or snowmelt, whereas in more temperate settings (i.e., winters characterized by less than approximately 1800 cumulated degree-days of frost), rain-on-snow events commonly interrupt this natural decline. The path followed by rain drops to reach headwater channels during winter depends on multiple factors including snowpack characteristics (snow water equivalent, temperature distribution, density distribution, etc.), vegetation characteristics (conifer, hardwood, grass, crops, etc.), and ground parameters (temperature distribution, porosity, layer thicknesses, etc.). After a runoff event and as cold air temperatures resume, Q is expected to follow a recession trend that can be exacerbated by ice production-induced flow depressions (or abstraction) of varying intensity and duration, e.g., [6–8].

The succession of cold and mild air temperature spells during winter, the latter potentially accompanied by rain that can generate river ice breakup events, produces hydrological pulses that can significantly impact aquatic parameters at varying time and space scales. The most significant pulse of the cold season is undoubtedly the spring melt hydrograph that triggers breakup. The parameters and thresholds that dictate river ice breakup chronology and intensity are multiple and very site (and sometimes winter) specific. Variations in Q caused by ice processes are usually significantly more sudden than those associated with open water conditions, especially when an ice jam (accumulation of ice blocs and floes that create a hydraulic restriction) releases, and this can impact aquatic species and increases mortality, e.g., [9]. The hydrological impacts of dynamic river ice formation and breakup events are synthesized in [7,10].

2.2. Ice Cover and Hydraulic Conditions

At the beginning of winter, the formation of a stationary ice cover necessarily generates an increase in water levels (Y). This rise, which can be estimated by different means, e.g., [11], is generally welcomed because it initially prevents shallow aquatic habitats from freezing. In turn, specific processes such as the production and transport of frazil (ice particles that form in the water column in turbulent flows), the accumulation of anchor ice (ice composed of frazil particles adhering to (and ice crystals that grow on) submerged surfaces; [12]), the formation of ice dams (composed of anchor ice and thermal ice [13]), the thickening of surface ice [14] and the formation of Aufeis (ice that freezes by layers as a result of a pressurized flow conditions [15]), combined with the Q recession, can be detrimental to aquatic species (e.g., freezing, suffocation, skin abrasion; [16]), even those sheltered in the substrate [17]. The type of ice cover and the potential occurrence of specific ice processes, in small to large channels and in steep to low-gradient reaches [18], significantly affect environmental parameters and the quality of aquatic habitats.

At freeze-up, the ice coverage (I$_c$) usually progresses gradually under moderately cold and stable air temperatures (by border ice lateral progression and by ice floes' juxtaposition), but under largely varying meteorological conditions, it can advance or regress quite dynamically. During winter, while I$_c$ remains stable (in the absence of runoff), hydraulic conditions evolve relatively smoothly, and the presence of an ice cover prevents the occurrence of most daily environmental parameter variations.

At breakup, ice and hydrological conditions can change rather suddenly, and I$_c$ may retreat significantly in a matter of minutes. Dynamic breakup events, and more specifically ice jams and ice runs (massive amounts of ice pieces flowing with the water at breakup), can cause direct (moving ice, rocks and woody debris, crushing individuals) and indirect (through hydraulic conditions alteration) impacts on aquatic habitats. Some species are known to prepare for breakup by finding shelters, e.g., [3], but this may not be enough and may certainly not be the case of all aquatic species, especially those of limited mobility.

2.3. Water Temperature

The water temperature (T$_w$) and its variations affect the metabolism, behavior and survival rates of multiple aquatic species [2,3,17]. Ice can only form in the water column if T$_w$ cools down to 0.0 °C, and this occurs in most cold regions' fluvial environments during winter. In addition, the occurrence of supercooling events (characterized by a T$_w$ slightly depressed under 0.0 °C), mostly taking place during freeze-up along turbulent streams, e.g., [19], as well as in lakes [20], is generally associated with the production of frazil and anchor ice that can be fatal to fish, at least at their young development stages [17].

During winter, T$_w$ mostly remains at 0.0 °C in the presence of an ice cover, and it cannot rise significantly unless the ice has been melted or flushed downstream. Nonetheless, specific spatiotemporal conditions can create favorable thermal environments for aquatic species' survival and even wealth. The formation of suspended ice covers along steep channels, e.g., [6], and the evolution of a floating ice cover into a free-spanning ice cover supported by the banks as Q decreases along narrow channels enable T$_w$ to rise above 0.0 °C for two reasons: the ice cover is no longer in contact with the flowing water, and groundwater heat cannot escape into the atmosphere. Similarly, T$_w$ may remain well above 0.0 °C in headwater channels, close to groundwater sources or at the confluence of small order channels, and this represents winter thermal refuges for mobile aquatic species, e.g., [16].

At breakup, the heat gained in open water leads can travel under an ice cover over great distances, which generates melting and additional heat absorption, e.g., [11]. At the end of winter, it is not surprising to measure abrupt changes in T$_w$, e.g., [21], and the aquatic environment can warm to 10 °C downstream of long open water sections and upstream of an ice jam [22].

2.4. Conductivity

Water conductivity (or the specific conductivity, Sp.C, corrected for T$_w$ variations) is a parameter that is often used to determine the contribution of groundwater during runoff events, e.g., [23]: usually, as Q increases, surface runoff generates a drop in Sp.C, and in turn, as Q declines, groundwater contribution dominates, and Sp.C increases. This explains why snowmelt events in the spring are normally associated with a decline in conductivity (or ionic concentration [24]). The presence and concentration of ions and contaminants has also been associated with Sp.C measurements in lake inlets, e.g., [25]. However, this parameter has not been widely measured under ice conditions in a freshwater environment.

2.5. Dissolved Oxygen

In the late 1990s, the exact link between an ice cover, biological activity and winter DO variations had been investigated, e.g., [26,27], but was still unclear. It was suggested [28] that a thick ice layer, covered with snow, would prevent most photosynthesis activity, which would explain why no variation in day-time and night-time DO could occur. Prowse [2] mentioned that the nature and rate of the winter DO depression depend on many factors, including "the quality and origin of source water comprising the flow, and various biochemical processes, such as decomposition and respiration, operating within the water column and channel bed". In the end, it appears that the first winter DO decline can be attributed to a drastic oxygen production decay combined with a sudden contact reduction between the water and the atmosphere, whereas the mid-winter DO decline would be due to the increasing dominance of poorly-oxygenated groundwater inflow. In turn, the sudden late-winter or spring rise in DO would be mostly associated with reaeration caused by river ice breakup and increasing turbulence. Note that the spring runoff can also generate a pronounced DO decline due to the resuspension of organic material [3,29].

In a temperate setting characterized by ice coverage variations in time and space, DO levels can behave differently from what has been reported for sub-arctic channels, and a winter DO depression may not occur or, at least, it could be less severe [2,26]. In fact, the low T_w promotes high absolute DO levels throughout winter (compared with summer DO levels), especially in organic-rich channels [2]. Nonetheless, a severe winter DO depression was measured in two tributaries of the St. John River, NB, Canada [30].

Above and beyond the harshness of winter and the downstream distance, it seems that a number of parameters affecting DO in streams and rivers has not been specifically investigated, including channel gradient, watershed land use, ice cover type and channel order. The actual technology enables the deployment of autonomous sensors that can measure DO levels on a continuous basis, which greatly facilitate environmental surveys in an aquatic environment and the quantification of spatiotemporal variations on a short time step.

2.6. Turbidity and Sediment Transport

The biotope of aquatic habitats is made of sediment and organic material that can be mobilized by natural forces, and sediment transport represents an important environmental fluvial process. The presence of stationary or fast-moving ice logically impacts the sediment transport capacity of a channel, e.g., [1,7,31–33], and it is not surprising that a number of studies has reported erosion and a redistribution of sediment in the presence of different forms of river ice, e.g., [34–36]. This winter process can either improve aquatic habitats, e.g., [37], or be detrimental to a number of species, including fish and their eggs, e.g., [17]. Some studies have also reported very low turbidity (Turb) measurements (or suspended load) during low winter flow conditions, e.g., [38]. The rate of sediment transport does not only depend on the transport capacity, but also on the supply of sediment, which can become very low during the cold season. Indeed, during winter and as Q declines, tributaries do not carry as much sediment; ice protects the banks and the bed (at grounded ice locations) from erosion; and the emerging portion of unstable banks is usually frozen and/or snow-covered, e.g., [3,39,40].

In turn, at breakup, the combined actions of the rising Q and ice abrasion are known to generate very high sediment transport rates, e.g., [41–43]. It is difficult to distinguish the sediment transport contribution of breakup from that of the spring freshet (spring snowmelt hydrograph) for multiple reasons: (1) they usually overlap, at least initially; (2) automated instruments can be damaged in the presence of moving ice; (3) water sampling is difficult and dangerous to perform in the presence of a deteriorating ice cover or during ice runs; (4) bedload measurements are virtually impossible to perform in the same conditions; and (5) the combined action of ground thawing and ice abrasion may increase sediment supply, e.g., [44], in such a way that sediment transport rates reach values that are well above those estimated from open water sediment transport rating curves, e.g., [3].

Dynamic processes such as ice jams and ice runs are known to generate sediment transport pulses, cause sediment accumulation on high banks, create pools and longitudinal scares in the channel and on the banks, damage the riparian vegetation and alter the surface armor of gravel bed channels, e.g., [3,35,45,46]. As a consequence, frequent and/or intense ice jamming and release events can destabilize channels, affect their morphology, e.g., [47], and therefore, disturb aquatic habitats, e.g., [3].

Techniques normally used to evaluate sediment transport in open water conditions (water samples, turbidity measurements, bedload traps) may underestimate sediment transport rates during winter because ice also represents a direct sediment transport vehicle. Anchor ice released from the bed and grounded ice floes lifted by the rising water level are referred to as "sediment rafts" that can transport significant amounts of material [48–50], including large rocks, e.g., [36,51,52]. The deposition of rafted sediments mostly depends on ice melting rates, and as a consequence, rafted sediment settling locations are often independent of prevalent hydraulic conditions. This means that these particles can disturb aquatic habitats and, in most cases, that they become available for subsequent, hydraulically-driven transport.

Sediment transport in cold region channels has been synthesized in review papers and reports [35,40,53].

3. Research Sites and Methodology

Both research watersheds, geographically presented in Figure 1, were selected for their comparable sizes, for their land use and morphology contrasts (Table 1), as well as for their proximity to Quebec City. The Montmorency (M) watershed is oriented north to south, and historically, its winter minimal Q has been two to three times less than its minimal summer Q. The Etchemin (E) watershed is inversely oriented, and historically, its winter and summer minimal Q have been comparable.

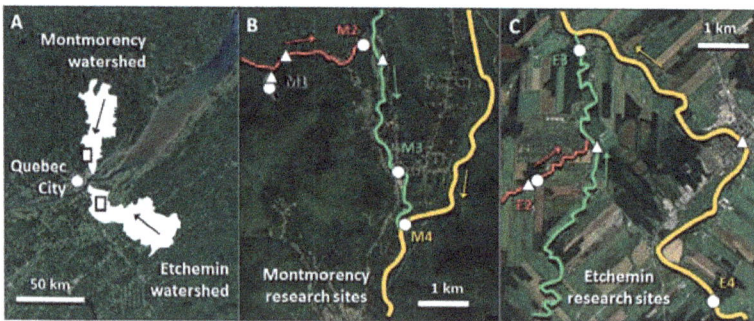

Figure 1. (**A**) Geographic location of the Montmorency (M) and Etchemin (E) watersheds located on both sides of the St. Lawrence River in Quebec City; (**B**) Research channels in the M watershed with instrumented sites identified by white circles; (**C**) Research channels in the E watershed with instrumented sites identified by white circles. Channel colors are representative of their approximate Strahler order. White triangles represent discharge estimation sites, and white circles indicate environmental parameter monitoring sites.

Table 1. Research channels with their respective Strahler order, code, watershed sizes and land use, as well as geographic and ice characteristics.

Channel. Name	Channel. Order	Channel. Code	Watershed Size (km^2)	Land Use (F: Forest; C: Crops)	Gradient (%)	Width (m)	Morphology	Ice Cover
Vallée Creek	1	M1	0.5	100% F	12	1	Cascades	Ice shells
Lépine Creek	2	M2	7	95% F	7	3	Step-pools	Suspended
De l'Île Stream	3	M3	90	95% F	1	20	Rapids	Suspended
Montmorency River	4	#M4	1100	95% F	1	60	Rapids	Suspended
Bélair-Sud Creek	2	E2	6	80% C /20% F	0.4	3	Artificially-confined ditch	Free-spanning snow
Le Bras Stream	3	E3	200	70% C /30% F	0.2	20	Meandering with few riffles	Confined surface ice
Etchemin River	4	E4	1100	35% C /65% F	0.3	60	Meandering with few rapids	Floating surface ice

It was initially believed that breakup events would be more severe on the northward flowing E River, as could be expected from what has been mentioned several times in the river ice literature. However, the authors experience over the years suggest that, at this scale (watersheds of about 1000 km² or less), the north to south T_{air} contrast is compensated by altitude, and that beyond watersheds orientation, their average gradient has a dominant influence on the river ice breakup scenario. This explains why both rivers have been affected by severe mechanical events on a regular basis, but with more frequent mid-winter breakup events on the E watershed. Table 1 also indicates that the ice cover along research channels depends on the local gradient or, more simply, on the morphology [18].

Table 2 presents the parameters that were measured on an hourly basis for 12 months (June 2011–June 2012), this paper focusing on winter results (November 2011–April 2012). The discharge (Q) was estimated using instruments and strategies adapted for each channel (white triangles in Figure 1 indicate Q estimation sites). Along the M1, M2 and M3 channels, specific hydraulic controls were only affected by ephemeral ice development because of a local, groundwater sources (see details about Q measurement in [6]). In this case, autonomous pressure sensors (HOBO U20 anchored to the bed using steel bars and weights) were deployed, and a Sontek Flow Tracker was sporadically used to confirm the stability of the local rating curve. Along the E2 and E3 channels, constant velocity and depth measurements (ISCO 2150 anchor to the channel bed using a PVC-covered steel weight) were performed to evaluate Q, and the Flow Tracker was used a few times (e.g., through holes in the ice cover) to facilitate the interpretation of the reach winter hydrological behavior. Finally, along the M4 and E4 channels, Q were estimated by the Quebec Provincial Government on a 15-min basis and converted into hourly-averaged data.

Table 2. Parameters measured or estimated during winter and instruments deployed into or along the different channels.

Parameter	Code	Units	Instrument	Acquisition Rate
Air temperature	T_{air}	°C	Onset HOBO U22-001	60 min
Discharge	Q	m³/s	● Onset HOBO U20 0–4 m ● YSI 6600 V2 ● ISCO 2150 ● Provincial Government ● Flow Tracker	60 min 60 min 60 min 15 min into 60 min Punctual
Ice coverage	I_c	%	Automated Canon 20D	60 min into 24 h
Water temperature	T_w	°C	YSI 6600 V2/YSI 6560	60 min
Specific conductivity	Sp.C	µs/cm	YSI 6600 V2/YSI 6560	60 min
Dissolved oxygen	DO	mg/L	YSI 6600 V2/YSI 6050 ROX	60 min
Turbidity	Turb	Nephelometric Turbidity Units (NTU)	YSI 6600 V2/YSI 6036 TRUB	60 min

The ice coverage I_c (converted into 24-h averaged data) was estimated by automated camera (Figure 2A) photographs' interpretation along reaches of several channel widths equivalent in length (at least 50). In addition to automated cameras (one per channel), about 30 field trips were completed from freeze-up to breakup in each watershed, and photographs were analyzed as objectively as possible. The same strategy was used to identify reach-specific ice processes that would affect environmental parameters.

Specific water quality monitoring sites (white circle in Figure 1) along each channel were first identified based on the position of nearby tributaries. Indeed, in order to attain the research objectives, instruments were anchored far downstream or some distance upstream of tributaries in order to minimize local water quality interferences that would not be representative of the reach characteristics. The instrument position was also selected based on accessibility and shelter from adverse phenomena. At each site (seven in total), a YSI 6600 V2 probe was placed in PVC tubes anchored to the riverbed

(Figure 2B) where inspections, calibration and battery changes would remain possible throughout winter (sometimes after tremendous amounts of the authors' calories spent breaking the ice cover; Figure 2C). At one site (E3), the channel depth and ice thickness imposed a more robust anchoring installation on a bridge pier (Figure 2D). Environmental parameters were also manually measured using a portable YSI Pro 2030 (T_w, Sp.C, DO) and a Lamotte 1979-EPA (Turb) several times during winter in order to confirm that YSI 6600 V2 sensors were not malfunctioning. In minor cases, data points were removed because of suspicious data of known (e.g., anchor ice) or unknown (e.g., possible fish or larvae using the sensor as a habitat or winter shelter; Figure 3) origin.

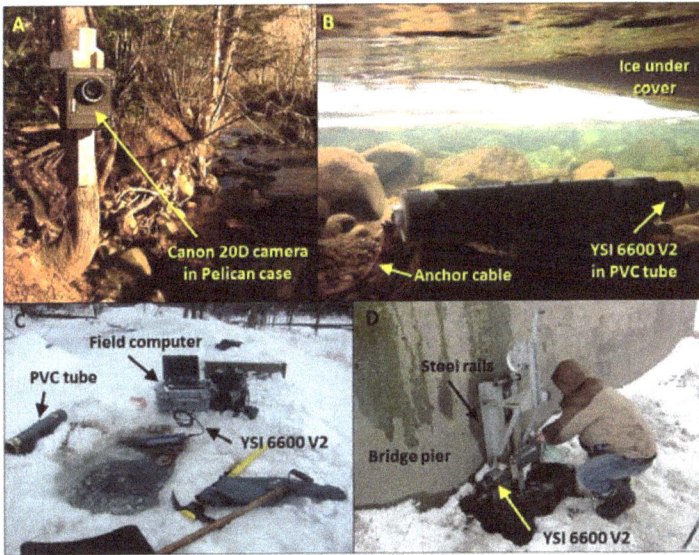

Figure 2. (**A**) Automated Canon 20D digital camera in adapted pelican case at Site M2; (**B**) aquatic view of the YSI 6600 V2 in its PVC tube at Site M3; (**C**) retrieving and downloading the YSI 6600 V2 at Site E4 in February 2012; and (**D**) retrieving and downloading the YSI 6600 V2 prior to breakup at Site E3.

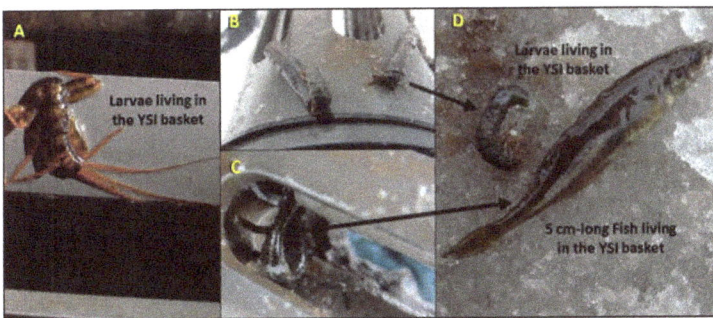

Figure 3. Larvae and fish adopting the YSI 6600 V2 basket as a wintering habitat at Site E2.

4. Results

4.1. Montmorency Watershed

This section presents the results of the 2011–2012 field campaign in the Montmorency (M) watershed. Overall results for each parameter and at each channel order are presented in Figure 4, and key observations, which were either surprising and/or relevant, are highlighted in the different subsections. Straight forward analyses are also presented, leaving more thoughtful interpretations and questioning for the discussion Section 5.

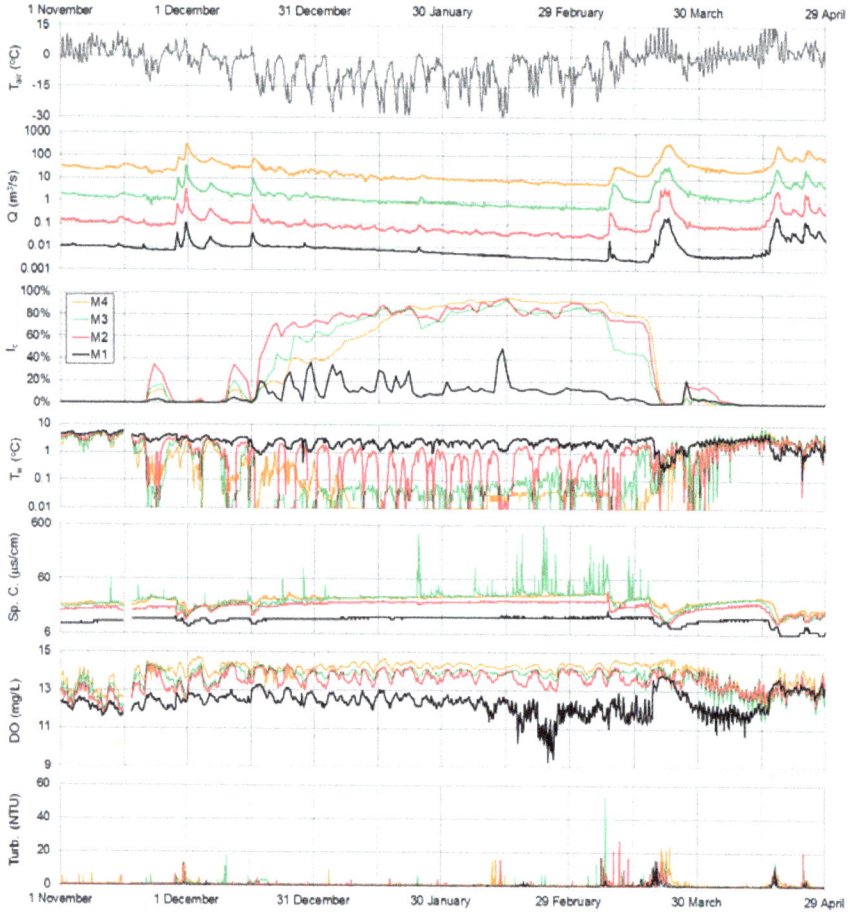

Figure 4. Hourly data (air temperature (T_{air}), discharge (Q), ice coverage (Ic), water temperature (T_w), specific conductivity (Sp.C), dissolved oxygen (DO) and turbidity (Turb)) from 1 November 2011 to 29 April 2012 at Sites M1, M2, M3 and M4.

4.1.1. Discharge

After the runoff event of 16 December (accompanied by a Q depressions documented in [6]), winter Q conditions were relatively stable with minor rain-on-snow events on 28 December and 24 January. Hydrological instabilities at Sites M2 (throughout winter), M3 (early winter) and M4 (first half of winter) most often represent Q depressions associated with ice production in upstream reaches and tributaries. A pre-freshet rain-on-snow event occurred on 8 March, and a very early and yet relatively thermal, multi-peak, snowmelt-dominated breakup scenario took place from 18 March–22 March.

The data revealed that, under open water conditions, the maximum Q of each runoff event was usually delayed by about 10 h between Sites M1 and M4, whereas, in the presence of snow on the ground and ice in the channels, runoff maxima were respectively delayed by 21 and 28 h on 24 January and 8 March. This winter alteration of the hydrological continuum is probably caused by the slow percolation of rain drops into the snow cover and by the slower runoff transit in the drainage system, partly associated with the presence of perforated ice dams, e.g., [13], that control the released Q until they cede or become overtopped.

4.1.2. Ice Coverage

Similarly to the 2010–2011 winter [5], the ice cover started forming first in the second order (M2) channel on 16 December by massive anchor ice formation and ice dam development [13]. Anchor ice and ice dams also formed about six days later (22 December) along Channel M3 and about 20 days later in the Montmorency River (M4; 6 January). The reason for this spatial pattern is due to local heat budget differences, but its exact origin still needs to be quantified. In the first order, groundwater dominated, channel (M1), the only ice observed were ice shells [18] that formed and melted in synchronicity with T_{air} variations. Openings in the suspended ice cover at Sites M2–M4 remained visible throughout winter, often downstream of low order tributaries.

The partial ice cover at Site M3 was mobilized during the pre-breakup runoff event of 8 March, mostly likely because ice dams were small (less than 0.3 m on average compared to about 0.6–1.5 m at Sites M2 and M4) and fragile along that specific reach. The exceptionally warm T_{air} from 18 March to 22 March (daily variations between 1 °C and 18 °C) caused an accelerated thermal breakup at Sites M2 and M4, and the only ice jam observed was located some 10 km downstream of Site M4.

4.1.3. Water Temperature

Figure 4 shows that the water temperature dropped, on average, with an increasing channel order. The groundwater heat explains the relatively high winter T_w at Site M1. The annual (July 2011–June 2012) average T_w at that site was 5.51 °C, which compares with the local annual T_{air} average of 4.6 °C. The coldest annual T_w (0.12 °C) at that site occurred during breakup in the presence of massive snowmelt runoff, but T_w barely dropped below 1 °C during the coldest winter nights.

At Site M2, anchor ice and ice dam development events were often initiated by slightly supercooled T_w associated with early-winter, mid-winter and even post-breakup cold spells. These short events were detected 12 times, and the minimum measured T_w was −0.04 °C (these data are not presented on the logarithmic axis in Figure 4 because the YSI 6600 V2 is not meant to measure T_w with such accuracy). As I_c progressed (increasing the channel insulation), freezing T_w at that site became less frequent and shorter in duration. As a consequence, relatively warm T_w (1 °C to 2 °C) transited to the third order channel during most of the mid-winter period.

The winter T_w behavior at Site M3 was comparable to that of Site M2, but winter absolute values were lower (about 0.08 °C) and freezing conditions more frequent and longer in duration. This is explained by the presence of floating ice cover sections (in contact with the flowing water) within the suspended ice cover-dominated reach and by the longer residence time in the drainage system, allowing for additional groundwater heat loss.

Finally, the surprisingly high T_w at Site M4 was probably caused by a local groundwater source that had not been identified before winter (the YSI 6600 V2 was sheltered in small pool along the bank [6]). At that site, measured $T_w = 0.00$ °C started with the massive formation of anchor ice and ice dams (6 January) and ended when the breaching of ice dams was over (9 February), leaving the ice cover suspended above the flowing water.

4.1.4. Conductivity

The Sp.C in the M watershed is normally close to that of pure water with a slight increase in the downstream direction, and this behavior persisted in the presence of ice. Sp.C values at Sites M1 and M2 were very low for 12 months (respective annual averages of 10 and 18 μs/cm), exhibiting an expected behavior of quick drops and gradual rises respectively taking place during and between runoff events. This indicates that the groundwater in the M2 sub-watershed is almost mineral-free.

Two unexpected results were observed on the Sp.C data at Sites M3 and M4. Along the third order channel, multiple Sp.C spikes were measured during winter. These spikes, reaching a maximum of 600 µs/cm compared with a mid-winter average of 30 µs/cm, either corresponded to the first cold T_{air} of winter or to snow melting conditions, especially after 6 February. They are probably respectively caused by the spreading of de-icing salt, e.g., [25], when T_{air} dropped below 0 °C and to the melting of salty snowbanks under the sun, during the second half of winter. Indeed, Site M3 was located a few meters downstream of a road bridge where sand and salt are used to improve tire adherence and to prevent ice formation on the road, and such Sp.C instabilities only occurred during the winter period.

At Site M4, intriguing Sp.C rises were measured between 16 December and 1 January. A publication by Turcotte et al. (Figure 8 in [6]) shows that these variations occur in synchronism with T_w variations, as well as with Q (and Y) depressions and were therefore associated with local groundwater flux pulses during the ice formation period. However, a similar 24-h rise in Sp.C at site M4 occurred at breakup (21–22 March) during high flow conditions, and the authors have not yet found a reasonable explanation for it.

4.1.5. Dissolved Oxygen

DO levels in the M watershed were usually high and increasing in the downstream direction during the entire monitoring period. Diurnal, open water DO fluctuations were detected at all sites before 10 November and after 30 March and are associated with T_w oscillations, as well as with aquatic biological activity. On the other hand, T_w variations after 10 November and before 30 March mostly concurred with cold and mild T_{air} spells that are normally longer than 24 h, and this explains, in part, why DO variations were so distinctive in the presence of ice and cold water.

At Site M1, DO levels were fairly high at the beginning of winter, but the dominance of poorly-oxygenated water (and potentially snow bridging across the channel) undoubtedly (e.g., [2]) caused these levels to drop prior to breakup. In turn, DO levels remained fairly high throughout winter between Sites M2 and M4 and were generally increasing with the channel order. This is due to I_c remaining lower than 100% (Figure 4), to the highly turbulent hydraulic conditions (steep morphologies; Table 1) and to the increasing cumulative duration of water contact with the atmosphere in the drainage system. Transient, lower-than-expected DO values at Site M4 after 16 December were measured at the same time as higher-than-normal Sp.C values, which reinforces the hypothesis that the monitored water at that site was momentarily affected by high groundwater concentrations.

4.1.6. Turbidity and Sediment Transport

Overall, very low turbidity (Turb) levels were measured at all sites during winter with average values oscillating between 0.1 Nephelometric Turbidity Units (NTU; Sites M1 and M2) to 0.5 NTU (Site M4). Isolated Turb spikes were associated with organic debris, ice pieces or aquatic animals transiting in front of optic sensors (a small fish was seen during winter in the instrumented small side pool at Site M4), whereas consistent, longer-lasting rises in Turb during runoff events were associated with an increase in suspended sediment transport. At Site M3, on 10 December and from 16 December to 20 December, high Turb measurements (respectively, 12 and 4 NTU) were probably caused by frazil transport interference during low T_w events (as observed). At breakup, often the most turbulent period of the year, steady values of 20 NTU were measured at Site M4, while remaining slightly lower at other sites (e.g., 15 NTU at Site M1). These values represent annual Turb maxima associated with the most prolonged suspended sediment transport event of the year. At breakup, Turb rises associated with runoff events generally peaked a few or many hours before Y (or Q; Figure 5). At Site M4 on 9 March and 23 March, a Turb peak seemed to occur when a "jave" (for ice jam release wave [10]) passed by, but this dynamic process (e.g., [45]) is difficult to confirm at an hourly data acquisition rate in a relatively small watershed.

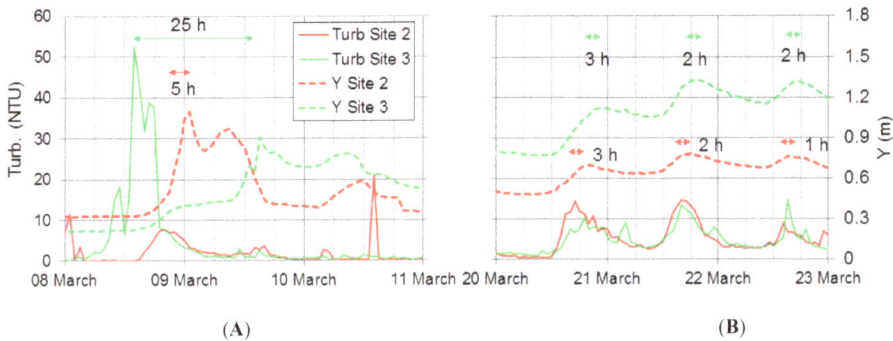

Figure 5. Hourly turbidity (Turb) and water depth (Y) data from Sites M2 and M3 (**A**) from 8 March to 11 March and (**B**) from 20 March to 23 March. The delay between turbidity and local Q peaks vary between 1 and 25 h.

It is important to note that sediment transport in steep gravel bed channels is dominated by bedload, a process that can hardly be measured with optic sensors. During winter 2011–2012, apart from anecdotal gravel and rock accumulations at Sites M1–M4 (that complicated the retrieval of aquatic sensors after the freshet period) bedload was not measured, nor estimated. This topic is addressed in the discussion.

4.2. Etchemin Watershed

This section presents the results of the 2011–2012 field campaign in the Etchemin (E) watershed. Overall results for each parameter and at every channel order are presented in Figure 6.

4.2.1. Discharge

The ice period officially started after the runoff event of 16 December. Mid-winter runoff (rain and/or thaw) events took place on 28 December, 2 January, 24 January, 17 February and 4 March. During the event of 24 January, Q was multiplied by two (E4) to five (E3) and seemed to replenish the phreatic storage for several subsequent weeks. The pre-breakup runoff event of 8 March was associated with a three- (E4) to 10-fold (E2) increase in Q, leading to the onset of breakup. Comparably to what occurred in the M watershed, an early, snowmelt-driven, breakup event took place in the E watershed from 18 March to 22 March, but, in this case, the resulting scenario was more mechanical.

Estimating Q at Sites E2 and E3 was challenging for different reasons, and despite the presence of multiple pressure sensors installed at different hydraulic controls, a certain degree of uncertainty remained. This is in part caused by the behavior of the ice cover during runoff events. An earlier research work [54] had revealed that the ice cover at Site E3 is often flooded during runoff events, and that the thickness of the ice cover and surface slush (i.e., the measured water level (Y)) increases as Q rises, but remains high despite Q declining afterward. Subsequent cold spells turn the surface slush into white ice with very limited impact on Y and on the prevalent backwater effect. It is also possible that pressurized flow conditions occurred at Sites E2 and E3, which made difficult the interpretation of the measured water velocity to estimate Q. Finally, the hydrological data later revealed that high water levels (corresponding to an open water Q of about 120 m^3/s) along the fourth order Etchemin River affected the rating curve at site E3, which complicated the post-winter estimation of high Q and breakup hydrological conditions.

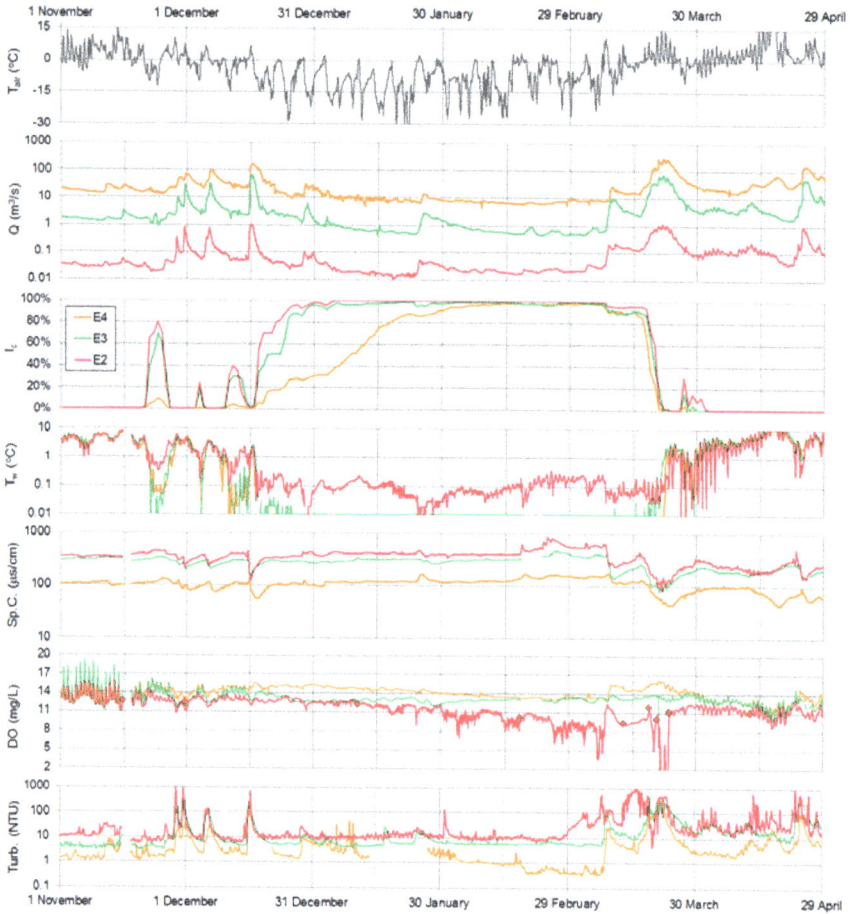

Figure 6. Hourly data (air temperature (T_{air}), discharge (Q), ice coverage (I_c), water temperature (T_w), specific conductivity (Sp.C), dissolved oxygen (DO) and turbidity (Turb)) from 1 November 2011 to 29 April 2012 at Sites E2, E3 and E4. The red diamonds in the DO graph represent punctual measurements at Site E2 with a portable instrument.

4.2.2. Ice Coverage

As presented in Table 1, the cryologic cover that insulates E2 and E3 channels is particular. Wind-blown snow across open fields is probably the main process that contributed to covering most instrumented and observation sites along Channel E2. This cover, dominated by free-spanning snow in narrow sections with thin (about 5 cm) surface cover made of snow ice at wider locations, extended along almost 100% of the channel in less than 15 days (after 16 December 2011). In Channel E3, the ice cover also formed swiftly, but this process was driven by frazil or snow slush bridging in meanders followed by surface interception and frontal progression in longitudinal segments [54]. As the ice cover thickened, bank and channel characteristics generated confinement, and when Q increased in mid-winter, the ice cover could not float freely at most observation sites. In turn, Channel E4 formed a more conventional, floating, surface ice cover quite gradually (over six weeks). This behavior was not associated with groundwater heat, as observed in the Montmorency watershed, but mostly with the presence of a dam (E4 instrumented site), by which the reservoir intercepted the frazil produced upstream. The downstream (observation) reaches were characterized by meanders with short riffles and bedrock-dominated rapids that were progressively covered by border ice migration, including ice-induced braided patterns (see Figure 9 in Turcotte and Morse [18]).

The snow and ice covers along the E channels were relatively intact on 17 March, before a massive amount of snowmelt water entered the drainage system. In five days, the average I_c dropped from 90% down to 10%, and ice jams formed at a few locations. The most significant observed jam was located upstream of Site E3 (Figure 1), precisely in the reach where Q was estimated (Figure 7A,B). The 1.0-m backwater effect caused by the intact ice cover (compared with the open water rating curve) on 17 March remained the same during the following days until the jam formed on 21 March at midnight. The backwater effect initially rose to about 1.8 m, but the jam started to melt in place. It released at 3 p.m. on the same day while Q was increasing.

Figure 7. (**A**) Measured water depth (Y) and estimated discharge (Q) at breakup on the third order channel E3 showing the signature of an ice jam; (**B**) ice jam (photograph taken on 21 March 2012) and (**C**) new gravel bar formed where the jam toe had been momentarily located (photograph taken on 1 August 2012).

4.2.3. Water Temperature

Globally, the water temperature (T_W) was lower and more stable during the ice season at all channel orders compared with the open water season. In the second order Channel E2, T_W remained at approximately 0.1 °C throughout winter, a condition explained by the dominance of groundwater heat combined with the free-spanning, insulating nature of the snow cover. Daily, small amplitude T_W variations were registered during the second portion of winter despite an I_c of 100%. These fluctuations began after the runoff event of 24 January and intensified after the event of 17 February. This indicates that each runoff event contributed to melting the underside of the free-spanning snow/ice cover (e.g., [54]), thus reducing the contact between the flowing water and cold surfaces. As a result, daily T_{air} variations could influence the air chamber temperature below the snow cover, thus affecting T_W.

In turn, T_W at both Sites E3 and E4 behaved as expected, dropping to 0.0 °C prior to freeze-up and maintaining this conditions until breakup. Supercooling events (not presented in Figure 6) were detected six times at Site E3 during freeze-up and once after breakup (−0.01 °C−−0.03 °C, keeping in mind the limited resolution of the YSI 6600V2). At Site E4, a similar degree of supercooling was measured only once during freeze-up (17 December) and once after breakup (27 March).

4.2.4. Conductivity

The Specific conductivity (Sp.C) in the E watershed (Figure 6) was one order of magnitude higher than that of the M watershed during winter 2011–2012 (Figure 4), a result that also extended to the open water period. The fact that Sp.C decreased with the channel order is probably due to the dominance of agricultural fields in the smaller watershed (Table 1). Indeed, the current agricultural practice involves the use of multiple organic and inorganic substances that modify soil properties, and the geology can also affect the groundwater and drainage system Sp.C.

As previously stated, the Sp.C signal is generally the mirror of Q. This logic was respected at all E sites prior to freeze-up (until 16 December), but mid-winter runoff events (e.g., 28 December,

24 January and 17 February) generated Sp.C rises rather than drops at Sites E2 and E4. This suggests that surface runoff during these events contained a higher concentration of ions and/or minerals than the standard groundwater. A this point, the authors could not identify the exact origin of these consistent high Sp.C levels, and further investigation should confirm if the agricultural practice, the decomposition of crops residues, the use of de-icing salt or another cause can be pointed out. A similar reasoning could also clarify why an important rise in Sp.C was measured at all sites during the second half of February, especially at Site E2, with values reaching 880 µs/cm (twice the value of Sp.C for a comparable Q during the summer of 2011).

4.2.5. Dissolved Oxygen

Aquatic measurements during the winter period indicate that the presence of a complete surface ice cover, a reduced biological activity and a stable T_w at 0.0 °C at Sites E3 and E4 erased diurnal dissolved oxygen (DO) variations. At site E4, DO gradually decreased during the winter period, as should be expected considering the reduced contact between the water and the atmosphere, the longer water residence time, as well as the dominance of poorly-oxygenated groundwater. At Site E3, a comparable decline was not detected, and DO remained fairly stable between 12 and 13 mg/L. It is possible that air pockets under the surface ice cover laying on the banks (confirmed during Q measurements) enabled gaseous exchanges between the atmosphere and the water, thus maintaining reasonably high winter DO levels.

The most relevant observation was made at Site E2 where a winter DO depletion was measured, reaching critically low values at the end of February. The representativeness of automated DO measurements was initially uncertain, but punctual DO measurements (see red diamonds in the DO graph, Figure 6) did not reveal any sign of continuous measurement errors. A number of publications have mentioned that low DO levels could be expected downstream of bogs [55] or industrial sources, e.g., [28,56]. It seems that agricultural practices could also temporarily deplete DO in small channels prior to and during breakup (e.g., [3]). The oxidation of highly concentrated solids and the presence of what appeared to be white, filamentary algae at Site E2 during the second half of winter would explain, at least in part, temporarily lethal DO levels.

4.2.6. Turbidity and Sediment Transport

Turbidity (Turb) values were generally low in the E watershed during winter, but high concentration events were monitored. At Sites E2 and E3, the winter Turb was less responsive to Q variations than what had been measured during open water conditions (Figure 8, based on averaged Turb values for determined relative Q increments), and values at very low Q were somewhat higher than what would have been expected based on open water conditions.

The Turb base value at Site E2 during winter was about 10 NTU, and intriguing daily variations (from 8 to 16 NTU) were detected during the entire cold period. About six days before breakup, while Q was relatively constant, very high daily bursts of Turb were measured. Unfortunately, the maximum YSI 6036 optical sensor range was 1000 NTU, not enough to fully monitor maximum levels. In turn, during the first days of massive snowmelt runoff, relatively low Turb levels were measured as if the sediment available for transport had been depleted (or was not available yet). Weeks after breakup, daily variations of significant magnitudes (50–400%) were again detected and concurred with Q variations (20–50%) and therefore with the sediment transport capacity unlimited by sediment supply. During breakup at Site E3, Turb variations occurred more frequently than Q variations, which suggests that dynamic ice processes generated Turb pulses (e.g., [45]). However, at a data acquisition rate of 1 h, no specific link could be made between detected ice jam release events (e.g., Figure 7) and Turb data.

Turb measurements at Site E4 revealed low winter levels, especially after mid-February and prior to the runoff event of 8 March (Figure 6). In turn, Turb values were higher than what would have been expected from the open water Turb-Q relationship (Figure 8), which is counter intuitive from the sediment supply and sediment transport capacity points of view, but not impossible considering the complexity of hydraulic conditions under an ice cover. The optical data between 16 December and 21 December were affected by anchor ice and were therefore removed from the dataset. A similar decision was made about the data between 13 January and 27 January because of thermal ice formation on the upstream face of the dam where the sensor was installed. At breakup (combined with the freshet), Turb levels reached values of 350 NTU, which is comparable to what had been measured in September 2011 for a Q that was 25% lower.

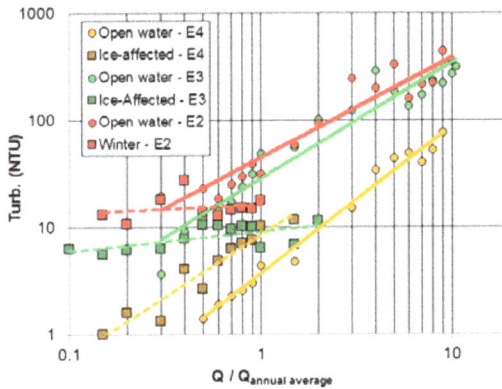

Figure 8. Data and power function interpolations between open water and ice-affected turbidity-Q relationships at Sites E4, E3 and E2. Q is made dimensionless by using the annual average Q at all sites.

The ratio of suspended vs. bedload transport in the E watershed is unknown, but it is most probably significantly higher than in the M watershed dominated by gravel-bed channels. In 2013, an attempt was made to link Turb with suspended sediment concentration (mg/L). Figure 9 presents the relationship at both Sites E3 and E4 (the maximum value of 1000 NTU had been reached too often at site E2 during the 12 months of environmental monitoring to perform this analysis). Keeping in mind that these relationships are approximate (e.g., they do not consider seasonal variations or the concentration of fine sediment in the ice), results suggest that respectively 0.2% (260 tons) and 0.5% (72 tons) of the annual suspended load transited through Sites E4 and E3 between 17 December and 7 March (21% of the year). In turn, between 8 March and 24 March (breakup and snowmelt runoff, 4% of the year), 22% of the annual suspended load transited through both sites (respectively 28,600 tons and 3400 tons).

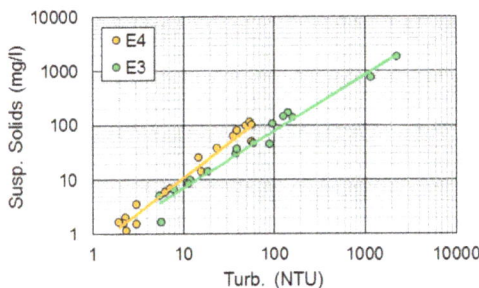

Figure 9. 2013 field data and interpolated relationship between suspended solids (mg/L) and turbidity (NTU) at Sites E3 and E4.

5. Discussion

5.1. Discharge

Discharge (Q) variations during winter 2011–2012 (Figures 4 and 6) did not directly generate extreme conditions from an environmental standpoint (e.g., there was no significant Q depression, only moderately low late-winter Q and no confirmed dynamic ice-induced Q instabilities associated with ice jam events). In turn, as expected, Q variations did affect other parameters. For example, when Q rose, I_c was reduced at most sites; T_w was reduced at sites where no floating ice cover was present (but the dominance of groundwater heat at site M1 was maintained despite 10-fold increases in Q); and Sp.C often rose unexpectedly (at sites E2 and M3) before dropping.

From a watershed hydrological continuum point of view, the ratio of Q per km^2 of the watershed generally remained fairly constant from M1 to M4, including during runoff events, which was apprehended because of the watershed topographical characteristics. On the other hand, this behavior was not observed in the flatter E watershed. It seems that Q could only be estimated with a limited level of confidence at E sites, a winter reality that applies to most cold regions' channels. Because Q directly affects most hydraulic, thermal, cryologic and water quality parameters (i.e., the entire environmental continuum), it appears that estimating Q in the presence of ice, despite being very challenging, is a necessary task that government agencies have often struggled to accomplish for several reasons, including resource and knowledge limitations. The velocity index method, e.g., [57,58], represents one avenue to address this issue. Other empirical approaches could be developed on the basis of site-specific knowledge that involves field data acquisition in a relatively dangerous environment [59].

5.2. Ice Processes

The data from Figures 4 and 6 indicate strong upstream to downstream ice cover formation chronologies in both E and M watersheds, the largest channel being the latest to become ice covered (despite generally lower gradients). At breakup, a combination of heat, Q variation and ice cover fragility usually dictated the spatial chronology of the I_c reduction. Therefore, understanding and monitoring the upstream to downstream ice dynamics (i.e., the cryologic continuum) appear crucial to identify the origin of specific environmental parameter fluctuations, such as T_w and DO.

A number of ice formation processes that were observed during winter 2011–2012 can negatively impact aquatic life. In low gradient channels, the formation of an ice cover at shallow locations (or when Q is low) and the downward migration of thermal ice in the substrate could both cause mortality by freezing. In steeper reaches, the formation of frazil and anchor ice, observed at many sites, could also directly (e.g., abrasion, isolation, freezing) and indirectly (e.g., suffocation, imposed migration to other sites and energy consumption) affect aquatic species (e.g., [2,3,17]). On the other hand, during the studied winter season, the presence of an ice or snow cover positively stabilized a number of environmental parameters, and no mid-winter breakup event was observed (although these processes are common in the region). The spring breakup scenario in the M watershed was relatively thermal, with limited dynamic impact on aquatic habitats; the moderately mechanical breakup in the E watershed generated ice jams that mostly melted in place; and no sign of major ice runs (only small shear walls) was observed.

During the following winters, the authors monitored four dynamic ice phenomena that were observed or assumed to affect aquatic life:

- In the spring of 2014 in the M watershed, a two kilometer-long ice jam was lifted and mobilized by an important jave [60]. Ice movements in secondary channels wiped large zones of riparian vegetation. After the event, fish were found swimming within isolated, shallow pools in the forest (formed by ice jam-induced high water levels), and various dead crayfish parts (probably crushed by moving rocks, woody debris and ice floes) were observed on newly-formed sandy bars (Figure 10A).
- In January 2015, under very low T_{air} (−25 °C), a jave was detected (water level acquisition rate of 5 min) in the Montmorency River at several sites along a 5 km-long reach. This "cold breakup",

an event that has rarely been documented, was probably caused by the release of an unstable ice dam that triggered a cascade effect. The jave celerity was not very high (5 km/h), and the wave amplitude was not significant (0.6 m), but it occurred under supercooling conditions, when species are the most vulnerable (e.g., less mobile). Although no mortality could be observed among the aquatic community, this result appears very likely.

- Mid-winter breakup events can be detrimental to aquatic life, especially when the rain is immediately followed by an intense cold spell. In January 2016, in the Ste. Anne River (fourth order gravel bed channel located near Quebec City, QC, Canada), a runoff event caused multiple ice runs and ice jams concurrently with massive frazil production and high frazil transport rates (T_{air} rapidly fell below $-10\,°C$ after the rainfall). This scenario and its outcome is probably comparable to the "cold breakup" described above, although its origin and suddenness are distinct.

- In December 2015, a snow storm generated a snow slush flow that travelled along a few kilometers of the Ste. Anne River. This dynamic event, comparable to a dynamic breakup, was probably caused by the release of a snow slush bridge under its own backwater pressure. Although the wave was not very high (about 1 m), it is still the most likely explanation for the observed mortality in the fish community (Figure 10B). A question arises regarding the ability of aquatic species to instinctively apprehend this type of snowfall-driven freeze-up consolidation event that our advanced society can hardly predict (e.g., [7,10]).

Overall, the grounded nature of moderate and intense ice runs (the water depth is comparable to the size of tumbling ice floes) in steep gravel bed channels such as the Montmorency and Ste. Anne Rivers represents a serious threat to aquatic species lying on or within the substrate. These ice runs can travel a significant distance downstream, rubbing the banks and bed with nowhere to hide.

(A) (B)

Figure 10. (**A**) Dead crayfish (8 cm in length) on a sandy bar along a secondary channel of the Montmorency River after the 15 April 2014 breakup and (**B**) dead fish (5–10 cm in length) found on a gravel bar of the Ste. Anne River after a snow slush consolidation event on 21 December 2012.

5.3. Water Temperature

Figures 4 and 6 revealed that T_w can remain above $0\,°C$ in headwater and/or steep channels, even during very cold spells, mostly because of a combination of groundwater heat and suspended ice (or free-spanning snow) cover insulation. Low order streams and confluences can represent a winter refuge for aquatic species including fish, e.g., [16], but questions arise regarding the impact of (1) relatively warm (and unstable) T_w and (2) the possible absence of ice cover on fish behavior, metabolism and predation, thus affecting survival rates, e.g., [17]. Furthermore, under relatively common circumstances, these channels are affected by transient supercooling events, as well as by massive anchor ice formation periods, and some reaches are not accessible to all species and individuals, partially because of the presence of ice, cascades and anthropic hydraulic structures that impede migration. Figure 11 presents a T_w dataset measured with a high resolution sensor (RBR Solo T) deployed in the Ste. Anne River during winter 2014–2015, substantiating that supercooling is very common in small rivers of the Quebec City region at freeze-up, as well as prior to and after breakup.

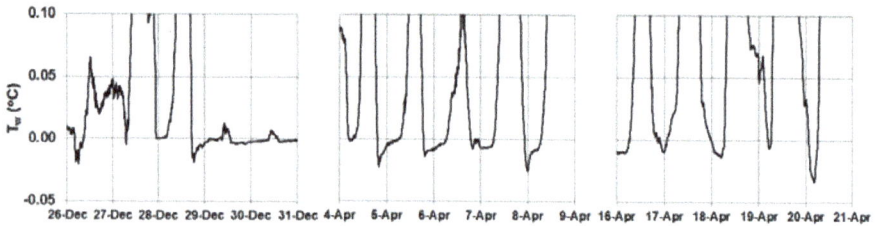

Figure 11. Supercooling events measured during freeze-up (**left**), prior to breakup (**middle**) and after breakup (**right**) in the Ste. Anne River watershed located northwest of Quebec City, QC, Canada, during winter 2014–2015.

In higher order and low-gradient channels located downstream, T_w most often remained close to 0.0 °C throughout winter, and most aquatic species are usually adapted to this environment, although they can suffer from long, cold winters, e.g., [17]. Prior to breakup, the rise in Q combined with a T_w of 0.0 °C could represent a limitation to overwintering aquatic species in preparation for a potentially dynamic breakup.

5.4. Conductivity

Results presented in Figures 4 and 6 suggest that the winter upstream-to-downstream Sp.C relationship in both M and E watersheds compares with what is normally monitored during the open water season (including the early November and late April periods). However, a number of site-specific signals that either faded or were not detected downstream suggest that humans can substantially modify the winter water quality, as interpreted through the specific conductance (Sp.C).

A publication [25] suggests that the use of de-icing salt on a highway in the Quebec City region was the main cause of the St. Augustin Lake's eutrophication. Transient salty spikes in streams (e.g., Figure 4 at Site M3) probably generated less ecological impacts, and dilution rather than accumulation can occurs downstream; but this should be confirmed by further investigation in a comparably pristine environment.

Furthermore, there are reasons to believe that the current agricultural practice in the E watershed can significantly modify the ion balance at all channel orders and therefore affect the Sp.C (e.g., [61]). The very high Sp.C values measured prior to breakup (e.g., Figure 6 at Site E2) are particularly intriguing. Further research should attempt measuring which ions (e.g., nitrates) or contaminants (e.g., fertilizer residues, de-icing salt) are responsible for this result, if the observed winter growth of algae can be linked to this parameter and if this can be directly or indirectly lethal to aquatic species. In this case also, the downstream site (E3) only registered a moderate late-winter rise in Sp.C, which confirms that contaminant dilution takes place downstream of confluences.

5.5. Dissolved Oxygen

Overall, the data presented in Figure 4 suggests that dissolved oxygen (DO) in steep (turbulent and partially ice-covered) channels located in forested settings is not a factor influencing the survival rate of aquatic species. In turn, aquatic species may have to choose between a relatively warm headwater channel with potentially low late-winter water depths and DO levels or a colder, deeper, higher order channel with a consistently DO-rich environment.

In a low-gradient setting (E watershed), the data in Figure 6 show that DO levels can remain surprisingly high during the entire winter season, despite the presence of a DO impermeable ice cover (e.g., [2]). At the opposite, in snow-covered headwater agricultural channels, lethal DO levels can be reached prior to and during breakup (e.g., Figure 6 at Site E2), which could be due to a high concentration of organic material [3]. Further research should confirm the origin, potential intensity and impacts of a multi-day "reduced DO wave" travelling downstream on aquatic life and if this phenomenon could impact higher order channels (which was not the case in the present study).

5.6. Turbidity and Sediment Transport

In the M watershed, turbidity (Turb, Figure 4) remained low (under 1 NTU) during the winter period, with measurable, but limited rises during runoff events, a behavior that compares with what was monitored during the open water period. In turn, at breakup and during the freshet event, Turb levels were multiplied by 10–50 for several days. Undoubtedly, a large ratio of the sediment transport along gravel bed channels in forested settings occurs in the form of bedload transport. The authors believe that the thermal nature of the March 2012 spring breakup event and the relatively low freshet runoff (about 330 m^3/s) in the M watershed did not generate a significant amount of bedload transport. However, in April 2014, a dynamic breakup event characterized by a succession of seven measurable javes [60] and a high Q (about 600 m^3/s) apparently moved a significant amount of sediment in the same watershed: two anchored instruments were lost (the local bed geometry had completely changed), and one instrument had been flipped over and buried under 200 mm-diameter stones. Further downstream, piles of gravel had been deposited on ice floes laying on the floodplain (Figure 12A), and a 300-mm rock was found in the thermal ice portion of an ice floe in the forest (Figure 12B). An observation from the Ste. Anne River also demonstrates that ice runs can mobilize 1.5-m boulders (Figure 12C). This highlights the fact that, beyond the complexity of estimating bedload transport during winter, quantifying ice rafting and ice pushing sediment transport is also very challenging, but necessary to understand how dynamic winter fluvial processes can impact the channel stability and aquatic habitats. It is known by river ice scientists and engineers that tributaries can trigger breakup [54] that can in turn generate bank and bed scour (e.g., [53]), which can directly cause mortality among the aquatic community. This highlights one of the potential impacts of the environmental continuum on aquatic species through sediment transport.

Figure 12. (**A**) Stones and mud found on an ice floe deposited in the floodplain after the 2014 breakup in the Montmorency River; (**B**) 300-mm stone trapped in thermal ice deposited on the floodplain after the 2014 breakup in the Montmorency River; and (**C**) 1.5-m boulder pushed by an ice run (1 April 2016) in the Ste. Anne River (left and right photographs respectively taken before and after winter at a similar discharge).

In the E watershed where a significant ratio of sediment transport takes place in the form of suspended load, Figure 6 and the data interpretation revealed that sediment transport rates were higher in smaller channels and that dilution tended through the drainage system. This could either be due to a decreasing proportion of agricultural land use (i.e., sediment supply; Table 1) or to a reduced sediment transport capacity (Site E4 being located in a small reservoir and Stream E3 presenting the lowest gradient). A change in land use from forest to agricultural (or logging) can substantially modify sediment transport modes and intensities in river systems by increasing fine sediment supply. Knowing that fine sediment can transport contaminants and obstruct gravel bed habitats [17], further investigation should reveal how this can impact the fluvial environment.

From a temporal point of view, Figure 6 shows that low sediment transport rates (although higher than expected; Figure 8) occurred in the presence of stationary ice, whereas a significant amount of suspended sediment transited through Sites E2–E4 during the moderately mechanical breakup event combined with the spring freshet. This result is consistent with other studies, e.g., [41,43], but additional winters of research should investigate how varying hydro-climatic conditions can impact the distribution of the annual sediment transport budget. For instance, in August 2011, the extratropical storm Irene generated an open water flow of 660 m^3/s at Site E4 (return period of 20–50 years), and a significant amount of suspended sediment was transported during the event (roughly 48% of the annual suspended load at Site E4 in three days), thus impacting the annual sediment transport budget distribution. Additional research should also investigate the net impact of a dynamic breakup on sediment transport and aquatic life. For instance, the newly formed gravel bar presented in Figure 7C was probably caused by significant water velocity variations under an ice jam.

5.7. Environmental Continuum Research Avenues

The data presented in Figures 4 and 6 were measured or estimated continuously during a single winter. Additional winters of field investigation at the same sites, data monitoring in distinct aquatic environments, as well as indirect approaches should expose how climate change (warmer T_{air}, shorter winters, more frequent mid-winter breakup events and consequent frazil production intensification) and anthropic activities (e.g., hydroelectric production, hydraulic structures, urbanization, agricultural practice, logging, etc.) can impact the hydrological, thermal, cryologic and morphological regimes of river systems and, more comprehensively, the environmental continuum. Their respective effect on the multiple parameters and factors that determine water and aquatic habitats quality should be documented.

From what has been learned through this study, in a climate change perspective involving warmer winters, more frequent rain-on-snow events and additional freeze-thaw cycles, the following winter environmental impacts could be expected at the watershed scale:

- Higher Q with more frequent runoff events in all channel orders;
- Lower I_c at all channel orders and a more fragile ice cover;
- Lower T_w in steep headwater channels (reduced I_c insulation) and warmer T_w in larger channels (reduced winter intensity and duration);
- The use of more de-icing salt that would potentially lead to more frequent Sp.C winter spikes downstream of roads and bridges;
- Higher sediment transport rates and more frequent sediment transport pulses in the drainage system that would eventually contribute in destabilizing cold region channels.

In turn, as an example of human impact, if the land use would change from forest (e.g., M watershed) to agriculture in a low gradient watershed (E watershed), the following winter environmental impacts could be anticipated:

- Higher runoff maximum Q at all channel orders (reduced response time in the absence of intercepting vegetation);

- Higher T_W in small channels (windblown [13] snow insulation);
- Potentially higher Sp.C (annual and) levels at all channel orders for reasons that would need to be identified (as measured at site E2);
- Potentially lower DO levels prior and during the breakup period (as detected at site E2);
- An increased sediment supply (absence of stabilizing vegetation) and transport capacity (consequent of higher Q) involving a change in channel bed characteristics and contaminant transport rates.

6. Conclusions

In cold and temperate regions river systems, winter may have been overlooked as a period during which aquatic life can be directly or indirectly impacted by cold air temperatures and by the consequent various forms of freshwater ice cover types and processes. This paper has presented a global portrait of environmental conditions in channels of different sizes in two distinct watersheds during the winter of 2011–2012 and made links from upstream to downstream conditions, as well as between various water quality parameters (discharge, ice coverage, water temperature, water conductivity, dissolved oxygen and turbidity) forming the watershed environmental continuum. The paper has also referred to key events, including some obtained during subsequent winters and in other rivers. The potential environmental impact of a number of hydrothermal events, ice processes and human activities on aquatic habitats has been highlighted and discussed.

Overall, this research has demonstrated that a multi-parameter, watershed scale, continuous environmental investigation campaign can provide information that facilitates the interpretation of specific water quality parameter variations and extremes. It furthermore proves that upstream to downstream, temporal and biophysical or biochemical interactions occur in watersheds during the winter period, that these interactions can directly or indirectly affect aquatic life and that streams and rivers are not as sleepy as they seem under their white cover. Finally, this paper suggests that including (representative) tributaries in freshwater aquatic investigation projects is necessary to obtain a comprehensive understanding of the aquatic habitat and species behavior.

Further data investigation, including the use of statistical tools, could reveal additional relationships among the various hourly datasets that were collected in this study (and subsequent studies). Although monitoring the aquatic environment in the presence of ice is challenging, today's technology enables the deployment of automated sensors that can monitor an increasing amount of parameters with suitable accuracy. However, preserving the instruments' integrity and reducing the occurrence of unexpected results or sensor readings will probably always imply the knowledge of river processes, as well as strategically planned presence in the field.

Acknowledgments: The authors would like to thank Shahrzad Bazri for the work done in the field in 2013 and for the analysis of environmental data. Thomas Simard-Robitaille provided field support in the Etchemin Watershed in 2011–2012, and Mathieu Dubé, as well as Félix Pigeon provided support in the Montmorency Watershed between 2011 and 2015. The reviewers and editor have provided comments that contributed to improving the content and structure of the paper. This research was made possible by the financial support of the Canada Foundation for Innovation (CFI) shared by François Anctil (Université Laval) and the Natural Sciences and Engineering Research Council of Canada (NSERC).

Author Contributions: The initial idea of this research was proposed by Brian Morse. Benoit Turcotte and Brian Morse did the research project planning; Benoit Turcotte completed the field work and analyzed the data; Benoit Turcotte proposed the structure of the paper and wrote each section with Brian Morse.

Conflicts of Interest: The authors declare no conflict of interest.

References

1. Prowse, T.D. River-ice ecology. I. Hydrologic, geomorphic, and water-quality aspects. *J. Cold Reg. Eng.* **2001**, *15*, 1–16. [CrossRef]
2. Prowse, T.D. River-ice ecology. I. Biological aspects. *J. Cold Reg. Eng.* **2001**, *15*, 17–33. [CrossRef]

3.   Prowse, T.D.; Culp, J.M. Ice Breakup: A Neglected Factor in River Ecology. In *River Ice Breakup*; Beltaos, S., Ed.; Water Resources Publications: Highland Ranch, CO, USA, 2008; pp. 349–376.

4.   Strahler, A.N. Quantitative Analysis of Watershed Geomorphology. *Trans. Am. Geophys. Union* **1957**, *8*, 913–920. [CrossRef]

5.   Turcotte, B.; Morse, B.; Anctil, F. Cryologic continuum of a steep watershed. *Hydrol. Process.* **2012**. [CrossRef]

6.   Turcotte, B.; Morse, B.; Anctil, F. The hydro-cryologic continuum of a steep watershed at freezeup. *J. Hydrol.* **2014**, *508*, 397–409. [CrossRef]

7.   Beltaos, S. Freezeup Jamming and Formation of Ice Cover. In *River Ice Formation*; Beltaos, S., Ed.; Committee on River Ice Processes and the Environment, Canadian Geophysical Union, Hydrology Section: Edmonton, AB, Canada, 2013; pp. 181–256.

8.   Prowse, T.D.; Carter, T. Significance of ice-induced storage to spring runoff: A case study of the Mackenzie River. *Hydrol. Process.* **2002**, *16*, 779–788. [CrossRef]

9.   Cunjak, R.A.; Prowse, T.D.; Parrish, D.L. Atlantic salmon in winter; the season of parr discontent. *Can. J. Fish. Aquat. Sci.* **1998**, *55*, 161–180. [CrossRef]

10. Jasek, M.; Beltaos, M. Ice-Jam Release: Javes, Ice Runs and Breaking Fronts. In *River Ice Breakup*; Beltaos, S., Ed.; Water Resources Publications: Highland Ranch, CO, USA, 2008; pp. 247–304.

11. Hicks, F.E. *An Introduction to River Ice Engineering for Civil Engineers and Geoscientists*; CreateSpace Independent Publishing Platform: Charleston, SC, USA, 2016; p. 159.

12. Malenchak, J.; Clark, S. Anchor Ice. In *River Ice Formation*; Beltaos, S., Ed.; Committee on River Ice Processes and the Environment, Canadian Geophysical Union, Hydrology Section: Edmonton, AB, Canada, 2013; pp. 135–158.

13. Turcotte, B.; Morse, B.; Dubé, M.; Anctil, F. Quantifying steep channels freezeup processes. *Cold Reg. Sci. Technol.* **2013**, *94*, 21–36. [CrossRef]

14. Ashton, G.D. Thermal processes. In *River Ice Formation*; Beltaos, S., Ed.; Committee on River Ice Processes and the Environment, Canadian Geophysical Union, Hydrology Section: Edmonton, AB, Canada, 2013; pp. 19–76.

15. Daly, S.F. Aufeis. In *River Ice Formation*; Beltaos, S., Ed.; Committee on River Ice Processes and the Environment, Canadian Geophysical Union, Hydrology Section: Edmonton, AB, Canada, 2013; pp. 159–180.

16. Power, G.; Brown, R.S.; Imhof, J.G. Groundwater and fish—Insights from northern North America. *Hydrol. Process.* **1999**, *13*, 401–422. [CrossRef]

17. Bergeron, N.E.; Enders, E.C. Fish Response to Freezeup. In *River Ice Formation*; Beltaos, S., Ed.; Committee on River Ice Processes and the Environment, Canadian Geophysical Union, Hydrology Section: Edmonton, AB, Canada, 2013; pp. 411–432.

18. Turcotte, B.; Morse, B. A global river ice classification model. *J. Hydrol.* **2013**, *507*, 134–148. [CrossRef]

19. Nafziger, H.; Hicks, F.; Thoms, P.; McFarlane, V.; Banack, J.; Cunjak, R.A. Measuring supercooling prevalence on small regulated and unregulated streams in New-Brunswick and Newfoundland, Canada. In Proceedings of the 17th CGU HSE CRIPE Workshop on River Ice, Edmonton, AB, Canada, 21–24 July 2013.

20. Daly, S.F.; Ettema, R. Frazil Ice Blockage of Water Intakes in the Great Lakes. *J. Hydraul. Eng.* **2006**, *132*, 814–824. [CrossRef]

21. Marsh, P.; Prowse, T.D. Water Temperature and Heast Flux at the Base of River Ice Covers. *Cold Reg. Sci. Technol.* **1987**, *14*, 33–50. [CrossRef]

22. Parkinson, F.E. Water temperature observations during break-up on the Liard–Mackenzie River system. In Proceedings of the 2nd Workshop on Hydraulics of Ice-Covered Rivers, Edmonton, AB, Canada, 1982.

23. Stewart, M.; Cimino, J.; Ross, M. Calibration of Base Flow Separation Methods with Streamflow Conductivity. *Ground Water* **2007**, *45*, 17–27. [CrossRef] [PubMed]

24. Hamilton, A.S.; Moore, R.D. Winter streamflow variability in two groundwater-fed sub-Arctic rivers, Yukon Territory, Canada. *Can. J. Civ. Eng.* **1996**, *23*, 1249–1259. [CrossRef]

25. Guesdon, G.; Santiago-Martin, A.; Raymond, S.; Messaoud, H.; Michaux, A.; Roy, S.; Galvez, R. Impact of Salinity on Saint-Augustin Lake, Canada: Remediation Measures at Watershed Scale. *Water* **2016**, *8*, 285. [CrossRef]

26. Schreier, H.; Erlebach, W.; Albright, L. Variations in water quality during winter in two Yukon rivers with emphasis on dissolved oxygen concentration. *Water Res.* **1980**, *14*, 1345–1351. [CrossRef]

27. Whitfield, P.H.; McNaughton, B. Dissolved-Oxygen Depressions under Ice Cover in Two Yukon Rivers. *Water Resour. Res.* **1986**, *22*, 1675–1679. [CrossRef]

28. Chambers, P.A.; Brown, S.; Culp, J.M.; Lowell, R.B.; Pietroniro, A. Dissolved oxygen decline in ice-covered rivers of northern Alberta and its effects on aquatic biota. *J. Aquat. Ecosyst. Stress Recovery* **2000**, *8*, 27–38. [CrossRef]

29. Hou, R.; Li, H. Modelling of BOD-DO dynamics in an ice-covered river in northern China. *Water Res.* **1987**, *21*, 247–251.

30. McBean, E.; Farquhar, G.; Kouwen, N. Predictions of ice-covered development in streams and its effect on dissolved oxygen modelling. *Can. J. Civ. Eng.* **1979**, *6*, 197–207. [CrossRef]

31. Demers, S.; Buffin-Bélanger, T.; Roy, A.G. Macroturbulent coherent structures in an ice-covered river flow using a pulse-coherent acoustic Doppler profiler. *Earth Surf. Process. Landf.* **2012**. [CrossRef]

32. Clark, S.P.; Peters, M.; Dow, K.; Malenchak, J.; Danielson, D. Investigating the effects of ice and bed roughness on the flow characteristics beneath a simulated partial ice cover. In Proceedings of the 23rd IAHR International Symposium on Ice, Ann Arbor, MI, USA, 31 May–3 June 2016.

33. Tsai, W.-F.; Ettema, R. Ice cover influence on transverse bed slopes in a curved alluvial channel. *J. Hydraul. Res.* **1994**, *32*, 561–581. [CrossRef]

34. Allard, G.; Buffin-Bélanger, T.; Bergeron, N.E. Fluvial and ice dynamics at a frazil-pool. *River Res. Appl.* **2011**. [CrossRef]

35. Ettema, R.; Daly, S.F. Sediment transport under ice. In *Cold Regions Research and Engineering Laboratory Report TR-04-20*; U.S. Army Engineer Research and Development Center: Hanover, NH, USA, 2004.

36. Sui, J.Y.; Wang, D.S.; Karney, B.W. Suspended sediment concentration and deformation of riverbed in a frazil jammed reach. *Can. J. Civ. Eng.* **2000**, *27*, 1120–1129. [CrossRef]

37. Ettema, R.; Zabilansky, L. Ice influences on channel stability: Insights from Missouri's Fort Peck reach. *J. Hydraul. Eng.* **2004**, *130*, 279–292. [CrossRef]

38. Tywonik, N.; Fowler, J.L. Winter measurements of suspended sediments. In Proceedings of the Banff International Symposium on the Role of Snow and Ice on Hydrology, Banff, AB, Canada, 1972.

39. Best, H.; McNamara, J.P.; Liberty, L. Association of ice and river channel morphology determined using ground-penetrating radar in the Kuparuk River, Alaska. *Arct. Antarct. Alp. Res.* **2005**, *37*, 157–162. [CrossRef]

40. Ettema, R. Review of alluvial-channel responses to river ice. *J. Cold Reg. Eng.* **2002**, *16*, 191–217. [CrossRef]

41. Beltaos, S.; Burrell, B.C. Suspended sediment concentrations in the Saint John River during ice breakup. In Proceedings of the Conference of the Canadian Society for Civil Engineering, London, ON, Canada, 7–10 June 2000.

42. Prowse, T.D. Suspended Sediment Concentration during River Ice Breakup. *Can. J. Civ. Eng.* **1993**, *20*, 872–875. [CrossRef]

43. Milburn, D.; Prowse, T.D. The effect of river-ice break-up on suspended sediment and select trace-element fluxes. *Nordic Hydrol.* **1996**, *27*, 69–84.

44. Gatto, L.W. Soil freeze-thaw effects on bank erodibility and stability. In *US Army Cold Regions Research and Engineering Laboratory Special Report 95-24*; USACE: Hanover, NH, USA, 1995.

45. Beltaos, S. Significance of javes in transporting suspended sediment during river ice breakup. In Proceedings of the 18th CHU-HS CRIPE Workshop on the Hydraulics of Ice Covered Rivers, Quebec City, QC, Canada, 18–20 August 2015.

46. Uunila, L.S. Effects of river ice on bank morphology and riparian vegetation along the Peace River, clayhurst to fort vermilion. In Proceedings of the 9th Workshop on River Ice, Fredericton, NB, Canada, 24–26 September 1997.

47. Hicks, F.E. Ice as the geomorphologic agent in an anastomosing river system. In Proceedings of the NHRI Workshop on Environmental Aspects of River Ice, National Hydrology Research Institute, Saskatoon, SK, Canada, 18-20 August 1993.

48. Beltaos, S.; Calkins, D.J.; Gatto, L.W.; Prowse, T.D.; Reedyk, S.; Scrimgeour, G.J.; Wilkins, S.P. Physical effect of river ice. In *Environmental Aspects of River Ice*; Prowse, T.D., Gridley, N.C., Eds.; Environmental Aspects of River Ice; National Hydrology Research Institute: Saskatoon, SK, Canada, 1993.

49. Kempema, E.W.; Reimnitz, E.; Clayton, J.R.; Payne, J.R. Interactions of Frazil and Anchor Ice with Sedimentary Particles in a Flume. *Cold Reg. Sci. Technol.* **1993**, *21*, 137–149. [CrossRef]

50. Kempema, E.W.; Ettema, R. Anchor ice rafting: Observations from the Laramie River. *River Res. Appl.* **2010**. [CrossRef]

51. Larsen, P.; Billfalk, L. Ice problems in Swedish hydro power operation. In Proceedings of the IAHR Symposium on Ice Problems, Delft, The Netherlands, 1978.

52. Martin, S. Frazil ice in rivers and oceans. *Annu. Rev. Fluid Mec.* **1981**, *13*, 379–397. [CrossRef]

53. Turcotte, B.; Morse, B.; Bergeron, N.E.; Roy, A.G. Sediment transport in ice-affected rivers. *J. Hydrol.* **2011**, *409*, 561–577. [CrossRef]

54. Turcotte, B.; Morse, B.; Anctil, F. Impacts of precipitation on the cryologic regime of stream channels. *Hydrol. Process.* **2012**, *26*, 2653–2662. [CrossRef]

55. Harper, P.P. Ecology of streams and high latitudes. In *Perspectives in Running Water Ecology*; Lock, M.A., Williams, D.D., Eds.; Plenum: New York, NY, USA, 1981; pp. 313–337.

56. Schallock, E.W.; Lotspeich, F.B. Low winter dissolved oxygen in some Alaskan rivers. In *Office of Ressources and Development*; U.S. Environmental Protection Agency: Corvallis, OR, USA, 1974.

57. Hicks, F.E.; Healy, D. Determining winter discharge based on modeling. *Can. J. Civ. Eng.* **2003**, *30*, 101–112. [CrossRef]

58. Morse, B.; Hamaï, K.; Choquette, Y. River discharge measurements using the velocity index method. In Proceedings of the 13th CRIPE Workshop on the Hydraulics of Ice Covered Rivers, Hanover, NH, USA, 15–16 September 2005.

59. Andrishak, R.; Hicks, F. Working safely on river ice. In Proceedings of the 18th CGU-HS CRIPE Workshop on the Hydraulics of Ice Covered Rivers, Quebec City, QC, Canada, 18–20 August 2015.

60. Pigeon, F.; Leclerc, M.; Morse, B.; Turcotte, B. Breakup 2014 on the Montmorency River. In Proceedings of the 18th CGU-HS CRIPE Workshop on the Hydraulics of Ice Covered Rivers, Quebec City, QC, Canada, 18–20 August 2015.

61. Ribeiro, K.H.; Favaretto, N.; Dieckow, J.; De Paula Souza, L.C.; Gomez Minella, J.P.; De Almeida, L.; Ribeiro Ramos, M. Quality of Surface Water Related to Land Use: A Case Study in a Catchment with Small Farms and Intensive Vegetable Crop Production in Southern Brazil. *Rev. Bras. Cienc. Solo* **2014**, *38*, 656–668. [CrossRef]

water MDPI

Article

Impacts of Climate Change on the Water Quality of a Regulated Prairie River

Nasim Hosseini [1],*, Jacinda Johnston [2] and Karl-Erich Lindenschmidt [1]

[1]   Global Institute for Water Security, University of Saskatchewan, 11 Innovation Boulevard, Saskatoon, SK S7N 3H5, Canada; karl-erich.lindenschmidt@usask.ca

[2]   Graduate Studies, University of Saskatchewan, Saskatoon, SK S7N 5C5, Canada; jacinda.johnston@usask.ca

*   Correspondence: nasim.hosseini@usask.ca; Tel.: +1-306-966-2825

Academic Editor: Richard Skeffington
Received: 29 November 2016; Accepted: 7 March 2017; Published: 10 March 2017

Abstract: Flows along the upper Qu'Appelle River are expected to increase in the future via increased discharge from Lake Diefenbaker to meet the demands of increased agricultural and industrial activity and population growth in southern Saskatchewan. This increased discharge and increased air temperature due to climate change are both expected to have an impact on the water quality of the river. The Water Quality Analysis Simulation Program (WASP7) was used to model current and future water quality of the upper Qu'Appelle River. The model was calibrated and validated to characterize the current state of the water quality of the river. The model was then used to predict water quality [nutrient (nitrogen and phosphorus) concentrations and oxygen dynamics] for the years 2050–2055 and 2080–2085. The modelling results indicate that global warming will result in a decrease in ice thickness, a shorter ice cover period, and decreased nutrient concentrations in 2050 or 2080 relative to 2010, with a greater decrease of nutrient concentrations in open water. In contrast to the effect of warmer water temperatures, increased flow through water management may cause increases in ammonium, nitrate, and dissolved oxygen concentrations and decreases in orthophosphate concentrations in summer.

Keywords: water quality model; climate change; WASP7; surface water; upper Qu'Appelle River; increased discharge

---

1. Introduction

Water demand in southern Saskatchewan has been increasing and is projected to continue to increase because of new and expanding mining (e.g., potash) industries, increased agricultural irrigation, and subsequent population increases [1]. Parsons et al. [1] undertook a study (upper Qu'Appelle Water Security Analysis, 2012) to assess the current water demands on the upper Qu'Appelle River (see Figure 1 for a map of the area) and what the future demands for water may be. Their findings indicate that an increased level of flow from Lake Diefenbaker into the upper Qu'Appelle River will be needed to accommodate the growing water demand in the region. This input of good quality water into the upper Qu'Appelle River would improve the water quality of the lower Qu'Appelle River downstream of Buffalo Pound Lake. However, more thorough assessments of how increased flow in the upper Qu'Appelle River will affect the overall water quality status of the river are necessary.

Climate change is one important factor that is known to affect ecosystems. The main impact of climate change on water quality is attributed to changing air temperature and hydrology [2]. Water temperature is directly affected by ambient air temperature [3] and is expected to increase as a result of global warming [4]. Variations in water temperature govern physico-chemical equilibriums (e.g., nitrification, mineralization of organic matter, etc.) in rivers and hence change transport and concentration of contaminants [2,3]. Increases in water temperature result in reduced oxygen solubility thus reducing dissolved oxygen (DO) concentrations and DO concentrations at which saturations occurs. Reduced DO concentrations will have an impact on the duration and intensity of algal blooms [3,4].

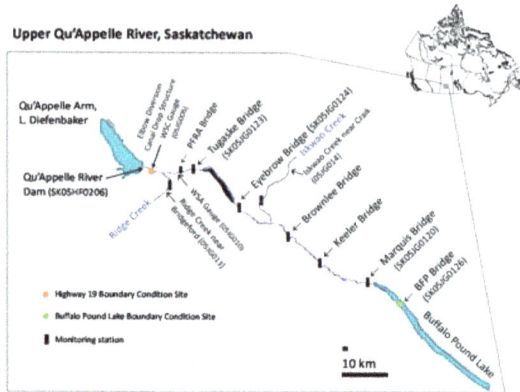

Figure 1. Map of the upper Qu'Appelle River. The lower Qu'Appelle River flows out of Buffalo Pound Lake.

Climate change is expected to alter the availability, seasonality and variability of flow in rivers [5–8]. These hydrological impacts of climate change are particularly pronounced in glacier-fed rivers [9,10]. A study on the hydrological impact of climate change on a river basin in Alberta, Canada [8], shows that high flow events due to climate change would be of a greater concern than low flow events in this region. Several studies point out that higher water temperatures and lower flow rates during summer may cause impairment to water quality in rivers [6,7]. For example, a review by Whitehead et al. [3] outlines how lower flow in summer may result in increases in phosphorus concentrations and biological oxygen demand (BOD) and decreases in DO concentrations in rivers [3] which, in turn, can lead to accelerated algal growth [4]. Under reduced flow in summer, ammonium concentrations decrease due to an increase in the nitrification rate with consequent increase in nitrate concentrations [3].

The flow in the upper Qu'Appelle River is regulated by the Qu'Appelle River Dam on the northeast arm of Lake Diefenbaker (Figure 1). Unlike many studies that have shown the impact of low flow due to climate change on the water quality of surface waters, this study was undertaken to assess the water quality (nutrient and dissolved oxygen concentrations) of the upper Qu'Appelle River due to increased air temperature resulting from climate change and increased flows to meet future water demand. The WASP7 program was used to characterize the water quality of the river and gain insight into how future increases in discharge and air temperature may affect water quality parameters. The periods of open-water (May through October) and ice cover (November through April) were compared to see how these changes would affect the water quality of the river seasonally. The results from this study provide valuable information on how water quality of other river systems throughout the world may change under the influence of increasing population and economic growth.

2. Materials and Methods

2.1. Water Quality Model

Descriptions of the WASP model are provided in the WASP7 manual and several other studies [11–15]. The WASP7 program was developed by United States Environmental Protection

Agency (USEPA) and has been improved from the original version, allowing greater flexibility to model water quality of different water systems (e.g., rivers, lakes, estuaries, etc.) [16–18]. WASP7 was designed to aid water resource management decisions by interpreting and predicting the responses of water quality to various factors such as natural phenomena and anthropogenic pollution. WASP7 was used in this study due to its robustness in simulation, high credibility acquired from many other successful applications in the past, and its application to other river systems in other arid and semi-arid river systems on the Canadian Prairies (e.g., South Saskatchewan River) [13,15].

The WASP7 model consists of several kinetic modules including sediment transport, eutrophication, toxicant transformations and fate, mercury methylation, and heat exchange. This study used the eutrophication module EUTRO and the heat module HEAT. The EUTRO module incorporates eutrophication parameters into the model, including several mass balance equations, to simulate nutrient transport and transformations, as well as phytoplankton and DO dynamics [17]. The HEAT module allows the simulation of processes influencing water temperature, such as surface heat exchange and ice formation and ablation [19]. The HEAT module simulates heat transfer based on both the conservation of water volume and heat. The processes of heat exchange include those between the atmosphere and the water column, and the water column and the bottom sediment and are based upon the U.S. Army Corps of Engineers CE-QUAL-W2 model formulations [20].

2.2. Model Set-Up

Water quality data were collected from several monitoring stations along the river, with locations shown in Figure 1. Water quality parameters used in this study include water temperature, dissolved oxygen (DO), orthophosphate (PO_4-P), nitrate (NO_3-N), ammonium (NH_4-N), dissolved organic phosphate (DOP), dissolved organic nitrogen (DON), and phytoplankton chlorophyll-α (Chl a). DON and DOP were not available in the database, therefore, DON was calculated as dissolved Kjeldahk nitrogen (DKN) minus NH_4-N, and similarly, DOP was calculated as total dissolved phosphorus (TDP) minus PO_4-P, as suggested by Tufford and McKellar [21].

The upper Qu'Appelle River was discretized into 165 longitudinal segments ranging in length from 600 to 800 m. The river morphology was surveyed at approximately 770 locations along the 97 km stretch, from the Qu'Appelle Dam at Lake Diefenbaker to Buffalo Pound Lake.

The average rate of discharge for the upper Qu'Appelle River is approximately 2.2 m^3/s (Jan 2010–December 2015), which is controlled at the Qu'Appelle River dam. Two naturally-flowing tributaries (Ridge Creek and Iskwao Creek) augment the river's flows. Daily flow rates were obtained from the Water Survey of Canada (WSC) gauges at the Elbow Diversion Canal (05JG006) and at Ridge Creek near Bridgeford (05JG013) (Figure 1). The flow rates are available from the WSC website (https://www.ec.gc.ca/rhc-wsc). Flow from Iskwao Creek was based on the seasonal historical flow data recorded at the WSC gauge at Iskwao Creek near Craik (05JG014). The flows for segments were simulated using 1-D kinematic wave routing [11]. Figure 2 shows the 2012–2015 flows for the upper Qu'Appelle River at 05JG006 and indicates when ice cover periods occurred on the river. Streamflow data attained from WSC identified with a "B" indicated estimated streamflow values under ice cover. During the 2012–2013 winter, a flow test was carried out by Lindenschmidt [22] on the upper Qu'Appelle River to determine the conveyance capacity of the river at higher flows under ice and how this flow increase should be regulated so that the risk of ice jamming at freeze-up is minimized [22]. The discharge was successfully increased in this study from 2.6 m^3/s in mid-November 2012, to 4 m^3/s by the end of January 2013, and then drastically reduced to about 0.8 m^3/s.

The Qu'Appelle River does not have any significant point loading sources (e.g., sewage or industrial effluent) that need to be accounted for; however, there are loadings from Ridge Creek and Iskwao Creek for which data were provided by the Saskatchewan Water Security Agency. The data were measured either biweekly or monthly and were available from April 2013 to December 2015. Also, there may be non-point loadings from agriculture and mining (potash) due to runoff. Extensive macrophyte growth was observed in the upstream portion of the river stretch between the PFRA Bridge and the Tugaske Bridge [23]. The large volume of macrophyte biomass affects aquatic ecosystems by emitting sufficient amounts of oxygen into the water to reach supersaturated concentrations [24,25]. In the Qu'Appelle River, this occurs from June to September, and sometimes into November, with July and August having the highest DO loadings [26]. The oxygen production from macrophyte biomass was considered by specifying in the model a dissolved oxygen loading of 350 kg/day between the PFRA and Tugaske bridges.

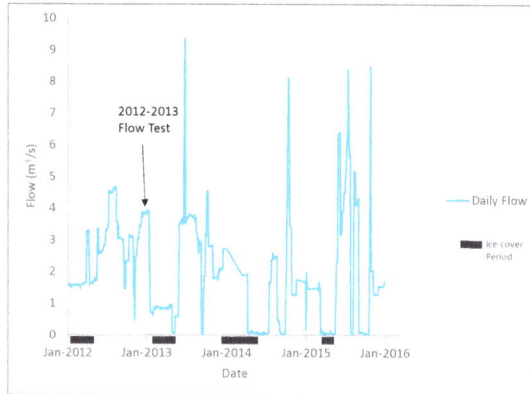

Figure 2. Upper Qu'Appelle River flow between 2012 and 2015 with indicated ice cover periods and 2012–2013 winter flow test.

Nutrient fluxes are important for lentic water under anaerobic conditions. The Qu'Appelle River is shallow (average depth is about 0.6 m) and due to aerobic conditions throughout the whole water column (minimum observed DO concentration was 4 mg/L), nutrient fluxes play a less important role and were not taken into account in this study.

The model boundaries were set at the most upstream and downstream monitoring stations SK05HF026 (Highway #19) and SK05JG0126 (Buffalo Pound Lake) (Figure 1). Figure 3 shows the range of concentrations occurring for water quality parameters NH_4-N, NO_3-N, DON, PO_4-P, OP, Chl a, and DO for the upstream boundary conditions used in the WASP7 model. These concentrations were measured during the 2012–2013 winter flow test and are summarized in Lindenschmidt [22].

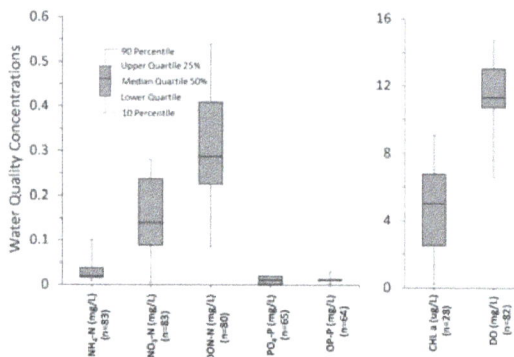

Figure 3. Range of concentrations occurring for water quality parameters at the upstream boundary condition used in the WASP7 model for the upper Qu'Appelle River. *n* indicates the number of observations.

Calibration and validation of the model were based on measured DO, NO_3-N, NH_4-N, PO_4-P, and Chl a concentrations that were collected at the monitoring stations for four years (2012–2015). Observed DON and DOP were not available in the database and were not considered in the calibration and validation. Note that only 15 measured concentrations were available for Chl a at these stations (Table 1). Calibration was obtained using the Dynamically Dimensioned Search (DDS) algorithm [27] through the OSTRICH (Optimization Software Tool for Research In Computational Heuristics) interface [28]. Water quality parameters were calibrated using 2014–2015 data from all the monitoring stations and then validated using 2012–2013 data. Table 1 lists the monitoring stations and the number of observations collected at each station for each parameter that were used for calibration and validation in this study. The water quality parameters were mostly collected bi-weekly at the Tugaske and Marquise bridges but not frequently at the other stations.

Table 1. Water quality monitoring stations and the number of observations used for calibration and validation.

Station Names	Number of Observations from January 2012 to December 2015				
	DO	NH_4-N	NO_3-N	PO_4-P	Chl a
Qu'Appelle River at Marquis Bridge	79	80	80	39	7
Qu'Appelle River at Keeler	5 *	5	5	5	5
Qu'Appelle River at Brownlee	53	2	2	2	2
Qu'Appelle River below Eyebrown Bridge	4	6	6	2	0
Qu'Appelle River at Tugaske Bridge	64	64	64	31	1

Note: * January 2013 to April 2013.

2.3. Climate Change Scenarios

The climate change scenarios used in this study were retrieved from the Pacific Climate Impacts Consortium (PCIC), University of Victoria (Pacific Climate Impacts Consortium, 2014). PCIC offers downscaled climate scenarios for the simulated period of 1950–2100 with a spatial resolution of 300 arc-seconds (about 10 km). The advantage of these downscaled scenarios over the North American Regional Climate Change Assessment Program (NARCCAP) and Atmosphere-Ocean Global Circulation Models (AOGCMs) is because of their better resolution. These data represent the average values of the region rather than a point quantity [29]. The downscaling scenarios stem from 12 climate models, each for three different greenhouse gas emission scenarios. The outputs include daily minimum and maximum air temperature and precipitation, which are based on Global Climate Model (GCM) projections [30] and historical daily gridded climate data for Canada [31,32].

For this study, four climate models (CanESM2-r1, GFDL-ESM2G-r1, HadGEM2-ES-r1, and MPI-ESM-LR-r3) from three Representative Concentration Pathways (RCPs) emissions scenarios (2.6, 4.5, and 8.5) were selected.

2.4. Water Managment

Increase in water demand may require a three-fold increase in the flow rate from approximately $2\ \text{m}^3/\text{s}$ to $6\ \text{m}^3/\text{s}$ during winter months (personal communication—with Saskatchewan Water Security Agency and see [1,22,23]). Great measures have been taken to test the flow capacity of the river in winter up to $4\ \text{m}^3/\text{s}$ [22] and a numerical model was used to test the ability of the river to accommodate a winter flow of up to $6\ \text{m}^3/\text{s}$ without ice jamming or overbank flooding [23]. The flow increase during the summer months is limited by the maximum conveyance capacity of the channel, which is $14\ \text{m}^3/\text{s}$ [22]. WASP7 allowed us to characterize the water quality of the river at these high flow rates ($6\ \text{m}^3/\text{s}$ in winter and $14\ \text{m}^3/\text{s}$ in summer) for more current conditions (2012–2015).

3. Results

3.1. Climate Change

3.1.1. Water Temperature and Ice Cover

The HEAT module was applied to our case study and the results were compared with the measured water temperature values at four locations along the river for four years (2012–2015) and ice thickness at Tugaske Bridge. The results are presented in Figures 4 and 5 which show that the model works well when simulating water temperature (with $R^2 = 0.88$) and ice thickness (with $R^2 = 0.93$).

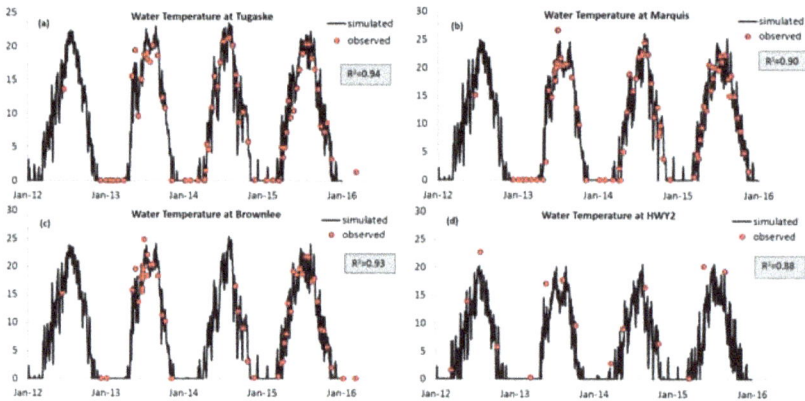

Figure 4. Simulated and measured water temperature at (**a**) Tugaske; (**b**) Marquis; (**c**) Brownlee; and (**d**) HWY#2.

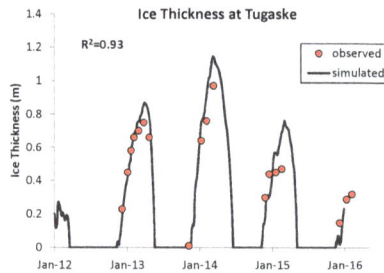

Figure 5. Daily simulated ice thickness at the Tugaske Bridge.

A local sensitivity analysis was applied to estimate the sensitivity of the water temperature and ice thickness to the forcing data, which included wind speed, cloud cover, air temperature, dew point, flow, and light extinction. The local sensitivity analysis is an effective and widely used method to determine the relative importance of parameters, although it does not account for parameter interactions [33–35]. The main advantage of this method is that it requires few model runs whereas global sensitivity analysis is computationally more expensive. Each model input is increased by a defined percentage (here 10%) while holding the other model inputs constant. The sensitivity ε was then assessed using

$$\varepsilon = \frac{\sqrt{\frac{1}{n}\sum_{i=1}^{n}(O_x - O_{base})^2}}{\Delta P} \qquad (1)$$

where O_{base} is the simulated result using the base input, O_x is the simulated result using the perturbed input, ΔP is the difference between base and perturbed input values (10% of the base parameter), and n is the total number of observations. The sensitivity analysis results indicated that air temperature has the most impact on both water temperature and ice thicknesses. Henceforth, for our modelling exercise when using climate models, daily air temperature from PCIC and the average daily historical flow discharge (from 2000 to 2016) seemed sufficient for our assessment. The other forcing data were excluded due to the lack of sensitivity of model results (water temperature and ice thickness) to these factors.

We ran the calibrated/validated model with the downscaled climate data for the same time frame (2012–2015) to verify similarity between model output from sampled data and model output from historical down-scaled climate data. The simulation results followed a similar trend as that shown in Figure 4a. There was good agreement between observed and simulated water temperature using daily maximum and minimum temperature from the CanESM2 RCP 2.6 model which resulted in R^2 values of 0.88 and 0.86, respectively. However, in winter 2012–2013, there was some lag between observed and simulated water temperatures.

Ice cover data were sampled with a coarse temporal resolution; therefore, ice cover duration was estimated based on the start and end date of 0 °C water temperature (i.e., the first and last day when the water temperature was 0 °C). Figure 6 illustrates the measured and simulated ice cover duration as well as the average of simulated values using the aforementioned four climate model scenarios. Ice cover periods were over-estimated by three weeks with the calibrated WASP7 model and were underestimated by about 2 weeks with the climate change models, due to the ice cover break up occurring too early.

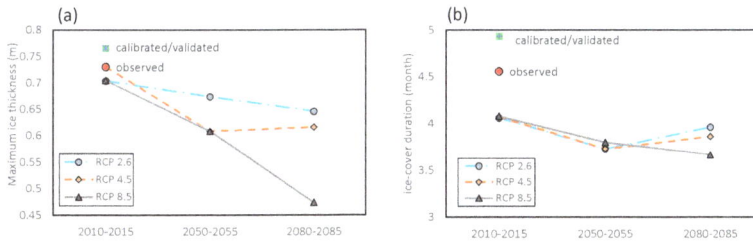

Figure 6. Climate change impact on (**a**) maximum ice thickness and (**b**) ice-on duration.

In Figure 6, the 5-year averaged annual maximum ice thickness and 5-year averaged mean ice-on duration are compared to the observed data. The maximum ice thickness will likely decrease considerably (up to 32.7%) as a result of global warming. The results suggest that the ice cover period will be shorter by up to 12 days in the future, though this is less than the error due to bias. The bias is not due to time of freeze-up but rather, the breakup being simulated early as a result of variability in flow discharge during the freshet. However, this only impacts the simulation of the aforementioned water quality parameters in March and April. Moreover, the water temperature data were collected bi-weekly to monthly rather than daily which may cause some degree of uncertainty in our estimated observed ice cover period.

3.1.2. Water Quality

Model calibration output for all parameters is shown in Figures 7 and 8. Calibration (2014–2015) produced a good fit to observed water quality data at the Tugaske and Marquis bridges, which was also the case in the validation (2012–2013). Comparing the simulated concentrations to the measured data yielded R^2 values ranging from 0.38 to 0.95. The low correlation coefficient value of 0.38 was found for DO at Marquis, although 90 percent of the data yielded a higher R^2 value of 0.78. There were little observed data to compare results for Chl a. Overall, calibration and validation were deemed successful when using the model to predict changes in water quality due to climate change.

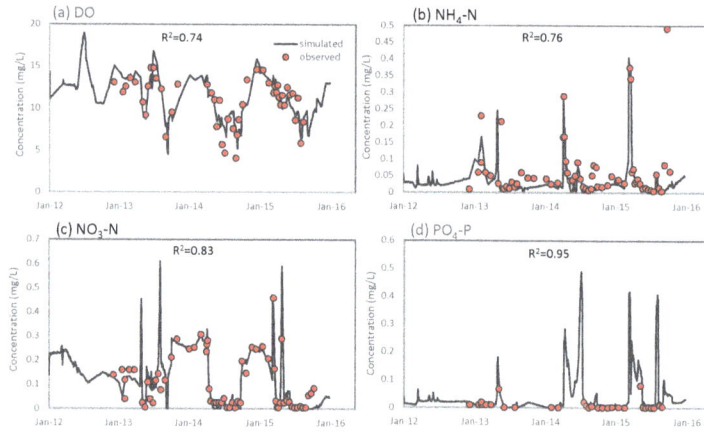

Figure 7. Calibration (2014–2015) and validation (2012–2013) results for simulated versus observed water quality data (**a**) DO (dissolved oxygen); (**b**) NH$_4$-N; (**c**) NO$_3$-N; and (**d**) PO$_4$-P at Tugaske Bridge.

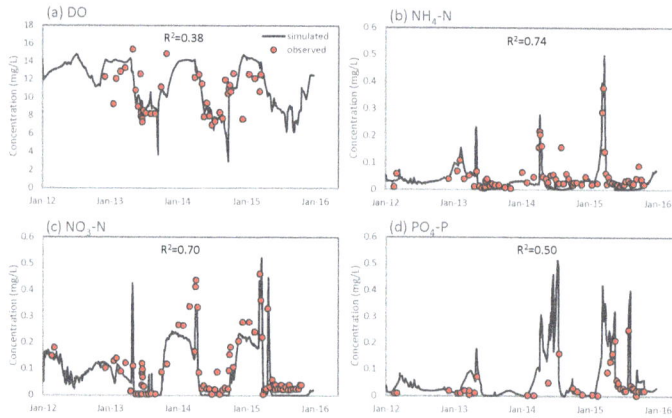

Figure 8. Calibration (2014–2015) and validation (2012–2013) results for simulated versus observed water quality data (**a**) DO; (**b**) NH$_4$-N; (**c**) NO$_3$-N; and (**d**) PO$_4$-P at Marquis Bridge.

Figure 9 compares the measured water quality concentrations and the outputs from the calibrated/validated WASP7 model and from climate change scenarios (i.e., the average of the outputs from all the climate change scenario models). The climate change scenario models were based on the simulated water temperature and ice cover period using historical down-scaled climate data. The 16-year averaged daily flow data at 05JG006 and 05JG013 flow gauges (Figure 1) were used for the climate change model. The 16-year averaged monthly historical concentrations of the water quality parameters including NH$_4$-N, NO$_3$-N, DON, PO$_4$-P, OP, Chl a, and DO were available at the SK05HF026 monitoring station (Figure 1) and 5-year averaged monthly water quality parameter concentrations (NH$_4$-N, NO$_3$-N, DON, PO$_4$-P, OP, and DO) were available at the Ridge Creek and Iskwao Creek monitoring stations (Figure 1) and were used for the climate change model. The mean simulated concentrations matched the measured mean concentrations relatively well.

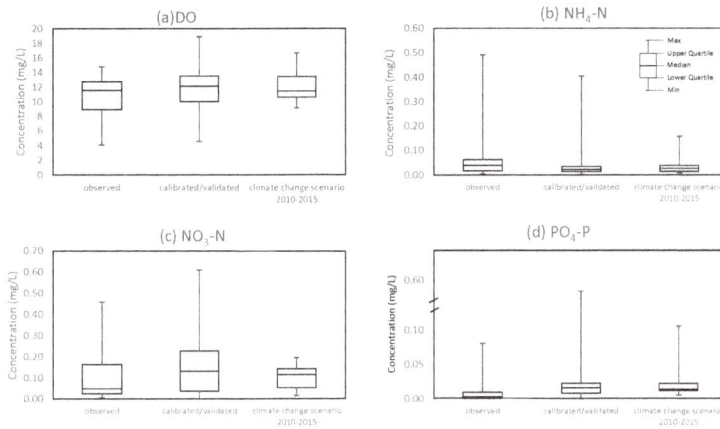

Figure 9. Measure and simulated water quality parameter concentrations using calibrated/validated model and climate change scenario models (**a**) DO; (**b**) NH$_4$-N; (**c**) NO$_3$-N; and (**d**) PO$_4$-P.

Figures 10 and 11 show the climate change impact on water quality parameters for 2050–2055 and 2080–2085. The outputs from the RCP 8.5 scenario, which represents the highest greenhouse gas emissions, indicate the largest change in water quality of the system. As might be expected, all the water quality parameters in the 2080–2085 period showed a bigger change than in the 2050–2055 period, except for a few cases (e.g., NH$_4$ in September using the RCP 2.6 scenario).

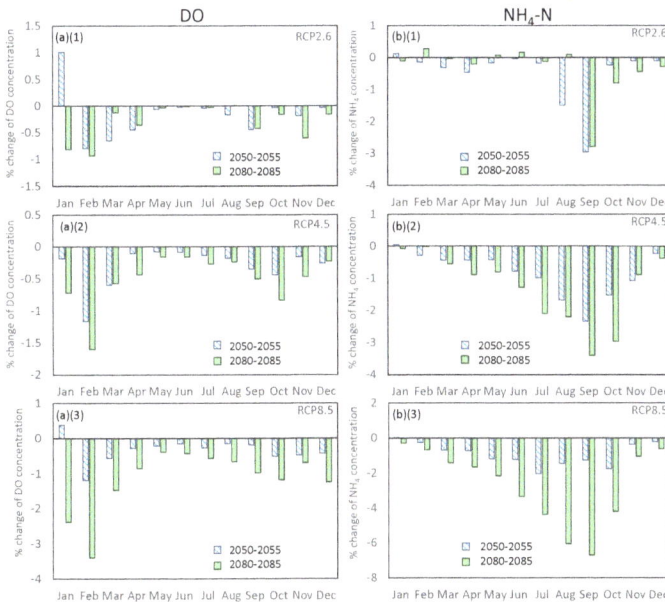

Figure 10. Percent change in water quality parameter concentrations for DO and NH$_4$-N, due to climate change scenarios (Representative Concentration Pathway) (**1**) RCP 2.6; (**2**) RCP 4.5; and (**3**) RCP 8.5 at the Tugaske location.

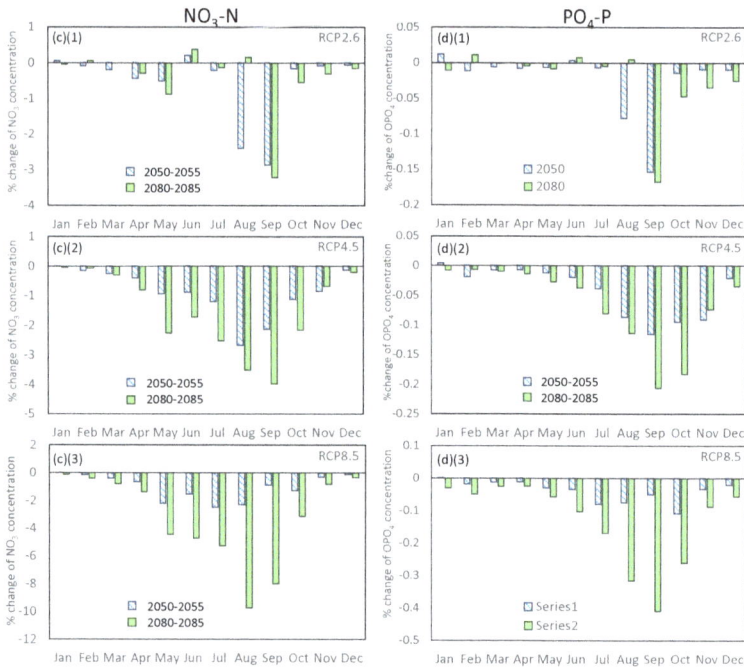

Figure 11. Percent change in water quality parameter concentrations for NO₃-N, and PO₄-P due to climate change scenarios (**1**) RCP 2.6; (**2**) RCP 4.5; and (**3**) RCP 8.5 at the Tugaske location.

In general, the mean monthly DO concentrations decrease due to climate change. The reduction in the DO concentrations is not significant in summer while it is pronounced during the ice cover. A minor increase in DO concentrations was predicted in January (about 1%).

Nutrient concentrations were expected to decrease with air temperature increase, although, the magnitude of the changes depends on the model scenarios and seasons. The greatest deviation was observed for nitrogen species with the maximum of 6.71% for NH_4-N and 9.79% for NO_3-N. In contrast, only a slight change was estimated for phosphorus with the maximum of 0.42%. The results suggest that the highest decrease in the mean monthly concentrations of nitrogen and phosphorus would occur in late summer (September and August).

3.2. Water Management

When flow rates during the ice cover season (November to April) were increased to 6 m³/s and to 14 m³/s in summer, the model output was significantly different from the original, calibrated/validated results (Figure 12). The results revealed that parameters are more sensitive to changes in flow than climate change. DO concentrations are increased in the ice cover periods (December–February), likely due to reaeration of the river, since flows are increased from the Qu'Appelle River Dam, and due to open water, allowing the exchange of gases between the river and atmosphere. However, DO is expected to decline in early spring and late fall. NH_4-N and NO_3-N generally increase in summer and fluctuate in winter. NH_4-N decreased by about 5% in the winter periods when winter flow was increased, while NO_3-N slightly increased (about 1.5%–9%) through the winter and into early spring. The phosphate concentrations are also expected to decrease in summer with the greater change in early spring during the freshet.

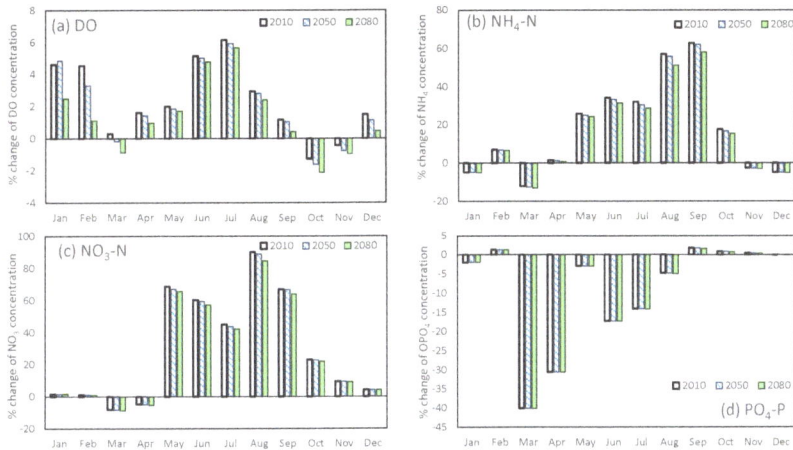

Figure 12. Changes in water quality parameters (**a**) DO; (**b**) NH$_4$-N; (**c**) NO$_3$-N; and (**d**) PO$_4$-P due to flow increase for the RCP 8.5 scenario at the Tugaske location.

4. Discussion

4.1. Impact of Increased Air Temperature on Water Quality

Many studies have concluded that water temperature will increase because of global warming [2–4,36]. Our modelling results suggest that water temperature will increase between 2010 and either 2050 or 2080 with a corresponding decrease in ice thickness due to warmer air temperatures. Ice cover duration may be up to 12 days shorter by the 2080s but this result is somewhat questionable due to uncertainty in our simulations (Figure 6b). The uncertainty in ice cover duration in this study is mainly attributed to the average flow rate used in our models. Ice breakup is strongly affected by when spring runoff occurs [37]. Historical trend analyses of river and lake ice covers by Magnuson et al. [37] showed similar results of later freeze-up of 5.8 days and earlier break-up of 6.5 days per 100 years.

A reduction in the concentrations of NH$_4$-N, NO$_3$-N, and PO$_4$-P was predicted as a result of warmer water temperatures due to increased air temperatures as a consequences of global warming (Figures 10 and 11). The most significant reduction in the nutrient concentrations were projected during the open-water period relative to the ice covered period. A similar study by Alam et al. [14] also concluded that nutrient concentrations in rivers decline when air temperature increases. The decreases in nutrient concentrations may be a product of increased phytoplankton growth, such that more algae are utilizing more nutrients, and therefore a lower concentration of nutrients remains in the water [14]. A declining trend in total phosphorus concentrations in winter during two decades (1992–2011) was found by Zhang et al. [38] in a shallow reservoir in China, which could be attributed to an increase in grazer abundance. Reduced ice cover duration and ice thickness as presented in the current study, would act favorably towards phytoplankton growth because of better light conditions in winter [39]. Dissolved oxygen concentrations were also projected in this study to decrease in all months, except in January when DO concentrations were predicted to increase slightly (Figure 10). Cox and Whitehead [40] also predicted a reduction in DO concentrations in the River Thames by the 2080s because of reduced saturation concentrations of DO and increased biological oxygen demand.

In contrast to rivers, some climate change studies on the water quality of lakes show that higher water temperatures and increases in oxygen demand promote the release of nutrients from sediments, resulting in more nutrient enrichment of the water column [2,4,39]. The upper Qu'Appelle River is shallow, hence the anoxic conditions which could result in remobilization of nutrients is not a major issue. However, we should note that as the upper Qu'Appelle River flows from Lake Diefenbaker

(Figure 1), impact of higher water temperatures on water quality of the lake would consequently affect the water quality of the river system. Such impact was not considered in this current study.

4.2. Impact of Increased Flow

The modelling results are based on water quality parameter concentrations predicted at Tugaske Bridge, at the point where nutrient concentrations and loadings are due to concentrations from Lake Diefenbaker and Ridge Creek. In summer when flow increased to 14 m^3/s from the Qu'Appelle River Dam, the increased flow led to an increase in NH$_4$-N and NO$_3$-N concentrations (Figure 12). This likely occurs because with increased flow in the upper Qu'Appelle River from the Dam relative to its base flow, the impact of NO$_3$-N and NH$_4$-N inputs from Ridge Creek was decreased. In contrast, the concentrations of PO$_4$-P decreased because of dilution (Figure 12). The contribution of NH$_4$-N, NO$_3$-N, and PO$_4$-P loads (kg/day) from the Ridge Creek in March and April are on average about 45%, 38%, and 73% of the loading at Tugaske Bridge, respectively. The increase in NO$_3$-N and NH$_4$-N concentrations in the summer period may also be due to higher nitrification and mineralization rates. Increases in nutrient concentrations have been reported in several studies as the result of drought in summer caused by climate change [6,41,42]. For example, increased NH$_4$-N concentrations in a catchment in central Greece were predicted as a result of a reduction in stream dilution capacity due to climate change [6]. Increased flow in the present study led to an increase in DO concentrations, which may be in response to the increased reaeration rate caused by the higher water velocity and flow rate.

Nutrient concentrations fluctuated in winter, when flow from the Qu'Appelle River Dam was increased from base flow to 6 m^3/s. The increase in NO$_3$-N concentrations and the decrease in NH$_4$-N in November–January may be due to the increase in DO concentrations which would cause NH$_4$-N oxidation to NO$_3$-N [43]. A small reduction in DO concentrations (about 2%) in October and November may be the result of macrophyte die-off. From the results of this study, it can be inferred that biological activities play a more important role in winter when flow is lower than in summer.

4.3. Future Studies

We assessed the potential impact of increased air temperature, caused by climate change, and increased flow rate, due to flow regulation, on the water quality of the upper Qu'Appelle River. Other factors such as land use change, water quality degradation of Lake Diefenbaker, and climate change impact on catchment nutrient loadings were not considered in this study. Other studies have assessed the impact of climate change on land use change and its consequences on nutrient loadings (e.g., [38,42,44–46]. Higher nitrogen and phosphorus loadings to the river may be expected due to increased precipitation caused by global warming especially from agricultural catchments [45,46]. Further studies on the Qu'Appelle River to consider such impacts on the long-term sustainability and security of this water source in Southern Saskatchewan would be beneficial.

Algal dynamics were not considered in this study due to the lack of sufficient measured data for Chl a for validation of the model results. To simulate algal dynamics in the river system, high-frequency and continuous sampling of Chl a would be necessary.

5. Conclusions

In this study, we assessed the future potential impacts of increased air temperature due to climate change and increased flow via water management on the concentrations of nutrients and dissolved oxygen in the upper Qu'Appelle River using the WASP7 model and PCIC climate change models. An important outcome of this study was to develop prediction capacity to assess how changes in flow management and climate change are anticipated to affect downstream changes in water quality. The results show that water quality parameters are highly sensitive to increased flow and air temperature. Warmer water temperatures caused a reduction in the concentrations of nutrients, with a greater decrease in the open water condition and a lower decrease in the ice cover condition.

Dissolved oxygen concentrations were predicted to decrease throughout the year except in January when DO concentrations will slightly increase. Based on our study, water quality parameters are more sensitive to flow changes than to climate warming. In summer, increased flow may cause an increase in NH_4-N and NO_3-N concentrations and a reduction in PO_4-P concentrations. Increased meso-eutrophication of Lake Diefenbaker due to climate change would have further impact on the water quality of the upper Qu'Appelle River.

From a water management perspective, the changes in the flow and nutrient regimes of the upper Qu'Appelle River will have implications for water quality in the downstream reservoir Buffalo Pound Lake, a key drinking water source for southern Saskatchewan. Our results indicate that a shift in the nutrient regime may occur primarily due to changes in flows, not climate change. Higher discharges can be brought about by the conveyance of more water from Lake Diefenbaker to Buffalo Pound Lake via the upper Qu'Appelle River. Since Lake Diefenbaker is phosphorus limited, larger transfers of its water downstream may increase the phosphorus limitation in Buffalo Pound Lake. The freshet also has a diluting impact on phosphorus loading in the system. Nitrogen concentrations follow an increasing trend with higher flows, particularly during the summer months due to the high loading from Lake Diefenbaker, which could ultimately enrich Buffalo Pound Lake with additional nitrogen. These conclusions are based on the assumption that the nutrient ratio in Lake Diefenbaker and the nutrient loading from the surrounding landscape in the upper Qu'Appelle Valley will not change substantially in the future. Further study is required to investigate the impacts of land-use changes and changes in the trophic status of Lake Diefenbaker on the upper Qu'Appelle River and ultimately on the aquatic ecosystem of Buffalo Pound Lake.

Acknowledgments: This research was funded by the Canada Excellence Research Chair (CERC) at the University of Saskatchewan's Global Institute for Water Security. The authors acknowledge the Saskatchewan Water Security Agency and the Saskatchewan Ministry of Environment for providing data used in this study.

Author Contributions: Nasim Hosseini, Jacinda Johnston and Karl-Erich Lindenschmidt conceived the idea and designed the modelling experiments. Nasim Hosseini conducted the simulations and processed the results. Nasim Hosseini and Karl-Erich Lindenschmidt analyzed the results. Nasim Hosseini and Jacinda Johnston wrote the initial draft, and Karl-Erich Lindenschmidt contributed to the writing.

Conflicts of Interest: The authors declare no conflict of interest.

References

1. Parsons, G.F.; Thorp, T.; Kulshreshtha, S.; Gates, C. *Upper Qu'Appelle Water Supply Project: Economic Impact Sensitivity Analysis*; Springer: Saskatchewan, SK, Canada, 2012; p. 113.
2. Delpla, I.; Jung, A.V.; Baures, E.; Clement, M.; Thomas, O. Impacts of climate change on surface water quality in relation to drinking water production. *Environ. Int.* **2009**, *35*, 1225–1233. [CrossRef] [PubMed]
3. Whitehead, P.G.; Wilby, R.L.; Battarbee, R.W.; Kernan, M.; Wade, A.J. A review of the potential impacts of climate change on surface water quality. *Hydrol. Sci. J.* **2009**, *54*, 101–123. [CrossRef]
4. Komatsu, E.; Fukushima, T.; Harasawa, H. A modeling approach to forecast the effect of long-term climate change on lake water quality. *Ecol. Model.* **2007**, *209*, 351–366. [CrossRef]
5. Whitehead, P.G.; Barbour, E.; Futter, M.N.; Sarkar, S.; Rodda, H.; Caesar, J.; Butterfield, D.; Jin, L.; Sinha, R.; Nicholls, R.; et al. Impacts of climate change and socio-economic scenarios on flow and water quality of the Ganges, Brahmaputra and Meghna (GBM) river systems: Low flow and flood statistics. *Environ. Sci. Process. Impacts* **2015**, *17*, 1057–1069. [CrossRef] [PubMed]
6. Mimikou, M.A.; Baltas, E.; Varanou, E.; Pantazis, K. Regional impacts of climate change on water resources quantity and quality indicators. *J. Hydrol.* **2000**, *234*, 95–109. [CrossRef]
7. Van Vliet, M.T.H.; Franssen, W.H.P.; Yearsley, J.R.; Ludwig, F.; Haddeland, I.; Lettenmaier, D.P.; Kabat, P. Global river discharge and water temperature under climate change. *Glob. Environ. Chang.* **2013**, *23*, 450–464. [CrossRef]
8. Kienzle, S.W.; Nemeth, M.W.; Byrne, J.M.; MacDonald, R.J. Simulating the hydrological impacts of climate change in the upper North Saskatchewan River basin, Alberta, Canada. *J. Hydrol.* **2012**, *412*, 76–89. [CrossRef]

9. Jasper, K.; Calanca, P.; Gyalistras, D.; Fuhrer, J. Differential imapcts of climate change on the hydrology of two alpine river basins. *Clim. Res.* **2004**, *26*, 113–129. [CrossRef]

10. Mote, P.W. Climate-driven variability and trends in mountain snowpack in western North America. *J. Clim.* **2006**, *19*, 6209–6220. [CrossRef]

11. Wool, T.A.; Ambrose, R.B.; Martin, J.L.; Comer, E.A. *Water Quality Analysis Simulation Program (WASP) Version 6.0 Manual. Draft User's Manual*; United States Environmental Protection Agency: Atlanta, GA, USA, 2006.

12. Cerucci, M.; Jaligama, G.; Amidon, T.; Cosgrove, A. The simulation of dissolved oxygen and orthophosphate for large scale watersheds using WASP7.1 with nutrient luxury uptake. *Water Environ. Fed.* **2007**, *12*, 5765–5776. [CrossRef]

13. Huang, J.; Liu, N.; Wang, M.; Yan, K. Application WASP Model on Validation of Reservoir-Drinking Water Source Protection Areas Delineation. In Proceedings of the 2010 3rd International Conference on Biomedical Engineering and Informatics (BMEI), Yantai, China, 16–18 October 2010.

14. Alam, A.; Badruzzaman, A.B.M.; Ali, M.A. Assessing effect of climate change on the water quality of the Sitalakhya river using WASP model. *J. Civ. Eng.* **2013**, *41*, 21–30.

15. Akomeah, E.; Chun, K.P.; Lindenschmidt, K.E. Dynamic water quality modelling and uncertainty analysis of phytoplankton and nutrient cycles for the upper South Saskatchewan River. *Environ. Sci. Pollut. Res.* **2015**, *22*, 18239–18251. [CrossRef] [PubMed]

16. Di Toro, D.M.; Fitzpatrick, J.J.; Thomann, R.V. *Documentation for Water Quality Analysis Simulation Program (WASP) and Model Verification Program (MVP)*; United States Environmental Protection Agency: Washington, DC, USA, 1983.

17. Ambrose, R.B.; Wool, T.A.; Connolly, J.P.; Schanz, R.W. *WASP4, a Hydrodynamic and Water Quality Model—Model Theory, User's Manual, and Programmer's Guide*; EPA-600/3-87/039; Office of Research and Development, United States Environmental Protection Agency: Washington, DC, USA, 1988.

18. Connolly, J.; Winfield, R. *A User's Guide for WASTOX, a Framework for Moeling the Fate of Toxic Chemicals in Aquatic Environments. Part 1: Exposure Concentration*; United States Environmental Protection Agency: Athens, GA, USA, 1984.

19. Ambrose, R.B.; Wool, T. *WASP7 Stream Transport-Model Theory and User's Guide: Supplement to Water Quality Analysis Simulation Program (WASP) User Documentation*; United States Environmental Protection Agency: Washington, DC, USA, 2009.

20. Cole, T.M.; Buchak, E.M. *CE-QUAL-W2: A Two-Dimensional, Laterally Averaged, Hydrodynamic and Water Quality Model, Version 2*; Instruction Report EL-95-1; Waterways Experiment Station, US Army Corps of Engineers: Vicksburg, MS, USA, 1995.

21. Tufford, D.L.; McKellar, H.N. Spatial and temporal hydrodynamic and water quality modeling analysis of a large reservoir on the South Carolina (USA) coastal plain. *Ecol. Model.* **1999**, *114*, 137–173. [CrossRef]

22. Lindenschmidt, K.-E. *Winter Flow Testing of the Upper Qu'Appelle River*; Lambert Academic Publishing: Saarbrücken, Germany, 2014.

23. Lindenschmidt, K.-E.; Sereda, J. The impact of macrophytes on winter flows along the Upper Qu'Appelle River. *Can. Water Resour. J.* **2014**, *39*, 342–355. [CrossRef]

24. Carpenter, S.R.; Lodge, D.M. Effects of submersed macrophytes on ecosystem processes. *Aquat. Bot.* **1986**, *26*, 341–370. [CrossRef]

25. Sand-jensen, K.; Prahl, C.; Stokholm, H. Oxygen release from roots of submerged aquatic macrophytes. *Wiley Behalf Nord. Soc. Oikos* **2016**, *38*, 349–354. [CrossRef]

26. Sereda, J. *Assessment of Macrophyte Control Options in the Upper Qu'Appelle Conveyance Channel: Nutrient Amendments and Harvesting*; Unpublished Report; Saskatchewan Water Security Agancy: Moose Jaw, SK, Canada, 2012.

27. Tolson, B.A.; Shoemaker, C.A. Dynamically dimensioned search algorithm for computationally efficient watershed model calibration. *Water Resour. Res.* **2007**, *43*, 1–16. [CrossRef]

28. Matott, L.S. *OSTRICH: An Optimization Software Tool; Documentation and User's Guide*; Univisity of Buffalo: Buffalo, NY, USA, 2008.

29. Murdock, T.Q.; Spittlehouse, D.L. *Selecting and Using Climate Change Scenarios for British Columbia*; Pacific Climate Impacts Consortium: Univisity of Victoria, Victoria, BC, Canada, 2011; Volume 39, p. 4.

30. Taylor, K.E.; Stouffer, R.J.; Meehl, G.A. An overview of CMIP5 and the experiment design. *Bull. Am. Meteorol. Soc.* **2012**, *93*, 485–498. [CrossRef]
31. McKenney, D.W.; Hutchiinson, M.F.; Papadopol, P.; Lawrence, K.; Pedlar, J.; Campbell, K.; Milewska, E.; Hopkinson, R.F.; Price, D.; Owen, T. Customized spatial climate models for North America. *Bull. Am. Meteorol. Soc.* **2011**, *92*, 1611–1622. [CrossRef]
32. Hopkinson, R.F.; Mckenney, D.W.; Milewska, E.J.; Hutchinson, M.F.; Papadopol, P.; Vincent, A.L.A. Impact of aligning climatological day on gridding daily maximum-minimum temperature and precipitation over Canada. *J. Appl. Meteorol. Climatol.* **2011**, *50*, 1654–1665. [CrossRef]
33. Tang, Y.; Reed, P.; Wagener, T.; van Werkhoven, K. Comparing sensitivity analysis methods to advance lumped watershed model identification and evaluation. *Hydrol. Earth Syst. Sci.* **2007**, *11*, 793–817. [CrossRef]
34. Razavi, S.; Gupta, H.V. What do we mean by sensitivity analysis? The need for comprehensive characterization of "global" sensitivity in earth and environmental systems models. *Water Resour. Res.* **2015**, *51*, 3070–3092. [CrossRef]
35. Saltelli, A.; Ratto, M.; Tarantola, S.; Campolongo, F. Sensitivity analysis for chemical models. *Chem. Rev.* **2005**, *105*, 2811–2827. [CrossRef] [PubMed]
36. Park, J.Y.; Park, G.A.; Kim, S.J. Assessment of future climate change impact on water quality of Chungju Lake, South Korea, Using WASP Coupled with SWAT. *J. Am. Water Resour. Assoc.* **2013**, *49*, 1225–1238. [CrossRef]
37. Magnuson, J.J.; Robertson, D.M.; Benson, B.J.; Wynne, R.H.; Livingstone, D.M.; Arai, T.; Assel, R.A.; Barry, R.G.; Card, V.; Kuuisisto, E.; et al. Historical trends in lake and river ice cover in the Northern Hemisphere. *Science* **2000**, *289*, 1743–1746. [CrossRef] [PubMed]
38. Zhang, C.; Lai, S.; Gao, X.; Xu, L. Potential impacts of climate change on water quality in a shallow reservoir in China. *Environ. Sci. Pollut. Res. Int.* **2015**, *22*, 14971–14982. [CrossRef] [PubMed]
39. Pettersson, K.; Grust, K.; Weyhenmeyer, G.; Blenckner, T. Seasonality of chlorophyll and nutrients in Lake Erken—Effects of weather conditions. *Hydrobiologia* **2003**, *506–509*, 75–81. [CrossRef]
40. Cox, B.A.; Whitehead, P.G. Impacts of climate change scenarios on dissolved oxygen in the River Thames, UK. *Hydrol. Res.* **2009**, *40*, 138. [CrossRef]
41. Murdoch, P.S.; Baron, J.S.; Miller, T.L. Potential effects of climate change on surface-water quality in North America. *J. Am. Water Resour. Assoc.* **2000**, *36*, 347–366. [CrossRef]
42. Wilby, R.L.; Whitehead, P.G.; Wade, A.J.; Butterfield, D.; Davis, R.J.; Watts, G. Integrated modelling of climate change impacts on water resources and quality in a lowland catchment: River Kennet, UK. *J. Hydrol.* **2006**, *330*, 204–220. [CrossRef]
43. Dodson, S.I. Introduction to Limnology. *J. N. Am. Benthol. Soc.* **2004**, *23*, 661–662. [CrossRef]
44. Tu, J. Combined impact of climate and land use changes on streamflow and water quality in eastern Massachusetts, USA. *J. Hydrol.* **2009**, *379*, 268–283. [CrossRef]
45. Arheimer, B.; Andreasson, J.; Fogelberg, S.; Johnsson, H.; Pers, C.B.; Persson, K. Climate change impact on water quality: Model results from southern Sweden. *AMBIO* **2005**, *34*, 559–566. [CrossRef] [PubMed]
46. Tong, S.T.Y.; Chen, W.L. Modeling the relationship between land use and surface water quality. *J. Environ. Manag.* **2002**, *66*, 377–393. [CrossRef]

water

MDPI

Concept Paper

Sustainable Ice-Jam Flood Management for Socio-Economic and Socio-Ecological Systems

Apurba Das [1,*], Maureen Reed [2] and Karl-Erich Lindenschmidt [1]

[1]   Global Institute for Water Security, University of Saskatchewan, 11 Innovation Boulevard, Saskatoon, SK S7N 3H5, Canada; karl-erich.lindenschmidt@usask.ca
[2]   School of Environment and Sustainability, University of Saskatchewan, 117 Science Place Saskatoon, SK S7N 5C8, Canada; maureen.reed@usask.ca
*   Correspondence: apurba.das@usask.ca; Tel.: +1-(306)-966-7797

Received: 20 November 2017; Accepted: 25 January 2018; Published: 31 January 2018

Abstract: Ice jams are critical components of the hydraulic regimes of rivers in cold regions. In addition to contributing to the maintenance of wetland ecology, including aquatic animals and waterfowl, ice jams provide essential moisture and nutrient replenishment to perched lakes and ponds in northern inland deltas. However, river ice-jam flooding can have detrimental impacts on in-stream aquatic ecosystems, cause damage to property and infrastructure, and present hazards to riverside communities. In order to maintain sustainable communities and ecosystems, ice-jam flooding must be both mitigated and promoted. This study reviews various flood management strategies used worldwide, and points to the knowledge gaps in these strategies. The main objective of the paper is to provide a framework for a sustainable ice-jam flood management strategy in order to better protect riverine socio-economic and socio-ecological systems. Sustainable flood management must be a carefully adopted and integrated strategy that includes both economic and ecological perspectives in order to mitigate ice-jam flooding in riverside socio-economic systems, while at the same time promoting ice-jam flooding of riverine socio-ecological systems such as inland deltas.

Keywords: ice jam flooding; sustainable flood management; socio-economic system; socio-ecological system

1. Introduction

Ice jams are a natural phenomenon in the rivers of cold regions that usually occur during spring ice cover breakup and in mid-winter in temperate regions. An ice jam forms when incoming ice floes along a reach are arrested by an obstacle (for example, existing ice blocks or river meanders), which constricts the channel, rendering it incapable of transporting the ice downstream. Hence, an ice jam resists the incoming river flow, raising water levels and overflowing the river banks [1]. Ice jams can be several metres thick and can extend for many kilometres along a river, often creating floods that are more severe than open water floods. Due to the high water levels, unusual velocities, and complex formation mechanisms of ice-induced floods, damage, injuries, and deaths to humans occur more frequently in these types of floods than in open-water floods. Ice-jam flood events are common in northern Canada, and have been documented around the world, such as for instance in the United States (US) [2,3], Finland [4], Iceland [5,6] Norway [7], Germany [8], Sweden [9], Russia [10,11], Japan [12], and China [13,14]. Most of these studies have focussed on the effects of ice-jam floods on riverine socio-economic systems; other studies have specifically investigated the impact of ice-induced floods on socio-ecological systems [15–21]. These latter studies show that the ice-induced flood is an important hydrological factor in maintaining sustainability in riverine socio-ecological systems in Canada.

Ice jams play a critical role in riverine communities and ecosystems in cold regions, as ice jam-related flooding can provide essential replenishment to perched lakes and ponds in inland deltas [15,22]. River ice jams can also cause devastating flood events, leading to detrimental impacts on aquatic ecosystems (for example, reduced fish habitat), property, and infrastructure (for example, damage to homes, bridges, roads, and businesses), and the loss of human life [23,24]. A previous study suggested that flow regulation and climatic variations increase the risk of extreme flood events to riverside communities [23], while at the same time reducing the frequency of ice jam-related flooding in inland deltas [15,25–28]. Hence, inland deltas have experienced prolonged dry periods and a reduction in wetland areas [16,17]. Therefore, ice-induced flood management requires the consideration of two measures: (1) the mitigation of ice-jam flooding in riverside urban communities, and (2) the promotion of ice-jam flooding in riverine ecosystems (for example, inland deltas), especially rivers, where both socio-economic and socio-ecological systems need to be equally maintained. A socio-economic system is a network of social and economic services that provides society with development, business, education, and homes, whereas a socio-ecological system is a network of interrelated processes that provide society with ecological services such as food, energy, water, and biodiversity. Determining an ice-jam flood management strategy could prove to be beneficial should the paucity of future ice-jam flooding or increases in prolonged drying in northern inland deltas occur due to climate change and regulation. The main objectives of this study are to: (a) provide a critical review of the current state of knowledge of the role of ice-jam flooding in sustaining river socio-economic and socio-ecological systems; (b) provide a framework for a sustainable ice-jam flood management strategy; and (c) identify challenges and knowledge gaps for sustainable ice-jam flood management strategies in cold regions.

2. The Role of Ice Jam Flooding in Systems

2.1. The Socio-Economic System

Ice-jam flooding is a considerable socio-economic concern for the communities, engineers, insurance companies, authorities, and government agencies in cold regions. Ice-jam floods are characterised by a quick increase in water levels and high flow velocities. As well as causing injury and death, these conditions often increase the size or height of river embankments, cause road erosion, damage infrastructure, and harm riverine ecosystems. A study indicated that the average annual damage in the United States from ice-jam formation is $120 M (USD) [2]. Ice-jam flooding interrupts the development of various river structures such as dikes, bridges, pipeline crossings, and pump stations [1]. These structures must be designed to manage the flow characteristics associated with ice-jam formation. Bridges and pipelines crossing a cold region's rivers are often designed to withstand more than the expected maximum ice-jam water levels. It is interesting to note that the dikes at Peace River in Alberta, Canada have been developed and upgraded several times to meet the maximum protection from ice-jam flood events [23]. The high flow velocities generated by the release of ice jams can also damage a river's navigation systems and gauge stations, and can pose a major risk to shipping and hydrological data losses. Another consequence of ice-jam formations along a river is economic losses from hydroelectric dam operations. Over 30% of the annual economic losses of hydro dams in Canada are due to ice-jam interference in hydropower operations [1].

2.2. The Socio-Ecological System

Floods play a key role in the physical, chemical, and biological processes of river ecosystems, and ice-induced floods can have significantly different impacts from open water floods on similar ecosystems [29]. Ice-induced flooding can create extremely high water levels that are capable of inundating elevated portions of a floodplain [15]. High water flows maintain the hydrological connectivity between the main channel and the perched channels, converting the diversion channel to the principal channel or creating a new channel [30]. Scientists have found that many of the deltas in northern Canada (for example, the Peace–Athabasca Delta and Slave River Delta) are drying up because ice-jam flooding has declined during the spring ice-cover breakup [17]. In these deltas, open water flooding is unable to flood perched lakes, while the ice jams can produce much higher water levels compared to open water floods and replenish the perched lakes. In these delta environments, many riparian vegetation are completely dependent on regular flood events for their reproduction, and regular flooding may also be necessary to maintain water balances [19]. Therefore, to prevent further problems, the deltaic environment must be maintained through regular flood events and a certain level of floodwater depth. Although ice-jam flooding is beneficial in many ways for river ecosystems, a large flood can adversely impact stream environments. High river flows and unusually high velocities can increase fish mortality and hydrological disturbances in the aquatic environment. During flood events, suspended sediments change the geomorphology of the channel, affecting aquatic organisms such as fish and invertebrates [31]. Suspended sediments change the chemical composition of the water column (for example, temperature, dissolved oxygen, and pH), and impede the oxygen process of fish habitats [29,32]. Therefore, consideration of an ecosystem's flow requirement is one of the essential tasks in any ice-jam flood management strategy and a moderate flood with a shorter return period should be maintained in order to avoid a major disturbance in the aquatic system [19].

3. Toward Sustainable Flood Risk Management

Sustainable flood management is a strategic approach that provides guidelines for maximum flood protection of the socio-economic and socio-ecological systems. According to the Flooding Issues Advisory Committee (FIAC) of Scotland, sustainable flood management is a strategy that "provides the maximum possible social and economic resilience against flooding, by protecting and working with the environment, in a way which is fair and affordable both now and in the future" [33]. According to this definition, the main objective of sustainable flood management is to incorporate strategies that benefit communities, reduce economic damage, protect environments, and meet the needs of future generations.

The key goal of sustainable flood risk management is to integrate all of the levels of actors and institutions into flood risk management processes, and ensure the maximum resilience of socio-economic and socio-ecological systems, both now and in the future. Therefore, a sustainable ice-jam flood risk management strategy should ensure maximum flood protection to the socio-economic systems and sustainable flow to the socio-ecological systems [34]. Maximum flood protection means that there is a minimal risk to any properties, infrastructure, and human beings exposed to any flood event. A sustainable amount of ecological flow to the socio-ecological systems refers to the maximum flow necessary for maintaining ecological integrity. However, human intervention (for example, flow regulation) and climatic conditions affect these systems, resulting in increasing flood damage and environmental deterioration around the world.

In recent decades, flood risk management experts have adopted new methods, tools, and technologies in an effort to transform policies and better manage and mitigate flood damage. Geographical information systems, remote sensing, and numerical modelling techniques have been widely introduced to map floodplains, quantify potential damages, and analyse flood risks [35]. Many countries have started to shift from flood damage and risk reduction strategies to damage and risk-management strategies. For example, in 2006, the US government integrated all of the flood management administrative authorities and federal and local agencies to form a national flood risk management program [36]. More recently, Emergency Preparedness Canada and the Institute Bureau of Canada identified a set of cultures in current flood-management approaches that contribute to the trend of increasing flood damage, and suggested policy changes in current flood management [37].

Despite all these efforts and the technological advancements in current flood management strategies, two main problems remain: first, national and regional flood management agencies do not collaborate as much as they could, and, second, environmental factors are not well-integrated into flood risk management approaches. To address these challenges and achieve sustainable flood management approaches, new strategies need to be adopted. These include the following: the engagement of all levels of stakeholders, communities, and institutions [38,39]; community or participatory-based flood management strategies [40,41]; non-structural measures and adaptation strategies [42]; and ecosystem-based flood management systems [19,43]. The development of such integrated strategies will require a transdisciplinary approach that combines the perspectives and individuals (e.g., scientists and stakeholders) from different disciplines related to water management (e.g., hydrology and ecology).

4. Ice-Jam Flood Management Strategies

Current ice-jam flood risk management strategies include structural and non-structural measures and flood risk assessment to reduce flood risk. Structural and non-structural measures are usually used to mitigate ice-jam formation and reduce damages during floods.

4.1. Structural Measures

Structural measures involve the construction of various types of physical structures such as dams, reservoirs, dikes, and channel modification to control the floodwater and ice-jam formation along a river. Structural measures prevent the overflow of the rivers to floodplain areas in many ways. For example, reservoirs can control water levels, while dams and levees confine the flow within the main channel, and floodways divert the excess flow. A study in northern Canada showed that an effective flow regulation by a reservoir can reduce the chances of ice-jam formation in ice-jam prone areas [44], and a certain level of flows can create favourable conditions for ice-jam flooding in an environment where seasonal ice-jam flooding is needed [25]. For example, in the winter of 1994–1995, an artificial ice jam was constructed using spray-ice techniques to cause the overtopping of remnants of the original rock-filled weir along a channel in the Peace-Athabasca Delta Although the project was partially successful—which was only because of low flow conditions along the channel [45]—it demonstrated the potential of the structural measures to replenish some of the perched basin. Structural measures are usually effective and suitable for controlling a large volume of ice and protecting the floodplain area from high water stages. In a river, a jam can extend for several kilometres, and a large volume of ice accumulated from upstream can create high shear forces and high-water stages. In this situation, a dyke system is very effective at keeping floodwaters and ice out of the flood prone area. However, structural measures are very costly, and may have long-term adverse environmental effects. Although the variability in ice-jam flood frequency along the Peace River has been shown to be statistically insignificant, and the evidence of the impact of dam operations on altering the ice-jam flood frequency is sparse [30], some previous studies have suggested that dam operations have a certain level of impact by altering the ice regime and reducing the ice-jam flood frequency downstream along the river, resulting in the reduction of wetlands that provide habitats for the aquatic environment [19,20,24]. Consequently, an environment friendly structural design or proper operational conditions for dams and reservoirs are necessary for the better protection of socio-economic systems, in order to avoid the long-term adverse impact of the measure on socio-ecological systems.

4.2. Non-Structural Measures

Non-structural measures can be categorised into two groups: (i) prevention and emergency measures, both of which involve spatial planning (for example, zoning regulation), public awareness programs, education, and research, and the provision of information and communication resources (e.g., flood forecast and early warning); and (ii) financial supports or penalties with respect to floodplain development [46]. Besides these measures, non-structural measures also include ice cutting, drilling, and blasting for emergency mechanical ice-jam breaking, and weakening winter ice covers. The main advantage of non-structural measures is that they are cost effective, environmentally friendly, and sustainable over time. For example, in the spring of 1996, a non-structural measure was employed along the PAD. Although a natural ice-jam was formed in the delta channel, a certain flow release from the reservoir assisted potential flooding with a 20-year return period [18]. Although non-structural measures reduced the catastrophic impacts of residual risks, they are unable to completely mitigate ice-jam flooding. Therefore, a combination of both structural and non-structural measures can be an effective way to optimally reduce the risks of flooding and better protect riverine ecosystems.

4.3. Flood Hazard Mapping

A flood-frequency or stage-probability curve is used to estimate the probability of ice-jam floods and produce the flood hazard map for a specific floodplain [23]. This curve is usually used to quantify the water level for ice-jam flood events. A long-term historical record of ice-jam stages and various statistical methods are traditionally applied to develop this curve. If historical data is inadequate, synthetic flood-frequency relationships are developed. This is an indirect approach to a stage-frequency curve using observation data or applying statistical analyses [47]. Additionally, various hydraulic one-dimensional (1D) and two-dimensional (2D) numerical modelling approaches are also widely used to synthesise ice-jam stages in northern regions [1,48]. Geographic information system (GIS) tools are usually used to develop event-based flood hazard maps and estimate the corresponding flood damage. A digital elevation model (DEM) is used to produce a flood hazard and risk map. A flood hazard map delineates the flood extent areas and provides information on flood characteristics (for example, depth and velocity). A flood risk map provides the information about the potential consequences of a flood event, including the number of inhabitants or infrastructures vulnerable to flooding and the potential economic damage for that specific event [49].

4.4. Flood Risk Assessment

Flood risk is defined as the probability of a flood event combined with the potential adverse impact of such an event on communities, the environment, and economic conditions [50]. Much research has already been conducted on ice-jam flood risk assessment [6,11,23,35,47,51,52], where the flood risks were assessed based on the potential for socio-economic and socio-ecological flood damage. An appropriate flood risk management strategy is adopted based on economic assessments. Such economic assessments incorporate only the direct flood damages, such as property and infrastructure damage, associated with a flood's stages. However, in many cases, indirect damages such as industrial production loss, emergency costs, and most importantly, damage to ecosystem services (which are used for livelihood, recreational, or spiritual purposes) are ignored in flood risk estimation [35]. Therefore, a sustainable flood management method should be introduced that can incorporate both direct and indirect flood damages in flood risk assessment processes.

5. Challenges

Ice-jam flood risk assessment is a challenging task compared to assessments of open water floods. Estimating water levels for ice-jam flood events is complex because ice-jam formation is site specific, and most often, historical data are unavailable [1]. Additionally, the freezing properties and high velocities of the flood water damage properties and structures. This damage is difficult to predict and evaluate during the assessment process.

Although a number of ice-jam prediction models have been developed to date, these models are very site specific, rely on adequate historical data input, and are unable to consider the full effect of climate change and flow regulations on ice-jam formation [47,53]. Research on the effects of climate change on ice regimes is sparse; therefore, little information is available on the impact of climate change on the frequency and the severity of ice jams.

Likewise, very little is known about how to incorporate environmental factors and conditions in flood management strategies [19]. Previous studies mainly considered the economic aspects of determining the flood risk because ecological damage is difficult to quantify in monetary terms. Therefore, future research should also concentrate on how to incorporate environmental damage into flood risk maps and management strategies.

Despite the advancements in scientific technologies of all aspects of ice-jam flood management strategies, national and international collaboration in this field is still very limited [54]. New sustainable ice-jam flood-management strategies should be more regional, integrated, and proactive.

6. Framework for Achieving Sustainable Ice-Jam Flood Risk Management

Sustainable ice-jam flood risk management strategies should deal with current and future influences of hydro-climatic and human interventions on river ice-jam formations, as well as with optimal flood risk reduction. The key idea of sustainable ice-jam management is to engage all kinds of people and organisations, such as governments, businesses, private organisations, scientists, and individuals, into flood risk assessment processes [38]. Some different ways of achieving the goal of sustainable ice-jam flood risk management are discussed below:

6.1. Integrating Structural and Non-Structural Measures

A sustainable ice-jam flood management strategy can be achieved by integrating structural components with non-structural components [55]. Many scientists argue that the construction of structural measures tends to increase flood risk, because perceptions of flood-risk reduction further increase development activities in floodplain areas [55,56]. However, integrating non-structural measures with structural measures can add floodplain development regulations that reduce the flood risk and subsequently increases economic and ecological sustainability in society. The main advantage of this measure is that it can simultaneously control river flows and riverine ecosystems by constructing eco-friendly flood control structures and adding necessary regulations to floodplain areas. For example, an artificial ice dam (structural measurement) with a certain discharge from the dam (non-structural measurement) along the Peace River can lead to favourable conditions for ice-jam flooding along the Peace–Athabasca Delta.

6.2. A Transdisciplinary Approach

A transdisciplinary approach to ice-jam flood management integrates all levels of individuals, such as stakeholders, practitioners, and indigenous peoples into flood management decision-making processes [56]. Such an approach connects practitioners from different disciplines and backgrounds, ranging from officials in public administration, to scholars in the social sciences, to experts in hydrology and river engineering. These practitioners share their own perceptions, as well as the common goal of developing a sustainable flood management strategy. The key objective of this approach is cooperative investigation among all of the disciplines to understand the effect of human activities and climate change on ice-jam formation, and their relation to flood hazards. For example, hydrologists and ecologists should investigate how the annual hydrological cycle and the functions of ecosystems are driven by human activities and hydro-climatic conditions, while institutions dealing with flooding should emphasise the need to better understand the consequences of these impacts on both the economy and the environment.

A transdisciplinary approach is an integrated management approach that can play a vital role in sustainable ice-jam flood management. The main objective of this approach is to ensure the participation of the general public, stakeholders' representatives, community groups, and local authorities in the decision-making processes. A suitable flood management strategy requires the incorporation of water resources, land use, and environmental strategies with flood management [56]. Therefore, incorporating these multi-sectoral strategies into flood management requires active participation from the community, stakeholders, government, local non-government organisations (NGOs), and politicians. They can influence decision-making processes because many decisions are vested completely in community and other participatory groups.

6.3. Incorporating Ecological Perspective in Ice-Jam Flood Management

The main sustainable goal is to preserve and protect natural resources. Riverine ecosystems provide natural resources such as water, energy, and habitats for aquatic communities. However, human activities and climate change are changing ice-jam flood regimes in cold regions, and potentially impacting the delta ecosystems [28,57]. Therefore, a sustainable ice-jam flood management strategy should add ecological perspectives to conventional flood management strategies [19]. To mitigate the impact of ice-jam flooding on riverine ecosystems, an environmental framework can be added to sustainable ice-jam flood management goals. This framework would then include an environmentally sensitive structural design, an environmental assessment in any flood management decision-making process, an environmentally sensitive economic analysis, and a regular monitoring program [56].

An environmentally sensitive structural measure can be designed by clearly understanding the morphology and ecology of a river, and how the river's floodplain and physical characteristics are driven by the flow regime. Environmental assessment is a tool to determine whether there is any significant environmental impact of a proposed flood protection measure or the existing flood protection scheme of a system. The economic value of losing ecosystem services due to any flood measures can also be incorporated in a flood management strategy to determine the feasibility of applying measures or finding alternatives. Continuous monitoring and evaluation of the environmental health and existing flood measures is necessary in flood management, because it helps to determine whether there is any long-term environmental effect of a current flood measure or if it is necessary to adopt any additional measures.

7. Summary and Conclusions

The preceding sections describe the various aspects of ice-jam flood management strategies, their limitations, and a framework for achieving sustainable ice-jam flood management strategies. To select an appropriate flood management strategy, a good understanding of ice-jam formation, growth, and release is critical. As the nature and characteristics of ice-jam formations vary from site to site and region to region, a detailed understanding of hydro-climatic conditions of the mitigation site is needed. To develop a stage-frequency curve requires adequate historical data or a synthetic approach to produce a stage-discharge relationship using numerical modelling tools. Structural measures are usually effective in extensive flood-prone areas, or where a hydropower operation is interrupted by an ice-jam formation; however, the initial investment and maintenance cost of structural measures is much higher than those of non-structural measures. To reduce the flood risk in the areas vulnerable to flood damage, both measures can be combined in a sustainable flood management strategy. Ice-jams impacting riverine ecosystems are widely visible, and much research has investigated the connection between ecology and ice jamming. Incorporating ecological perspectives in ice-jam flood management strategies is a pressing need. Such integration will enable scientists to better manage important natural resources and sustainability in riverine ecosystems.

A sustainable flood management approach should be able to simultaneously protect people and socio-economic activities, as well as secure natural ecosystems functioning through disturbances created by floods. Incorporating both structural and non-structural measures can help ensure maximum flood protection and mitigation, as well as the promotion of flooding in deltaic environments. However, combining both measures is not always an easy task, and requires comprehensive study to determine the feasibility of both measures working simultaneously. Scientists and other individuals working together on such an approach would enhance multi-perspective analyses with consideration of the impact of flooding on both society and on the environment. Ecological perspectives are key to a transdisciplinary approach to ice-jam flood management strategies, as they optimise the resiliency of riverine ecosystems (Figure 1). The ecological perspective can be used as a flood management tool in decision-making processes, policies, and flood mitigation and promotion measures.

Finally, to apply the required flood management measures, a clear understanding of ice-jam processes in the geographic area under study is vital, because river ice processes vary from site to site depending on local climatic conditions. In addition to this, measures must be carefully balanced in order to mitigate ice-jam flooding in riverside socio-economic systems, and at the same time, promote ice-jam flooding in riverine socio-ecological systems such as inland deltas.

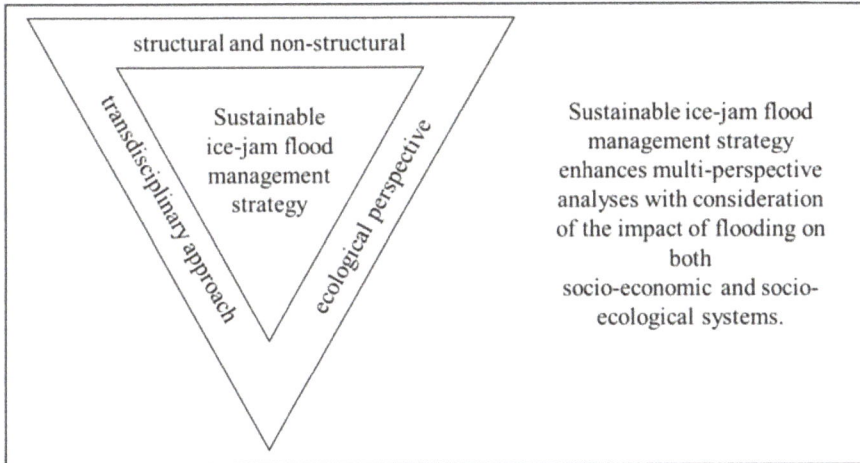

Figure 1. The summary of the sustainable ice-jam flood management strategy.

Acknowledgments: We thank the Natural Sciences and Engineering Research Council of Canada (NSERC), the Saskatchewan Government and the University of Saskatchewan for their funding support of this research.

Author Contributions: Apurba Das and Karl-Erich Lindenschmidt developed the concept and Apurba Das drafted the texts in the manuscript. Maureen Reed helped to improve the concept of the paper and Karl-Erich Lindenschmidt and Maureen Reed reviewed the final version of the manuscript.

Conflicts of Interest: The authors declare no conflict of interest.

References

1. Beltaos, S. *River Ice Jams*; Water Resources Publication: Littleton, CO, USA, 1995.
2. White, K.D.; Tuthill, A.M.; Furman, L. Studies of Ice Jam flooding in the United States. In *Extreme Hydrological Events: New Concepts for Security*; Springer: Houten, The Netherlands, 2006; pp. 255–268.
3. White, K.D.; Eames, H.J. *CRREL Ice Jam Database*; DTIC Document; Defense Technical Information Center: Fort Belvoir, VA, USA, 1999.
4. Centre for Economic Development, Transport and the Environment. *The Preliminary Flood Risk Assessment in Tornionjoki-Muonionjoki River Basin*; Centre for Economic Development, Transport and the Environment: Helsinki, Finland, 2010.
5. Eliasson, J.; Gröndal, G.O. Development of a river ice jam by a combined heat loss and hydraulic model. *Hydrol. Earth Syst. Sci. Discuss.* **2008**, *5*, 1021–1042. [CrossRef]
6. Pagneux, E.; Gísladóttir, G.; Jónsdóttir, S. Public perception of flood hazard and flood risk in Iceland: A case study in a watershed prone to ice-jam floods. *Nat. Hazards* **2011**, *58*, 269–287. [CrossRef]
7. Lier, Ø.E. Modeling of Ice Dams in the Karasjohka River. In Proceedings of the 16th IAHR International Symposium on Ice, Dunedin, New Zealand, 2–6 December 2002.
8. Carstensen, D. Ice conditions and ice forces. In Proceedings of the Chinese-German Joint Symposium on Hydraulic and Ocean Engineering, Darmstadt, German, 24–30 August 2008.
9. Ahopelto, L.; Huokuna, M.; Aaltonen, J.; Koskela, J.J. Flood frequencies in places prone to ice jams, case city of Tornio. In Proceedings of the CGU HS Committee on River Ice Processes and the Environment, 18th Workshop on the Hydraulics of Ice Covered Rivers, Quebec City, QC, Canada, 18–20 August 2015.

10. Buzin, V.A.; D'yachenko, N.Y. Forecasting the intrawater ice formation and ice jams in the Neva River. *Russ. Meteorol. Hydrol.* **2011**, *36*, 770–775. [CrossRef]

11. Frolova, N.L.; Agafonova, S.A.; Krylenko, I.N.; Zavadsky, A.S. An assessment of danger during spring floods and ice jams in the north of European Russia. *Proc. Int. Assoc. Hydrol. Sci.* **2015**, *369*, 37–41. [CrossRef]

12. Shen, H.T.; Liu, L. Shokotsu River ice jam formation. *Cold Reg. Sci. Technol.* **2003**, *37*, 35–49. [CrossRef]

13. Fu, C.; Popescu, I.; Wang, C.; Mynett, A.E.; Zhang, F. Challenges in modelling river flow and ice regime on the Ningxia–Inner Mongolia reach of the Yellow River, China. *Hydrol. Earth Syst. Sci.* **2014**, *18*, 1225–1237. [CrossRef]

14. Rao, S.; Yang, T.; Liu, J.; Chen, D. Characteristics of Ice Regime in the Upper Yellow River in the Last Ten Years. In Proceedings of the 21st IAHR International Symposium on Ice, Dalian, China, 11–15 June 2012.

15. Peters, D.; Prowse, T.; Marsh, P.; Lafleur, P.; Buttle, J. Persistence of Water within Perched Basins of the Peace-Athabasca Delta, Northern Canada. *Wetl. Ecol. Manag.* **2006**, *14*, 221–243. [CrossRef]

16. Peters, D.; Prowse, T.; Pietroniro, A.; Leconte, R. Flood hydrology of the Peace-Athabasca Delta, Northern Canada. *Hydrol. Process.* **2006**, *20*, 4073–4096. [CrossRef]

17. Prowse, T.D.; Conly, F.M. A review of hydroecological results of the Northern River Basins Study, Canada. Part 2. Peace-Athabasca Delta. *River Res. Appl.* **2002**, *18*, 447–460. [CrossRef]

18. Prowse, T.D.; Peters, D.; Beltaos, S.; Pietroniro, A.; Romolo, L.; Töyrä, J.; Leconte, R. Restoring Ice-jam Floodwater to a Drying Delta Ecosystem. *Water Int.* **2002**, *27*, 58–69. [CrossRef]

19. Peters, D.; Caissie, D.; Monk, W.; Rood, S.; St-Hilaire, A. An ecological perspective on floods in Canada. *Can. Water Resour. J.* **2016**, *41*, 288–306. [CrossRef]

20. Lindenschmidt, K.-E.; Das, A. A geospatial model to determine patterns of ice cover breakup along the Slave River. *Can. J. Civ. Eng.* **2015**, *42*, 675–685. [CrossRef]

21. Wolfe, B.B.; Karst-Riddoch, T.L.; Vardy, S.R.; Falcone, M.D.; Hall, R.I.; Edwards, T.W.D. Impacts of climate and river flooding on the hydro-ecology of a floodplain basin, Peace-Athabasca Delta, Canada since A.D. 1700. *Quat. Res.* **2005**, *64*, 147–162. [CrossRef]

22. Timoney, K. A dying delta? A case study of a wetland paradigm. *Wetlands* **2002**, *22*, 282–300.

23. Lindenschmidt, K.-E.; Das, A.; Rokaya, P.; Chu, T. Ice-jam flood risk assessment and mapping. *Hydrol. Process.* **2016**, *30*, 3754–3769. [CrossRef]

24. Church, M. *The Regulation of Peace River*; John Wiley & Sons, Ltd.: New York, NY, USA, 2014.

25. Beltaos, S. Numerical modelling of ice-jam flooding on the Peace-Athabasca delta. *Hydrol. Process.* **2003**, *17*, 3685–3702. [CrossRef]

26. Beltaos, S.; Prowse, T.D.; Carter, T. Ice regime of the lower Peace River and ice-jam flooding of the Peace-Athabasca Delta. *Hydrol. Process.* **2006**, *20*, 4009–4029.

27. Beltaos, S. The role of waves in ice-jam flooding of the Peace-Athabasca Delta. *Hydrol. Process.* **2007**, *21*, 2548–2559. [CrossRef]

28. Beltaos, S. Comparing the impacts of regulation and climate on ice-jam flooding of the Peace-Athabasca Delta. *Cold Reg. Sci. Technol.* **2014**, *108*, 49–58. [CrossRef]

29. Prowse, T.D.; Culp, J.M. Ice breakup: A neglected factor in river ecology. *Can. J. Civ. Eng.* **2003**, *30*, 128–144. [CrossRef]

30. Timoney, K.P. *The Peace-Athabasca Delta: Portrait of a Dynamic Ecosystem*; University of Alberta: Edmonton, AB, Canada, 2013.

31. Milner, A.M.; Robertson, A.L.; McDermott, M.J.; Klaar, M.J.; Brown, L.E. Major flood disturbance alters river ecosystem evolution. *Nat. Clim. Chang.* **2013**, *3*, 137–141. [CrossRef]

32. Waters, T.F. *Sediment in Streams: Sources, Biological Effects, and Control*; American Fisheries Society: Bethesda, MD, USA, 1995.

33. Scottish Executive. *Final Report of the National Technical Advisory Group on Flooding*; Scottish Executive: Edinburgh, UK, 2005.

34. Hooijer, A.; Klijn, F.; Pedroli, G.B.M.; Van Os, A.G. Towards sustainable flood risk management in the Rhine and Meuse river basins: Synopsis of the findings of IRMA-SPONGE. *River Res. Appl.* **2004**, *20*, 343–357. [CrossRef]

35. Burrell, B.; Huokuna, M.; Beltaos, S.; Kovachis, N.; Turcotte, B.; Jasek, M. Flood Hazard and Risk Delineation of Ice-Related Floods: Present Status and Outlook. In Proceedings of the 18th CGUHS CRIPE Workshop on the Hydraulics of Ice Covered Rivers, Quebec City, QC, Canada, 18–20 August 2015.

36. Wood, M.D.; Linkov, I.; Kovacs, D.; Butte, G. Flood Risk Management. In *Mental Modeling Approach*; Springer: Berlin, Germany, 2017.
37. Shrubsole, D. Flood management in Canada at the crossroads. *Glob. Environ. Chang. Part B Environ. Hazards* **2000**, *2*, 63–75. [CrossRef]
38. Sayers, P.; Yuanyuan, L.; Galloway, G.; Penning-Rowsell, E.; Fuxin, S.; Kang, W.; Yiwei, C.; Le Quesne, T. Flood Risk Management: A Strategic Approach. The United Nations Educational, Scientific and Cultural Organization: Paris, France, 2013.
39. Akter, T.; Simonovic, S.P. Aggregation of fuzzy views of a large number of stakeholders for multi-objective flood management decision-making. *J. Environ. Manag.* **2005**, *77*, 133–143. [CrossRef] [PubMed]
40. Buckland, J.; Rahman, M. Community-based Disaster Management During the 1997 Red River Flood in Canada. *Disasters* **1999**, *23*, 174–191. [CrossRef] [PubMed]
41. Simonovic, S.P.; Akter, T. Participatory floodplain management in the Red River Basin, Canada. *Annu. Rev. Control* **2006**, *30*, 183–192. [CrossRef]
42. Kundzewicz, Z.W. Non-structural flood protection and sustainability. *Water Int.* **2002**, *27*, 3–13. [CrossRef]
43. Poff, N.L. Ecological response to and management of increased flooding caused by climate change. *Philos. Trans. R. Soc. Lond. A* **2002**, *360*, 1497–1510. [CrossRef] [PubMed]
44. Jasek, M.; Friesenham, E.; Granson, W. Operational river ice forecasting on the Peace River–managing flood risk and hydropower production. In Proceedings of the 14th Workshop on the Hydraulics of Ice Covered Rivers 2007 CGU HS Committee on River Ice Processes and the Environment, Quebec City, QC, Canada, 19–22 June 2007; pp. 19–22.
45. Prowse, T.D.; Demuth, M.N. Using ice to flood the Peace–Athabasca Delta, Canada. *Regul. Rivers Res. Manag.* **1996**, *12*, 447–457. [CrossRef]
46. Thampapillai, D.J.; Musgrave, W.F. Flood damage mitigation: A review of structural and nonstructural measures and alternative decision frameworks. *Water Resour. Res.* **1985**, *21*, 411–424. [CrossRef]
47. Beltaos, S. Assessing Ice-Jam Flood Risk: Methodology and Limitations. In Proceedings of the 20th IAHR Inernational Symposium on Ice, Lathi, Finland, 14–17 June 2010; pp. 14–17.
48. Tuthill, A.M.; Wuebben, J.L.; Daly, S.F.; White, K.D. Probability distributions for peak stage on rivers affected by ice jams. *J. Cold Reg. Eng.* **1996**, *10*, 36–57. [CrossRef]
49. Aaltonen, J.; Huokuna, M. Flood mapping of river ice breakup jams in River Kyrönjoki delta. In Proceedings of the CGU HS Committee on River Ice Processes and the Environment, 18th Workshop on the Hydraulics of Ice Covered Rivers, Quebec City, QC, Canada, 18–20 August 2015.
50. Directive, E. 60/EC of the European Parliament and of the Council of 23 October 2007 on the assessment and management of flood risks. *Off. J. Eur. Union L* **2007**, *288*, 186–193.
51. Beltaos, S.; Burrell, B.C. *Extreme Ice Jam Floods along the Saint John River, New Brunswick, Canada*; Iahs Publication: Oxfordshire, UK, 2002; pp. 9–14.
52. Mahabir, C.; Hicks, F.; Fayek, A.R. Forecasting ice jam risk at Fort McMurray, AB, using fuzzy logic. In Proceedings of the 16th IAHR International Symposium on Ice, International Association of Hydraulic Engineering and Research, Dunedin, New Zealand, 2–6 December 2002; pp. 91–98.
53. Beltaos, S.; Prowse, T.D. Climate impacts on extreme ice-jam events in Canadian rivers. *Hydrol. Sci. J.* **2001**, *46*, 157–181. [CrossRef]
54. White, K.; Hicks, F.; Beltaos, S.; Loss, G. Ice jam response and mitigation: The need for cooperative succession planning and knowledge transfer. In Proceedings of the 14th Workshop on River Ice, Quebec City, QC, Canada, 20–22 June 2007; pp. 20–22.
55. Yu, X.; Huang, Y. Sustainable Flood Risk Management: Lesson from Recent Cases. *GeoRisk* **2011**. [CrossRef]
56. Bonacci, O. Environmental aspects of integrated flood management. *Gospodarstvo i Okoliš* **2007**, *15*, 146–159.
57. Prowse, T.D.; Conly, F.M. Effects of climatic variability and flow regulation on ice-jam flooding of a northern delta. *Hydrol. Process.* **1998**, *12*, 1589–1610. [CrossRef]

MDPI
St. Alban-Anlage 66
4052 Basel
Switzerland
Tel. +41 61 683 77 34
Fax +41 61 302 89 18
www.mdpi.com

Water Editorial Office
E-mail: water@mdpi.com
www.mdpi.com/journal/water

www.ingramcontent.com/pod-product-compliance
Lightning Source LLC
Chambersburg PA
CBHW051848210326
41597CB00033B/5817